PLANET
MANAGEMENT

PLANET MANAGEMENT

GENERAL EDITOR

Dr Michael Williams

New York
OXFORD UNIVERSITY PRESS
1993

CONSULTANT EDITOR
Professor Peter Haggett, University of Bristol

John D. Baines, Chorleywood, UK
Italy and Greece, Central Europe

Dr Terry Cannon, Thames Polytechnic, UK
China and its neighbors

Dr Mike Clark, University of Southampton, UK
Canada and the Arctic

Professor Susan L. Cutter, Rutgers – University of
New Jersey, USA
The United States

Dr John P. Dickenson, University of Liverpool, UK
South America

Professor Ezekiel Kalipeni, Colgate University, New York, USA
Central Africa

Dr Richard Knowles, University of Salford, UK
The Nordic Countries

Drs H. Meijer, Utrecht, The Netherlands
The Low Countries

Dr Nick Middleton, University of Oxford, UK
The Middle East, The Indian Subcontinent

Dr Andrew Millington, University of Reading, UK
Northern Africa

Dr Lesley Potter, University of Adelaide, Australia
Southeast Asia

Dr David Simon, Royal Holloway and Bedford New College,
University of London, UK
Southern Africa

Dr J. Tuppen, Lyon Graduate School of Business, France
France and its neighbors

Dr David Turnock, University of Leicester, UK
Eastern Europe

Dr Tatyana Vlasova, Academy of Science, Russia
Northern Eurasia

Dr David Watts, University of Hull, UK
Central America and the Caribbean

Dr Allan Williams, University of Exeter, UK
Spain and Portugal

Dr Michael Williams, University of Oxford, UK
*Managing the World's Environments, The British Isles
Australasia, Oceania and Antarctica*

Dr M. E. Witherick, Southampton, UK
Japan and Korea

AN EQUINOX BOOK
Copyright © Andromeda Oxford Limited 1993

Planned and produced by
Andromeda Oxford Limited
9-15 The Vineyard, Abingdon
Oxfordshire, England OX14 3PX

Published in the United States of America by
Oxford University Press, Inc.,
200 Madison Avenue,
New York, N.Y. 10016

Oxford is a registered trademark of
Oxford University Press

Library of Congress
Cataloging-in-Publication Data

Planet management / edited by Michael Williams.
      p.   cm.
"Planned and produced by Andromeda (Oxford)
Limited"--T.p. verso.
Includes bibliographical references and index.
ISBN 0-19-520945-1
1. Environmental protection.   2. Human ecology.
I. Williams, Michael. 1935 June 06-   II. Andromeda
Oxford Ltd.
TD170.P57   1992
363.7--dc20
                                         92-34497
                                         CIP

Volume editor	Victoria Egan
Assistant editors	James Bates, Lauren Bourque
Designers	Isobel Gillan, Chris Munday
Cartographic manager	Olive Pearson
Cartographic editor	Clare Cuthbertson
Picture research managers	Alison Renney, Leanda Shrimpton
Picture researcher	David Pratt
Editorial director	Graham Bateman
Project editor	Susan Kennedy
Art editor	Steve McCurdy

ISBN 0-19-520945-1

Printing (last digit): 9 8 7 6 5 4 3 2 1

Printed in Singapore by C.S. Graphics Ltd

INTRODUCTORY PHOTOGRAPHS
Half title: *Tree planting, Niger (Panos pictures/Ron Giling)*
Half title verso: *Antarctica's fragile wilderness (OSF/Kim Westerskov)*
Title page: *Dump site, New Mexico (Rob Badger)*
This page: *Rainforest destruction, Borneo (Magnum Photos/Peter Marlow)*

Contents

PREFACE

WHEN HUMAN POPULATION NUMBERS WERE LOW AND PEOPLE HAD little energy at their disposal except their own muscles and those of animals, they lived in some sort of harmony with nature. Of course, whenever they cultivated the soil, cut down the forest for fuel and timber or dammed a river for irrigation, they were altering the environment for their use, but the impact was usually minimal and the scars soon healed. With industrialization, and the application of technology and science to farming and manufacturing processes, we acquired the ability to lift ourselves out of a dependence on the organic world with its limits to production. By the beginning of the 20th century, domination of the environment and the pursuit of progress had become goals for all nations as they strove to emulate the countries of Europe and North America.

However, since 1945 the global population has risen from 2.5 to over 5.5 billion today, and a doubling of that number by 2020 is confidently forecast. Awareness that increasing population and production levels were depleting resources and damaging the environment began to grow as river quality deteriorated, urban air became unbreathable, valued landscapes were destroyed and the application of agrochemicals caused the disappearance of familiar animals, insects and flowers. By the mid 1960s a new environmental lobby was making itself felt, arguing that government legislation was necessary to stop pollution, control production and protect species and landscapes, even if it meant reduced or no further increase in material well-being. In the developing world the attempt to raise standards of living through rapid industrialization put a huge strain on the environment.

Later, environmental concern took on a new dimension with rising awareness that problems are not confined within national boundaries but are global. The seas in one part of the globe are affected by pollution in another, and greenhouse gas emissions may influence climates, sea levels and the quality of life everywhere. The environment is now everyone's business as we recognize the interdependence of the world's economic and natural systems.

This book sets out to show how the major environmental issues of dwindling resources, pollution, waste disposal, habitat and species destruction, and climate change have come to the attention of the world, and what management strategies are being employed in the attempt to halt, minimize and rectify the worst excesses of environmental degradation by political parties, action groups, national governments and international organizations. The regional studies show that not all environmental problems affect all people equally, and that there are different degrees of willingness and ability to tackle them.

Dr Michael Williams
UNIVERSITY OF OXFORD

People power Protesters against toxic waste incineration, California

Responding to a crisis Gulf War oil pollution, Kuwait (*overleaf*)

MANAGING THE WORLD'S ENVIRONMENTS

The Gathering Pace of Change

E VER SINCE THE START OF HUMAN HISTORY, people have been changing their environment, making use of what was locally available in nature to provide food and shelter, clothing, tools and other possessions. At first, however, their impact was small and localized. The small groups of nomadic hunter–gatherers who inhabited the Earth more than a million years ago caused only minor disturbances to animal populations and the vegetation. Change has come in the last 10,000 years – gradually at first, and at an ever accelerating pace in modern times – as a result of technological advances in three main areas of human activity: agriculture, industry and medicine.

The first major turning point came with the development of settled forms of farming in Southeast Asia, China, the Middle East and Central and South America from about 8000 BC. Large areas of natural vegetation were lost as land was plowed up for crops or used to graze livestock; forests were cut down for timber and fuel. Hillsides were terraced to conserve soil moisture, and water was diverted from rivers to irrigate croplands.

These innovations produced food surpluses that sustained the rise of urban societies. As agricultural techniques improved over the next thousands of years, world populations slowly increased and the human impact on the environment spread and intensified. After 1500 AD European colonization brought new areas of land into cultivation throughout North and South America, Australia and New Zealand. The area of land given over to crops doubled from 265 million ha (655 million acres) in 1700 to 537 million ha (1,327 million acres) in 1850.

Industrialization, starting in western Europe about the mid 18th century and spreading from there around the world, brought environmental change on an unprecedented and ever-rising scale. The Earth's mineral resources – fossil fuels and metal ores – were dug out of the ground and transported by sea and rail for processing into energy or manufactured products.

Industry and manufacturing created unwanted byproducts, many of them toxic, which were discharged into the air or rivers, or dumped on the ground. Urban populations rapidly expanded, leading to sprawling, pollution-choked conurbations. Agricultural methods became more intensive and environmentally damaging in order to grow more and cheaper food. Increasing use of motor vehicles raised levels of pollutants in the atmosphere.

The population factor

The environmental impacts of modern advances in medicine have been as far-reaching as those brought about by the

Highway construction (*left*) in the United States. The rise of the automobile has transformed the 20th century environment. Large areas of land in town and country have been taken for roads and parking, and levels of air pollution have soared, especially in the cities.

The impact of farming (*below*) A hillside in East Africa has been cleared of forest for cultivation; grazing livestock on the valley pastures prevents trees and shrubs regenerating. Over the centuries, farming has played a crucial role in shaping our "natural" landscapes.

agricultural and industrial revolutions. Control of epidemic disease, more widely available medical treatment and improved nutrition have reduced mortality rates and increased life expectancies in all societies. Between 1950 and 1990, the world population more than doubled from 2.5 billion to 5.3 billion; it is expected to reach 8 billion by 2020 and possibly 9–10 billion by the end of the 21st century. The greatest population increases will take place in the developing world, which is expected to contain five-sixths of the world population by 2020.

It is the affluent, high energy-using industrial nations that produce the bulk of the world's pollutants and wastes – those that contribute to increasing levels of carbon dioxide in the atmosphere and cause acid rain. On the other hand, the quest for food and fuelwood in the countries of the developing world leads to cultivation of marginal land and deforestation, causing land erosion and degradation. As the scale of human impact on the environment grows ever greater, the need for fairer and better management of the Earth's resources grows ever more urgent.

Controlling Nature

THOUSANDS OF YEARS AGO PEOPLE OBSERVED that some wild plants produced larger grains and withstood drought and wind stress better than others. This led them to select certain varieties to sow for next year's food supply, giving rise to the domestication of wheat and other cereal crops. Similarly, the observation that certain crops grew better on land that was inundated each year with river floodwater led to early experiments in irrigation techniques. The basic human need to increase and maintain supplies of food and water has been the principal motivation throughout history for discovering new ways of controlling nature.

At first results were achieved through trial and error, and passed on as traditional folk wisdom. As scientific understanding and engineering technology grew during the 18th and 19th centuries, intervention became more systematic and largescale. Nevertheless, relatively low-tech changes to the environment – such as the terracing of sloping land – have had enormous cumulative effect on the landscape over the centuries.

Water management is one of the most widespread and most important ways by which humans alter nature. Water can be diverted from rivers and lakes by dams, irrigation channels and canals, and also pumped from natural underground reserves (aquifers). In drier areas, water management mainly aims to increase water supplies and reduce waste; in low-lying wet areas the goal is to remove excess water by drainage, and so increase land for cultivation and development.

Most of the crops and domestic animals that feed us today bear little resemblance to their ancestors. Careful selection of

Burning the savannas (*above*) Hundreds of thousands of years ago hunter–gatherers used fire to improve pastures for game; it is still used by pastoral farmers to encourage a new growth of grass.

A string of lakes (*right*) fills the valley behind a hydroelectric dam on the Waitiki river in South Island, New Zealand. Some 310,000 sq km (120,000 sq mi) of land worldwide has been flooded by dam building.

desirable traits (such as size, strength, hardiness or yield) has changed them. After World War II, selective breeding gave way to scientific cross-breeding to produce hybrid varieties that combine the best features of the parent stock.

Artificial pesticides and herbicides have been developed to deal with unwanted plant pests and weeds. Natural predators are also controlled through the deliberate introduction of disease or other predators. The viral disease myxomatosis, for example, decimated rabbit populations in the 1950s, and the minute insect *cactoblastus* is used to devour the prickly pear cactus, a pest to farmers in Australia. Genetic engineering should theoretically give humans more control over other species than they have ever had before. Experiments producing pest or disease resistant plants and higher yielding livestock are already well advanced.

Knock-on effects

Collections of animals and plants that interact with each other and their non-living environment (such as climate, soils and water) form communities that are known as ecosystems. These may range from a rock pool to a tropical rainforest, but whatever their size their interdependence means that alteration of one element in the ecosystem has a knock-on effect on all the others. Human interven-

DAMS – A RISKY FORM OF MANAGEMENT

One of the simplest methods of water management is the building of a dam, or barrier, across a river to control the flow of floodwaters and create a holding lake from which water can be diverted for irrigation and other purposes. The Egyptians are known to have made use of this technique about 5,000 years ago. Advances in civil engineering in the 19th and early 20th centuries allowed bigger and bigger dams to be built, often linked to hydroelectricity projects, such as the Hoover Dam on the Colorado river, completed in 1936.

In the 1960s a number of developing countries embarked on major dam-building schemes, undertaken to enhance national prestige as well as to increase the supply of water for irrigation and provide power for industry. It is now recognized that the environmental costs of schemes such as the Aswan Dam on the river Nile in Egypt and the Akosombo Dam on the Volta in Ghana have outweighed their gains.

The adverse effects of building large dams are many. The flooding of vast areas of land destroys wildlife and displaces thousands of people. The reservoirs behind the dams silt up, decreasing the river's load of nutrients and making land downstream less fertile; the faster river flows increase the rate of coastal erosion, and valuable delta wetlands are lost. Despite these and other known risks, dam-building schemes continue. More than 2 million people will be displaced if the Three Gorges Dam on the Chang river in China – planned as the world's biggest dam – goes ahead.

tion can have disastrous effects on the delicate balance of ecosystems.

Higher yielding rice varieties – developed in the 1960s to boost agricultural production in developing countries – drain the soil of nutrients, forcing farmers to use expensive fertilizers. Synthetic nonspecific insecticides such as DDT (now banned in most countries) leave indestructible residues that work their way up through the food chain, killing fish, birds and other predators. Pest control of one species can result in an epidemic of another by destroying its natural predator. These and many other examples are a sobering reminder that the forces of nature are frequently stronger than our attempts to control them.

The Rise of Environmental Concern

Hugging the trees (*above*) The Chipko movement in India began as a nonviolent resistance movement of Himalayan villagers – many of them women – who joined together to halt the government-aided logging schemes that were threatening their way of life.

Close encounter (*right*) In a protest against whaling, Greenpeace protesters sail their inflatable boat close to the bows of a Japanese factory ship in the southern Pacific. Such direct action has been successful in influencing governments to change their policies.

DIFFERENT CULTURES AROUND THE WORLD have different concepts of their relationship with nature. Western philosophies have traditionally stressed that humans are separate from and superior to their natural environment. This attitude has been criticized as an arrogant misunderstanding of the interdependence of people and nature, leading to mutual ill-effects. Some of the worst environmental damage in the westernized world has occurred under Marxist political regimes in which responsibility for the environment has been sacrificed in order to attain production goals at any cost.

Many economically primitive societies have a more harmonious view of nature. Australian Aborigines, the Inuit of the Arctic and other hunter–gatherers including many African peoples see themselves as part of the land and its produce, all of which is sacred and has to be cared for. Some non-European cultures such as Buddhism or Taoism emphasize noninterference with the natural world and favor the idea of the unity of humans and nature. Nevertheless, the stresses of population growth and the desire to develop economically mean that Asian countries are as liable as Western ones to damage their environment through activities such as deforestation or overgrazing.

The modern movement

In 1864 the American scientist, politician and diplomat George Perkins Marsh (1801–82) published *Man and Nature*, in which he challenged the prevailing view in Western societies that the Earth provides unlimited resources for human growth and expansion. Much of what he said was ignored or forgotten until the 1960s when the modern environmental movement began. In 1962, the publication of another book, *Silent Spring*, by Rachel Carson, revealed the extent of the damage caused to the environment by increasing and indiscriminate use of pesticides and herbicides.

The book aroused unprecedented international concern. It appeared at a time when public anxiety was mounting over the environmental dangers of nuclear bomb testing, when pollution of the seas and marine life by oil tankers was first becoming an issue, and when there was increasing debate about the relationship between rising population numbers, the exhaustion of nonrenewable natural resources and the continued health of the environment. The airing of these and other issues in the 1960s changed the environmental argument, and the nature of conservationist lobbying, for all time.

In the century or so between the publication of Marsh's and Carson's books, many pressure groups had been formed to protect specific landscape features in national parks or particular animal or plant species and their environments. But after the 1960s policymakers, scientists and environmental campaigners mounted a series of more concerted efforts to study the effects of human activity on the planet. The United Nations Conference on the Human Environment held in Stockholm in 1972 resulted in the creation of the United Nations Environmental Program (UNEP) to monitor global change, and gave rise to many national environmental agencies.

Many activist and pressure groups, such as the Sierra Club in the United States, Greenpeace and Friends of the Earth, put pressure on politicians to include environmental protection in their legislation. The 1980s saw Green parties in one western European country after another achieve a proportion of the popular vote (ranging from 1 percent in Britain in 1983 to 5.6 percent in the Netherlands in 1984) in national elections, forcing the major parties to incorporate environmental measures in their own political programs. Environmental groups, too, had a role in bringing about the political changes that led to the demise of the ruling communist parties in the Soviet Union and Eastern Europe.

GREENPEACE – PUTTING THE MESSAGE ACROSS

Greenpeace was formed in 1971 by American and Canadian environmentalists who wanted to stop the United States carrying out atomic tests in the Aleutian Islands off the Alaska Peninsula. It was the most influential of a number of direct action environmental groups that came into being in the 1970s and 1980s, born of frustration at the slow rate of change that was being achieved through legislation by conventional political means.

Now a worldwide organization (it had formed groups in the Soviet Union even before its dissolution in 1990), Greenpeace has been criticized for being too militant in its actions, putting its campaigners into potentially dangerous situations and directly confronting those it opposes, such as whalers or seal-cullers. However, its officials describe its work as nonviolent passive resistance, and it skillfully uses the media to focus public attention onto environmental issues. The sinking of the Greenpeace boat *Rainbow Warrior* (which was attempting to stop French atomic tests in Mururoa Atoll in French Polynesia) in Auckland Harbor, New Zealand, in 1985 received widespread condemnation and drew world attention to its activities.

Greenpeace has been responsible for some notable successes. By publicizing information about the dumping of radioactive wastes and toxic wastes at sea, it has been influential in forcing governments to change their policies. It has played a major role in halting commercial whaling and seal hunting.

Dwindling Resources

Natural resources are anything that we obtain from the Earth's physical environment. Some, such as soils, water and vegetation, are renewable: if properly cared for, they will last forever. But others – material resources such as fossil fuels (coal, gas, and oil) and metal and mineral ores – are nonrenewable: once supplies are used up, they cannot be replaced.

There is a direct relationship between nonrenewable resource consumption and per capita income. Between 60 and 80 percent of the world's supplies of aluminum, copper, lead, nickel, tin and zinc, and 55 percent of its iron ore, are consumed by only nine countries in the developed world: the United States, Japan, the former Soviet Union, Britain, Canada, Italy, France, Germany and South Korea. The same countries consume 65 percent of world production of coal, gas and oil; if all Europe, Australia and South Africa are included, the figure rises to 77 percent. The only country in the less-developed world that consistently enters the "Top Ten" of resource consumption is China, because of its sheer size.

In 1988, taking the current rate of consumption as a guide, predictions were made about how long the known recoverable reserves of these resources would last. The most abundant material, aluminum, had a predicted life of 224 years, followed by iron ore with 167 years, nickel with 65 years and copper with 41 years. Lead, mercury, tin and zinc were due to run out within 21 to 22 years.

Fears that resource scarcity will limit future economic development have – so far – usually been unfounded. Scarcity of a particular commodity in a free market causes its price to rise, and it may therefore be replaced by a less expensive material, such as aluminum for iron or plastic for wood. Resources may be conserved by more efficient use: car engines may be designed to consume less fuel, buildings insulated to conserve energy, and scrap metal recycled. Finally, new reserves may be opened up.

The most accessible, and therefore the cheapest, reserves of any resource will always be used first. As supplies become scarcer and prices rise, it becomes economically viable to tap reserves that were previously thought too expensive or too marginal to exploit. Rich countries will usually pay whatever prices are necessary to secure supplies. The losers will be the poorer countries, who will not be able to afford these prices. Ultimately, however, the world will have to face the fact that some resources have been exhausted or are too expensive to exploit.

The loss of the forests

Wood is a resource that is renewable provided it is not used up more quickly than it can be replaced. In the northern hemisphere, with careful management, the quick-growing coniferous forests that supply softwood are more or less self-sustaining: extraction rarely exceeds growth. But the slow-maturing tropical hardwood forests of Southeast Asia and Brazil are being rapidly depleted – the rate of clearance far outstrips the ability of the forest to regenerate.

The countries where these forests are located are mostly poor and have soaring population growth. Much logging takes place so that the timber can be sold to

A vast woodpile (*above*) in Xinjiang autonomous region in the far west of China, where wood is the principal domestic fuel. Severe land degradation results from the widespread felling of trees.

A global problem (*right*) The areas of acute fuelwood shortage extend right around the world. Commercial logging is often made the culprit for deforestation, but half the trees felled each year are burned as fuel.

wealthy nations to earn cash to buy imports, but even more of the forests are being cleared to create land to grow food. In this case, timber is regarded not as a resource but as an obstacle and is burnt on the spot. Projections suggest that Southeast Asia will be logged out by 2020.

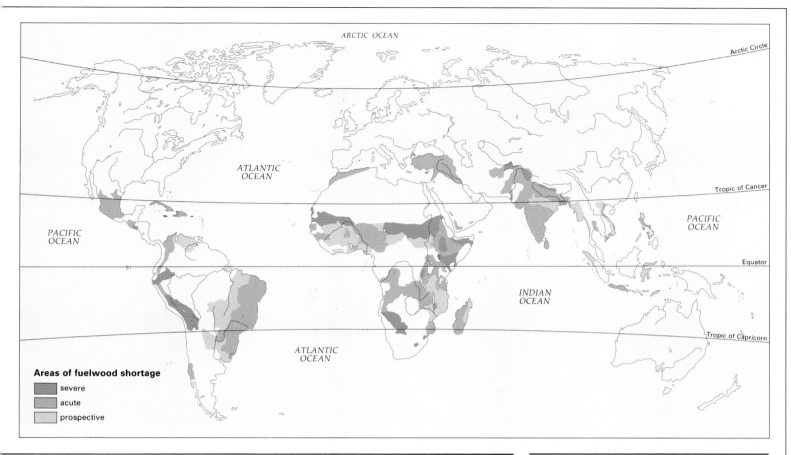

Areas of fuelwood shortage

- severe
- acute
- prospective

THE FUELWOOD CRISIS IN THE THIRD WORLD

In many countries, very large numbers of people – probably about 2 billion in all – rely on wood as a fuel for warmth and cooking. This demand has produced an acute scarcity of fuelwood in parts of Andean South America, the Caribbean islands, most of the Indian subcontinent (particularly Nepal), and Africa. Here the situation is particularly alarming – in some places wood provides up to 90 percent of all energy requirements and forested areas, particularly around the cities, are being stripped of their trees to provide it. Up to half the family budget may be spent on fuelwood, and up to four-fifths of the working year taken up with scouring the surrounding countryside for wood or woody fiber.

In 1990 just over half of all wood extracted from the world's forests was used for fuelwood. The demand had doubled over the previous 20 years, and with the predicted increase in population is likely to double again by 2020. Community forestry or tree-planting schemes have had local success in restoring the loss in some countries. However, these schemes barely scratch the surface of the problem, and the need for sustainable forestry or alternative energy sources in poorer nations is now of global importance.

Acquired value Uranium, being mined here at Rössing, Namibia, had no particular use until scientists in the 1930s, exploring nuclear fission, discovered its potential as an energy source, and its value soared.

Pollution in the Atmosphere

EVERY YEAR THOUSANDS OF MILLIONS OF tonnes of pollutants enter the atmosphere. Some air pollution is caused naturally. Eroded soil may be whipped up by strong winds into the atmosphere to create dust storms, and huge clouds of gas and particles of dust are released in volcanic explosions. But it is the growth in human industrial activity that has caused the massive increase in the scale, intensity and variety of air pollution over the last 200 years.

Almost 99.9 percent of the atmosphere is made up of two stable, nonreactive gases: nitrogen and oxygen. It is the complex chemical reactions that take place when other gases and particulate matter (droplets of liquids or tiny particles of solids such as soot, ash or metals) are released into the atmosphere that cause air pollution. The changes that occur to them once they are in the atmosphere depend on climatic factors such as temperature, wind, cloud and solar radiation, as well as the other gases that are present.

Some gases emitted into the atmosphere (for example, carbon dioxide and carbon monoxide) are known as "greenhouse gases" – that is, they act like glass in a greenhouse, contributing to global warming by trapping heat radiated from the surface of the Earth that would otherwise be radiated back into space. Other gases (such as sulfur dioxide and nitrogen oxide) cause acid rain. All of these gases – released when fossil fuels are burned in power plants, factories and domestic fires – are the largest source of air pollution. Another major source, particularly in urban areas, are motor vehicle exhausts, which contain high levels of carbon monoxide, carbon dioxide, hydrocarbons and nitrogen oxide. In addition, car exhausts emit particulate matter such as lead, which in high concentrations can kill plants and animals and cause mental retardation in young children.

Choking cities
Many pollutants cause direct damage to human health – various respiratory diseases, eye and skin irritations, nervous disorders and brain damage can all be attributed to high concentrations of gases and particulate matter in the atmosphere. The effects are obviously most severe in congested cities and areas of intense industrial activity where there are inadequate pollution controls.

Most large urban–industrial areas suffer to some degree from smog – the noxious fog that forms when sunlight acts on nitrogen oxides and hydrocarbons (given off in vehicle exhausts) in the lower atmosphere to produce a mixture of gases. The most prominent of these is ozone, which causes eye and lung irritation and damages vegetation. The gases

Going up in smoke More than 240 million tires are discarded every year in the United States. Many end up on dumps such as this one in North Carolina. Fires start easily and can smolder for months, releasing pollutants such as hydrocarbons into the atmosphere – just one environmental hazard of the automobile.

ACID RAIN

When industrial and motor vehicle pollutants, mainly sulfur dioxide, nitrogen oxide and hydrocarbons, react with sunlight and water vapor in the atmosphere they produce sulfuric and nitric acids. These acids descend to the Earth through wet deposition (rain, snow, fog, mist, dew) or dry deposition (in gases and particulate matter). All these forms of deposition come under the general heading of acid rain.

Acid rain has far-reaching, damaging effects. It raises the acid levels in lakes and rivers, killing fish and other aquatic life. It reduces soil fertility and releases heavy metals such as aluminum, copper and cadmium into the soil, which hinder the growth of trees and crop plants. It also attacks buildings and can corrode railroad tracks.

Acid concentrations are particularly high over northern Europe, eastern Canada and the northeastern United States, but are found as far afield as China and South America, where it is destroying forests and lakes. Because the sources of pollution are often many thousands of miles away, acid rain has become a contentious political issue. It is estimated, for example, that 20,000 lakes in Sweden are acidified, but most of the pollution comes from elsewhere in Europe. In 1984 Sweden and Norway formed the "30 Percent Club", which aimed to reduce emissions by at least that figure by 1993. Although many countries joined, four of the largest producers of acid rain – the United States, Britain, Spain and Poland – refused to do so.

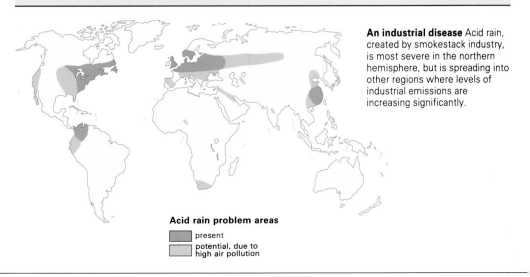

An industrial disease Acid rain, created by smokestack industry, is most severe in the northern hemisphere, but is spreading into other regions where levels of industrial emissions are increasing significantly.

Acid rain problem areas
- present
- potential, due to high air pollution

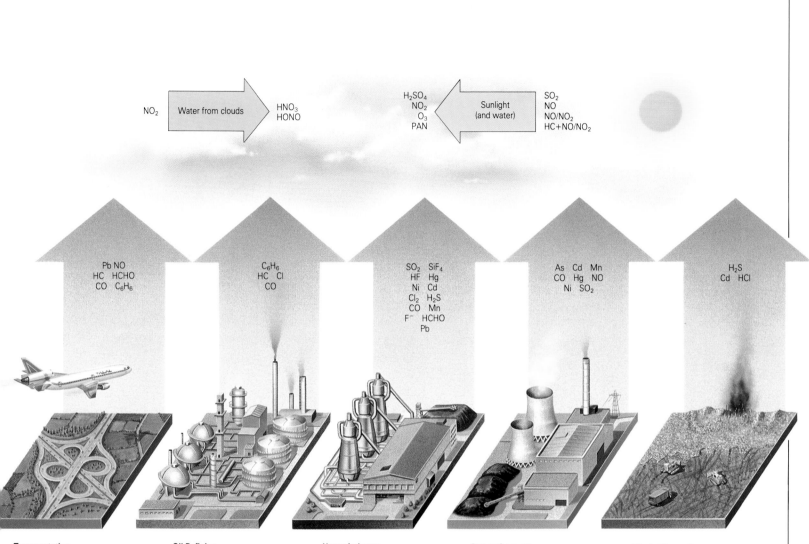

Transportation · **Oil Refining** · **Heavy Industry** (smelting/chemicals) · **Power Generation** (coal- and oil-fired) · **Waste Disposal** (incinerators, sewerage plants)

Principal sources of air pollution Modern industrial processes release primary pollutants – gases such as carbon monoxide (CO); sulfur dioxide (SO_2) and nitric oxide (NO); volatile compounds such as hydrocarbons (HCs); and suspended particulate matter such as lead (Pb), arsenic (As) and cadmium (Cd) – into the atmosphere. Many of these react with sunlight and water vapor to create secondary pollutants such as ozone (O_3) and sulfuric acid (H_2SO_4). It can be very difficult to trace chemical releases to a particular source, so that persistent offenders have often escaped detection.

build up during the day, but if the temperature drops at night, the air nearer the ground cools more rapidly than the air above, and the layer of warm air prevents the smog from dispersing. Wind eventually clears the smog, but in cities such as Los Angeles, Mexico City and São Paulo, which are surrounded by mountains, the lethal cocktail of gases can be trapped for months on end.

The frequency and severity of smog thus depends on a combination of factors including local climate and topography. Its incidence can be reduced by limiting vehicle use and controlling industrial emissions, particularly from the burning of fossil fuels.

AIR POLLUTANTS THAT DAMAGE HUMAN HEALTH

Pollutant	Effects
As (arsenic)	Lung and skin cancer
C_6H_6 (benzene)	Leukemia
Cd (cadmium)	Kidney and lung damage; bone weakening
Cl_2 (chlorine)	Irritation of mucous membranes of mouth and nose
CO (carbon monoxide)	Reduces absorption by body of oxygen; heart damage
F^- (fluoride ion)	Children's teeth become mottled
HCHO (formaldehyde)	Irritation of eyes and nose
HCl (hydrogen chloride)	Irritation of eyes and lungs
HF (hydrogen fluoride)	Irritation of eyes, skin and mucous membranes of mouth and nose
Hg (mercury)	Nervous disorders
HNO_3 (nitric acid)*	Respiratory diseases
HONO (nitrous acid)*	Respiratory diseases
H_2S (hydrogen sulfide)	Nausea; irritation of eyes
H_2SO_4 (sulfuric acid)*	Respiratory diseases
Mn (manganese)	Contributes to Parkinson's disease
Ni (nickel)	Lung cancer
NO (nitric oxide)	Easily forms NO_2
NO_2 (nitrogen dioxide)*	Bronchitis; reduces resistance to influenza
O_3 (ozone)	Irritation of eyes; aggravates asthma
PAN (peroxyacetyl nitrate)	Irritation of eyes; aggravates asthma
Pb (lead)	Brain damage; high blood pressure; growth impairment
SiF_4 (silicon tetrafluoride)	Irritation of lungs
SO_2 (sulfur dioxide)	Hinders breathing; irritation of eyes

(* Substances produced in the atmosphere from pollutants released at ground level.)

Heating Up the Earth

ENERGY FROM THE SUN IS ABSORBED BY THE Earth as heat and is then radiated back again. However, most of the heat given off by the Earth is prevented from escaping into space by carbon dioxide and other gases such as carbon monoxide and methane in the lower atmosphere. These allow sunlight to enter but prevent some of the heat from escaping. Without the shield provided by these "greenhouse gases", it is thought that the average world temperature would be $-18°C$.

Carbon dioxide accumulates naturally in the atmosphere, but over the last 100 years concentrations have increased dramatically from 250 ppm (parts per million) to 370 ppm, mostly as a result of the burning of fossil fuels for energy and the largescale destruction of the tropical rainforests: vasts amounts of carbon dioxide, carbon monoxide and methane are given off when forests are cleared and burned for cultivation. During the same period of time, average global surface temperatures have increased by 0.5°C. Most scientists now agree that these two changes are linked.

If present trends continue, carbon dioxide levels could reach 500 ppm by the middle of the 21st century. Scientists predict that this could raise temperatures by between 3°C and 5.5°C. The Earth has not experienced such a change since the ending of the last ice age between 18,000 and 10,000 years ago, when temperatures rose by 5°C. But the latest predicted rise would take place 10 to 100 times more quickly than that.

The effects on sea and land

It has been suggested that if this happens the polar ice caps and glaciers will melt, causing the sea level to rise by between 0.8 and 1.8 m (2.6 and 6.0 ft) flooding some of the world's most densely populated low-lying areas. Global warming could also change ocean currents, causing stormier seas and increased rainfall or flooding in some regions. The degree of warming is expected to be far greater near the poles than at the Equator. This would have profound consequences for the world's ecology, changing tundra to forest, for example, provided animal and plant life could adapt and shift quickly enough. Many of today's most productive farmlands could be affected by drought, though increased rainfall closer to the Equator could improve growing conditions in the tropics.

These scenarios are conjectural – there are many unknown factors in possible climate change, and climatologists disagree about all the consequences. The role of the oceans is not fully understood. It is known that the seas absorb up to 50 percent of the carbon dioxide created by humans, and this may help to counteract global warming. A warming sea might encourage greater phytoplankton growth, which would trap and absorb even more carbon dioxide.

The need for action

Although there is disagreement about the consequences of global warming, there is a consensus that the emission of greenhouse gases must be slowed down. Achieving it is more difficult.

Developed countries – which produce three-quarters of the world's pollution – have the technical and financial resources to adopt energy-saving measures and clean up industrial emissions. But the consequences will be higher prices and a lower standard of living, and politicians hesitate to take stringent action. In contrast, developing nations, faced with huge population growth, cannot afford to slow down or postpone industrialization, and lack the resources to adopt clean technology. They need to earn currency from their forests, and cannot afford largescale replanting schemes.

Global warming is a global political issue. The multinational agreement signed at the Rio Earth Summit in 1992 to reduce greenhouse gas emissions by the end of the century to the level that they were in 1990 may not go far enough, but is an indication that the world is beginning to take action.

Industrial laboratory (*above*) Oil and coal-burning industries are major contributors to global warming. One way that industrialized countries can help reduce it is by taxing fossil fuels and using the revenues to develop "cleaner" energy sources.

The ozone hole (*below*) On this satellite-generated map the hole in the ozone layer above Antarctica shows as a purple oval. The hole has grown steadily since it was first detected in 1979.

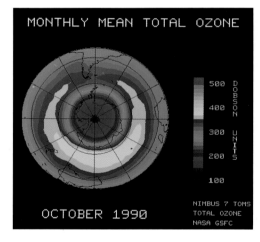

THE OZONE LAYER

The ozone layer lies about 15–20 km (9–12 mi) above the Earth's surface in the stratosphere. It protects the Earth from harmful ultraviolet (UV) radiation, particularly the UV-B radiation that can cause sunburn, skin cancer and eye cataracts.

Ozone is destroyed by chlorofluorocarbons (CFCs) released into the atmosphere from aerosol sprays, refrigerator coolants and plastic foams such as hamburger cartons. Methane and nitrous oxide, byproducts of agriculture, also contribute to the depletion of the ozone layer. In 1984 scientists discovered a massive hole in the ozone layer over Antarctica. Although some scientists have argued that this is a natural phenomenon, most now agree that it has been exacerbated by human interference.

This view is being reinforced by the strong probability that another hole is opening up in the atmosphere above the industrialized countries of the northern hemisphere, from New England to north central Europe. As a result, the production of CFCs has been banned in most developed countries. However, many Third World countries continue to rely on CFCs for cheap refrigerator production.

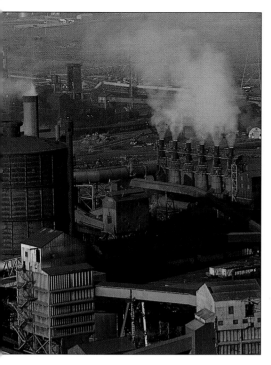

THE GREENHOUSE GASES

Greenhouse gas	Total contribution to global warming (%)	Sources and contribution to global warming (%)	Lifespan (years)
Carbon dioxide (CO_2)	50	Energy from fossil fuels (35) Deforestation (10) Agriculture (3) Industry (2)	500
Methane (CH_4)	16	Energy from fossil fuels (4) Deforestation (4) Agriculture (8)	7–10
Nitrous oxide (N_2O)	6	Energy from fossil fuels (4) Agriculture (2)	140–190
Chlorofluorocarbons (CFCs)	20	Industry (20)	65–110
Ozone (O_3)	8	Energy from fossil fuels (6) Industry (2)	Hours to a few days

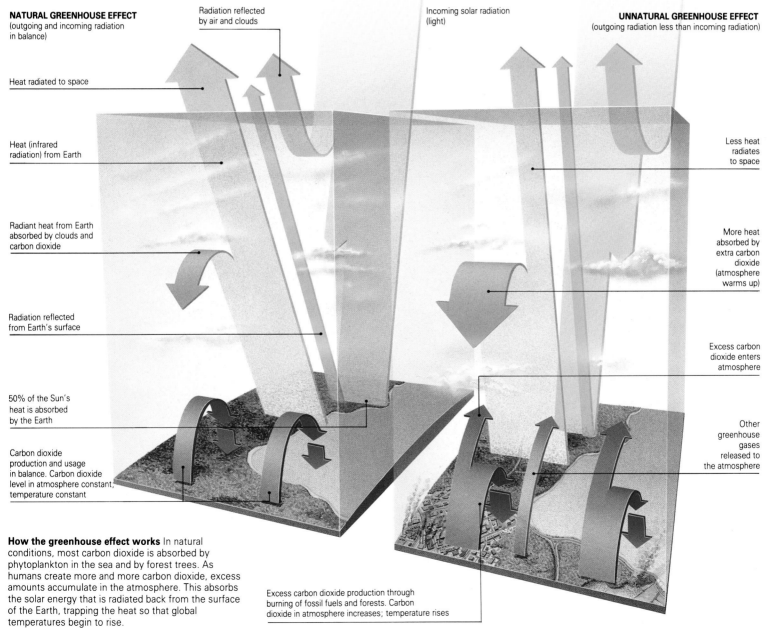

NATURAL GREENHOUSE EFFECT
(outgoing and incoming radiation in balance)

Radiation reflected by air and clouds

Incoming solar radiation (light)

UNNATURAL GREENHOUSE EFFECT
(outgoing radiation less than incoming radiation)

Heat radiated to space

Heat (infrared radiation) from Earth

Radiant heat from Earth absorbed by clouds and carbon dioxide

Radiation reflected from Earth's surface

50% of the Sun's heat is absorbed by the Earth

Carbon dioxide production and usage in balance. Carbon dioxide level in atmosphere constant; temperature constant

Less heat radiates to space

More heat absorbed by extra carbon dioxide (atmosphere warms up)

Excess carbon dioxide enters atmosphere

Other greenhouse gases released to the atmosphere

How the greenhouse effect works In natural conditions, most carbon dioxide is absorbed by phytoplankton in the sea and by forest trees. As humans create more and more carbon dioxide, excess amounts accumulate in the atmosphere. This absorbs the solar energy that is radiated back from the surface of the Earth, trapping the heat so that global temperatures begin to rise.

Excess carbon dioxide production through burning of fossil fuels and forests. Carbon dioxide in atmosphere increases; temperature rises

The Problems of Clean Water Supply

WATER IS THE MOST POORLY MANAGED OF all the Earth's resources, yet it is essential to all living organisms and a vital commodity for agriculture and industry. Although water covers 71 percent of the Earth's surface, only a tiny proportion of this (just over 0.25 percent) is available for human use as fresh water from lakes and rivers. Fresh water is constantly renewed by rain and snow, but it is distributed unequally around the globe. In deserts and semiarid areas, where there is no surface water, supplies have to be pumped up from natural reservoirs, or aquifers, under the ground. Even in humid regions of the world, however, droughts can occur and water supplies falter.

Access to a water supply is only part of the problem – there is also the question of water quality. Normally, organic substances such as decaying vegetation, and animal and human excreta, degrade in water, but if excess amounts accumulate then eutrophication (overenrichment by nutrients and minerals) occurs. This encourages algal growth, which reduces the water's oxygen content and kills fish and other aquatic life.

Organic contamination of the water supply can also be triggered either by deforestation, which causes nutrient-rich silts to accumulate in rivers, or by over-irrigation, which increases salt levels in the water. Chemical pesticides and herbicides, used in intensive agriculture, seep into rivers, killing fish and aquatic life. Nitrogen, phosphate and potassium from chemical fertilizers drain off from farmland to cause algal blooms and eutrophication in rivers and lakes, and increased levels of all these substances affect the quality of drinking water.

Industry is a major source of water pollution. Most countries have strict bans against the direct discharge of industrial wastes into streams and rivers, but controls are not always observed. Moreover, toxic industrial wastes are frequently buried in landfill sites. Chemicals seep

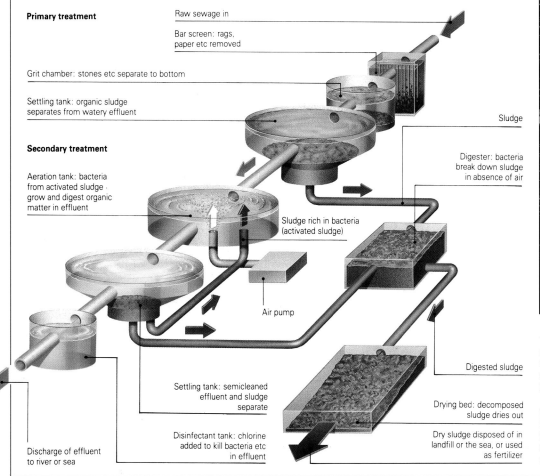

Primary treatment

Raw sewage in

Bar screen: rags, paper etc removed

Grit chamber: stones etc separate to bottom

Settling tank: organic sludge separates from watery effluent

Sludge

Secondary treatment

Aeration tank: bacteria from activated sludge grow and digest organic matter in effluent

Digester: bacteria break down sludge in absence of air

Sludge rich in bacteria (activated sludge)

Air pump

Settling tank: semicleaned effluent and sludge separate

Digested sludge

Drying bed: decomposed sludge dries out

Disinfectant tank: chlorine added to kill bacteria etc in effluent

Dry sludge disposed of in landfill or the sea, or used as fertilizer

Discharge of effluent to river or sea

The detritus of modern city life (*above*) forms a thick scum on the surface of the river at Long Beach, California. Garbage such as this kills aquatic life, destroys coastal wetland habitats and turns resort beaches into ugly eyesores and potential health risks.

Sewage treatment (*left*) Solid wastes are separated from liquid effluent in filtering and settling tanks. The former are broken down by bacteria and dried to form sludge. The latter is disinfected to remove bacteria and discharged into waterways.

chain from minute aquatic invertebrates to fish, birds, and eventually humans. The World Health Organization (WHO) estimates that about 1 million people are poisoned by pesticides every year, causing up to 20,000 deaths.

Water that contains large amounts of human excreta (common in developing countries, where sewage is untreated before being discharged) carries a high risk of contamination from life-threatening micro-organisms, such as those responsible for cholera, typhoid and dysentery. As many as a quarter of urban dwellers in the developing world have an unsatisfactory water source, and 35 percent an inadequate sanitation service. The situation in rural areas is much worse.

The buildup of chemicals and heavy metals in drinking water can be highly injurious. Many pollutants are known to cause cancers and tumors, for example, though pollution is rarely, if ever, given as a cause of death. With over 1,000 new chemical substances entering the market annually, there is cause for concern.

Problems of water pollution are most acute in the newly industrializing and less-developed countries of the world, particularly in areas that rely on underground water reserves, since – without atmospheric oxygen to purify the supply – the water quality progressively deteriorates unless the source of pollution is eradicated. Whatever form the supply takes, however, pollution is much better tackled at source rather than controlled by building purification plants or increasing water supplies, as these merely put extra pressure on an already scarce resource.

through the soil into groundwater and so travel underground to drain into rivers, often many miles away, before eventually entering drinking supplies. Toxic heavy metals such as lead or cadmium enter rivers in the same way and accumulate in sediments on the bottom, where they kill fish and other aquatic life. The muds cannot be dredged and spread over the land as they would then pollute drinking water sources and food crops.

The risks to health

Pollution of the water supply affects the health of all the living organisms – plants, animals and humans – that depend on it. Pesticides, which concentrate in the tissue of living organisms, pass up the food

Waste and Waste Disposal

TODAY'S MODERN CONSUMER SOCIETIES PRO-
duce vast amounts of waste. Products
are packaged in convenient and attractive
throwaway materials; when they break
they are discarded and replaced rather
than repaired. Most domestic waste consists of paper, glass, plastics, metals and
food matter.

The rate at which domestic waste is
produced corresponds directly with standards of living. Each citizen of New York,
for example, throws away on average 1.8
kg (4 lbs) of waste every day, while at the
other end of the scale, each citizen of
Calcutta discards 0.5 kg (1 lb). Multiply
these daily figures to give an annual total
for the whole population and New York is
seen to generate over 8 million tonnes of
waste each year, Calcutta nearly 3 million
tonnes. A billion people live in cities of
over 100,000 people, producing more than
340 million tonnes of waste annually. The
United States alone is responsible for
more than half of this.

Domestic waste is rarely toxic (capable
of causing death or serious injury) but is

Time bomb (*above*) Most of the world's toxic wastes
are dumped in landfill sites, such as this one in Britain.
These are covered with soil, with longterm risks of
chemical seepage and buildup of gases. Housing
developments are often built above landfill sites.

Second time around (*left*) In a salvage yard in Atlanta,
Georgia, metal from discarded household appliances
and used cars is sorted into different types and
compacted before being sent for recycling. The re-use
of scrap metal makes huge energy savings.

frequently hazardous (causing ill health).
Industrial wastes, on the other hand, are
often highly toxic. In the late 1970s, after
the residents of Love Canal in upper
New York State began complaining of
unpleasant smells and skin irritations, it
was revealed that their homes had been
built on top of an old canal excavation
where 50,000 tonnes of toxic chemicals
had been dumped in steel drums between
1942 and 1953. As time passed, the steel
drums corroded, allowing the chemicals

estimated 375 million tonnes of industrial waste that are produced each year (140 million tonnes of them in the United States) increasingly seems the solution. It can reduce the volume of waste by 90 percent, and if the incinerator is equipped with a "scrubber", almost all toxins are destroyed. However, small amounts of toxins and particulate matter are still released into the atmosphere, and the contaminated ash has to be efficiently disposed of in some way.

Waste reduction and recycling

Manufacturing processes in industries such as mining can be altered to reduce the amounts of waste produced. More and more companies will be forced to do this as the costs of disposing of waste products mount. Even greater benefits are to be had from recycling waste. For example, much less energy is needed to re-use scrap metals than to process metals from ore: in the aluminum industry, which requires large amounts of electricity for smelting, energy savings of as much as 96 percent can be made through recycling waste metal.

Recycling also cuts down air and water pollution and reduces the need for scarce landfill sites. It is particularly suitable for domestic waste. All paper and glass, and nearly all metal and plastics, can be recycled. Food waste can be processed to make humus, fertilizer or animal feed; and rotting vegetation used to produce methane (biogas) for heating or power.

to seep into the water, soil and building foundations of the surrounding area. Health surveys later revealed an unusually high incidence of cancer, miscarriages, and birth defects among local residents, who had to be relocated.

Over 90 percent of the world's waste is dumped in landfills – holes in the ground that are eventually covered over with soil. This was long considered a safe means of waste disposal, as it was thought that natural processes would disperse and dilute the waste, eventually rendering it harmless. It took the Love Canal episode, and others like it, to show that over a period of time chemicals create leachate, a concentrated toxic liquid that percolates into groundwater.

Dumping at sea has similar drawbacks. Waste containers can disintegrate or be washed ashore, polluting marine life and thus the food chain. Incineration of the

NUCLEAR WASTE

By far the most difficult disposal problem of all is what to do with the waste from nuclear power production. Nuclear waste disposal carries special risks and responsibilities because of the extreme and protracted dangers to the environment and all plant and animal life from exposure to radioactivity. Low-level wastes have a relatively short half-life (the amount of time it takes for radioactivity to subside to a nondangerous level), but high-level wastes have a half-life of at least 10,000 years.

The spent fuel rods that are periodically withdrawn from the reactor core are counted as high-level wastes. The cumulative amount of spent fuel in the world is already 90,900 tonnes, not counting the unknown amounts produced in the former Soviet Union. Spent fuel is usually reprocessed to

recover a small proportion of the uranium and plutonium for new fuel, but most is discarded as nonreusable waste. Moreover, the liquids that are created through reprocessing increase the amount of radioactive material by nearly 60 times.

In the past, low-level wastes were dumped in steel containers in the sea or stored in landfills, but both these methods have now been banned. All nuclear wastes are currently stored in liquid form in steel tanks protected by concrete or lead, or buried deep underground as solids encased in glass. No method currently used is 100 percent safe, however, and after disasters such as the explosions at the Chernobyl nuclear reactor in 1986, public concern over nuclear power and the disposal of wastes remains high.

Accumulation in the Oceans

OCEANS COVER SOME TWO-THIRDS OF THE Earth's surface. They are of vital importance to the working of the hydrological cycle – the constant circulation of water through the Earth's environments: moisture that is evaporated by sunlight from the surface of seas, lakes and rivers condenses into clouds in the atmosphere, eventually falling back to the Earth's surface as precipitation, to drain into rivers and the sea again. The oceans are also a major source of food, supplying 14 percent of animal protein in the world diet. Yet nowhere is human insensitivity to nature more apparent than in our attitude toward the oceans. They are treated as vast dumping grounds for every kind of human detritus.

All around the world's coasts sewage is discharged into the sea, frequently untreated. Some countries, such as the United States and Britain, use the sea as a place to dispose of excess urban sewage in the form of sludge. In the past, it was felt that the oceans could quite easily dilute and disperse organic wastes, but it is now recognized that too much sludge dumping encourages the growth of algal blooms (eutrophication), which rob the seas of oxygen and kill marine life. Eutrophication is a particular problem in shallow, sheltered coastal waters or lagoons where there is little or no tide to disperse the pollutants. An algal bloom 1,000 km (620 mi) long that formed in the almost enclosed Adriatic Sea off northeastern Italy in the late 1980s was believed to have been caused by nitrate fertilizers draining from farmlands in the fertile north Italian plain and carried via the river Po.

When sewage sludge originates from a heavily industrialized area it often contains toxic heavy metals, such as arsenic, lead and mercury. These can enter the food chain and eventually cause serious harm – and even death – to humans. Hazardous industrial wastes may also be dumped directly into the sea, particularly as landfill sites become scarcer and more expensive. Other sources of pollution include the incineration of toxic wastes at sea, the dumping of low-level radioactive wastes (both of which have now been banned) and oil pollution, either deliberately discharged from tankers at sea, or accidentally spilt during collisions.

Polluted coastlines

Great amounts of pollution enter the sea from industrial plants such as mines, pulp and paper mills and oil refineries located in estuaries and around the coasts. Mud spoil dredged from harbors and estuaries (usually to be dumped at sea) is frequently high in heavy metals and other toxins. Because tides wash material discharged into the sea back toward the land, marine pollution is highest along the world's shallow continental shelves, where the greatest quantities of fish are to be found, and along the shoreline.

Coastal wetlands, such as tidal marshes and mangroves, are particularly vulnerable to pollution. They are among the world's most productive wildlife habitats. Beaches, too, support a wide diversity of natural life and are commonly used by

Fire at sea (*right*) An incinerator ship burns toxic waste at sea. Although this disposal method reduces contamination risks on land, pollutants in the soot and smoke fall into the sea, where they add to the cocktail of substances already there.

Areas of marine pollution (*below*) Tides and currents push marine pollution toward the shore to mix with the pollution discharged directly from the land. A permanent slick of oil is left by tankers using the world's major shipping routes.

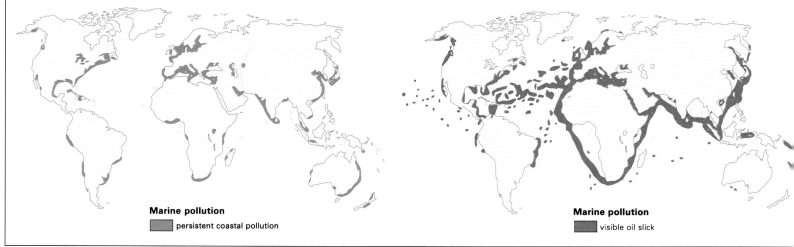

Marine pollution
█ persistent coastal pollution

Marine pollution
█ visible oil slick

Green slime (*above*) washed up along Italy's Adriatic coast is the stinking remains of an algal bloom caused by fertilizers draining from farmland in the Po valley. A series of blooms such as this in the 1980s damaged the local tourist and fishing industries.

Sources of oil pollution in the sea

Other

From rivers

Leakage from offshore oil rigs

Coastal refineries

Runoff from towns and cities

Industrial wastes

Natural discharges

Bilge and fuel oils

From atmosphere; originally from industry and motor vehicles

Tanker accidents

Dumped municipal waste

Operational discharges from oil tankers, eg washing out oil tanks

people for recreation. Unacceptably high levels of sewage and other pollution pose a direct threat to health, as well as being aesthetically displeasing. Many once-popular resort beaches have now been declared unfit for bathing.

Many shallow, enclosed areas of water near heavily populated areas, such as the Inland Sea in Japan, Chesapeake Bay in the eastern United States and the Baltic and North Seas in northern Europe, are showing marked deterioration in quality. The discovery of significant numbers of fish and crustaceans with liver cancers, ulcers and finrot is normally the first sign

that pollution is exceeding the sea's ca-pacity to absorb it. Cooperative agree-ments between states bordering sensitive areas – such as the 1990 Nicosia Charter, dealing with the Mediterranean – have led to new legislation that ensures the monitoring and regulation of discharges into shallow marine waters, but these efforts may be too little and too late for some seas and their marine life.

Sources of oil pollution (*right*) Surprisingly, tanker accidents account for only an eighth of marine oil pollution. By far the largest amounts are discharged by tankers operating at sea or are released from dumped municipal wastes.

Deforestation and Erosion

L ARGE TRACTS OF TEMPERATE DECIDUOUS forests once covered much of the northern hemisphere. Over hundreds of years they have been felled to provide land for farming, fuelwood and timber. Traditional farming practices did not degrade the humus-rich forest soils, and seminatural countryside has often been left, supporting the original species. If clearing stops the forests are able to grow back to something like their previous state. The same is not true of tropical forests. Tropical soils are not naturally fertile – the forests' fertility is provided by nutrients that are locked up in the vegetation and recycled through decay and regrowth. Once a patch of forest is cleared the cycle is broken; the land quickly becomes sterile and vulnerable to erosion.

Once felled, the tropical forests may be lost for ever.

The destruction of the tropical forests of South America, Southeast Asia and Central Africa may be the most serious environmental issue facing the world today. They play a crucial role in the hydrological cycle – their dense vegetation controls runoff after tropical rainstorms by holding and gradually releasing water like a sponge. Once a forest is cleared, there is a greater danger of rivers drying up or of flash floods occurring. Moisture is released into the air through transpiration and evaporation from the leaves' surface. If large areas of forest are felled, precipitation declines and droughts can occur. On a world scale, destruction of the rainforests contributes to global warming.

ATLANTIC OCEAN

PACIFIC OCEAN

Catastrophic deforestation
(*left*) The hillsides of Haiti, in the Caribbean, have been stripped of their once abundant tree cover. Today severe soil erosion and land degradation add to the massive problems that this impoverished country is having to face.

Erosion and desertification
Deforestation is also one of the major causes of soil erosion. Although soil erosion occurs naturally when topsoil is carried away by water, wind, ice or landslides, the process is greatly accelerated by the removal of the natural vegetation. Other human activities that contribute to erosion are the overgrazing of grasslands; the overplowing of cultivated land, which breaks up the soil structure, allowing it to blow away; and badly managed irrigation systems that can result in salinization and waterlogging, both of which ultimately reduce plant cover.

The consequences of soil erosion are various. Gullying occurs – stripped of the vegetation that hold them together, the soils are washed away by rainfall, and deep valleys are formed, dramatically reducing the area of land for cultivation. The eroded soil silts up rivers, waterways, dams and irrigation channels, adding to problems of water supply. Deprived of topsoil, the land becomes less productive and ultimately crops fail.

In arid and semiarid areas soil erosion can result in desertification. It occurs in

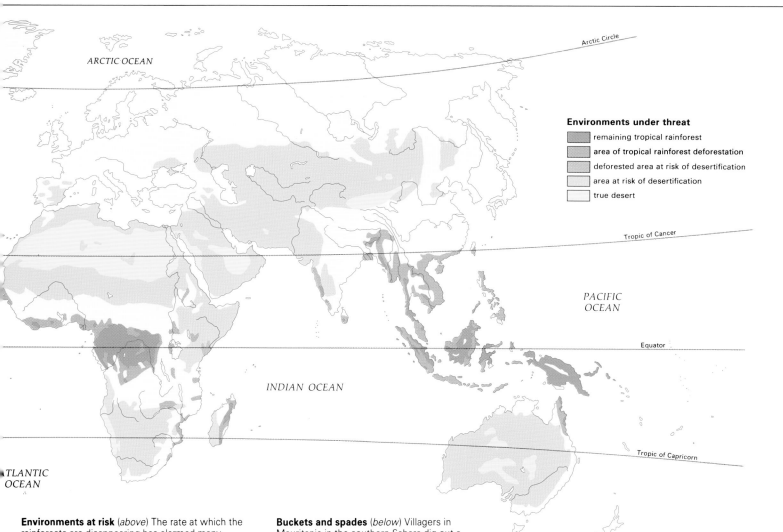

ARCTIC OCEAN

Arctic Circle

PACIFIC OCEAN

Tropic of Cancer

Equator

INDIAN OCEAN

Tropic of Capricorn

ATLANTIC OCEAN

Environments at risk (*above*) The rate at which the rainforests are disappearing has alarmed many environmentalists. Desertification affects even larger areas of the world, with consequent loss of productive land and widespread risk of famine.

Buckets and spades (*below*) Villagers in Mauritania in the southern Sahara dig out a house engulfed by sand. Overgrazing helps to loosen light soils, and windstorms blow them until they pile up round an obstruction.

marginal areas of land where cultivation and grazing rangelands merge into each other: light eroded soils are blown by the wind to form dunes that swallow up productive land. Population pressures – and the need to grow more food – lead to overgrazing of pastureland and the shortening and abandonment of fallow periods in crop cultivation. This increases pressure on already fragile lands, pushing them toward desertification.

Areas of particular concern around the world are the broad belt across the southern fringes of the Sahara from west to east Africa, western and central Asia, and considerable parts of the western United States and Australia. Desertification is also taking place in areas that are not on the fringes of deserts, such as parts of Mediterranean Europe. The United Nations Environmental Program (UNEP) estimates that 48 million sq km (18.6 million sq mi) – one-third of the world's surface – are affected to some degree by desertification. It predicts that if desertification is allowed to continue at the present rate, it could threaten the livelihoods of at least 850 million people.

Upsetting the Balance

Frrom the very beginning of the Earth's history, climatic or geological changes have altered its physical landscapes, and the plant and animal species that inhabit them have evolved, adapted or become extinct. In the comparatively short time that humans have inhabited the Earth they have contributed to this process of change by exploiting the natural resources available to them. However, over the past 300 years more species have become extinct than ever before, and surviving wilderness areas are dwindling to an alarming extent.

Swamps are drained, forests cleared and grasslands plowed up for land to grow more crops and graze more animals for food. Urban and industrial growth swallows up wildlife habitats beneath buildings, airports, pipelines and roads, and pollutes fragile ecosystems such as tundra or wetlands. The worldwide expansion of the tourist and recreation industries brings the impact of human activities to previously untouched areas. Hundreds of rare – and not so rare – species face extinction if their natural habitats are destroyed any further.

Humans have always hunted animals for their flesh, fur, feathers and other products. Many – such as the Passenger pigeon and the quagga (a type of zebra) – are now extinct as a result. Although hundreds of species are threatened by overhunting, it is the plight of the big mammals such as the rhinoceros, elephant, tiger and whale that arouses most public concern. Whales, for example, have been hunted for centuries for their meat, blubber and oil, but the introduction of factory ships after 1914 led to their slaughter in untold millions. The world stock of 2.2 million whales at the beginning of the century had been reduced to about 625,000 by the 1970s, and some species, such as the Blue and Humpback, have all but disappeared.

The decline of one species can have dramatic effects on others. For example, when wolves were eliminated from the forest of the Kaibab Plateau in Arizona, the deer population surged so much that eventually they overgrazed their habitat. They became diseased through overcrowding and the population crashed. Wolves were later reintroduced. The introduction of alien species of plants and animals also upsets nature's balance. Throughout the Americas, Australia and New Zealand, the introduction of horses,

cattle and sheep, to say nothing of pests, weeds and diseases, has produced numerous booms and crashes in the populations of indigenous species.

Why conserve?

There are compelling reasons to try to prevent or slow down the human destruction of other species and their habitats. Human interference with the complex and delicate balance of ecosystems often has unforeseen and unwelcome results. We do not yet know the potential value

Birds for sale (*right*) Gray-headed parakeets on a market stall in Bangkok, Thailand. Efforts by organizations such as CITES have failed to end the international trade in live animals and animal products – bringing some species close to extinction.

Massive interference (*below*) Napalm is used to clear native bush in New Zealand for replanting with Monterey pine. These fast-growing trees are a valuable source of timber, but biodiversity is lost through large planting schemes of exotic species

for food and medicine of countless plants and animals that seem unimportant to us now: by maintaining the diversity of species we are passing on a priceless resource to future generations. There is above all a strong ethical argument for conservation: to destroy in perhaps 20 years a species that has taken 20 million to evolve cannot be acceptable.

Since 1950 the number of protected areas has risen from about 500 (totaling 80 million ha/198 million acres) to 3,510 (totaling 428 million ha/1,056 million acres) worldwide. However, protected areas still account for only 3.2 percent of the Earth's surface. While the developed countries can afford the luxury of allowing land to remain unexploited for farming or industry, poorer countries cannot when the bulk of their populations are struggling to make a living and produce enough food. Under such circumstances the arguments for conservation are more

difficult to justify, though increased food aid or financial incentives from the richer countries not to develop wild areas could be part of the solution.

Land for leisure Rainforest in Bali, Indonesia, has been cleared to make way for a new golf course. The greens will be planted with alien grasses, and the large amounts of fertilizers needed to maintain them will be washed off by the rain into watercourses.

GENETIC ENGINEERING

DNA (deoxyribonucleic acid) is the genetic material present in the nucleus of every living cell that determines heredity. The discovery in 1973 of recombinant DNA (DNA with new genetic characteristics that has been produced under laboratory conditions) opened up vast new possibilities for human control over the evolution of other species. Genetic engineering, as this branch of science is called, enables scientists to select and cross-fertilize desired traits within the embryo, thus altering the genetic characteristics of a species and eliminating the chance mutations of traditional breeding.

Genetic engineering is still largely experimental and remains highly controversial. Its proponents point out that it has the potential to solve the world's food problems by producing higher-yielding varieties of crops and livestock, as well as drought or disease resistant strains. However, critics fear that the deliberate or accidental release of genetically engineered plants or animals into the wild could have unpredictable and possibly disastrous environmental consequences. The reduction of natural genetic variations within species could eventually lead to a loss of species diversity. Whatever view they hold, most people are agreed that genetic engineering is now too advanced to be brought to a halt, but that strict controls are necessary.

Alternative Sources of Energy

I**N RECENT DECADES, THE GOVERNMENTS OF** most developed countries have sought to reduce their reliance on expensive and polluting fossil fuels and seek alternative, more efficient sources of energy. After World War II the development of nuclear technology seemed to hold out the best prospect of supplying cheap, sustainable energy. It was especially attractive to countries such as France that lacked substantial indigenous coal reserves or great potential for hydroelectricity. However, although nuclear energy accounts for about 18 percent of world electricity production, the expected massive expansion of nuclear power has not taken place. One reason is the mounting cost of implementing such a program. Concern over safety and the problems of disposing of nuclear waste have also played a part.

The Earth's forces

Increased attention is now being focused on harnessing the Earth's natural energy sources: flowing water, tides, wind, the Sun and heat from the Earth's molten core (geothermal energy). Hydroelectricity – utilizing the power from falling or fast-flowing rivers to turn turbines to generate electricity – is the most widely used of these alternative energies, supplying over 20 percent of the world's electricity. Most of the hydroelectric capacity of Europe and North America is already being exploited, but it remains largely underdeveloped in other parts of the world. However, large plants are in operation in Brazil, China and India. Largescale schemes are efficient and relatively cheap to operate, but these gains need to be balanced against their longterm environmental disadvantages.

The daily movement of tides is a source of energy that can be harnessed by building a barrage across a bay or estuary and allowing the water level behind it to rise at high tide and fall through turbines at low tide. At present the world's only commercial tidal power station is at La Rance in Brittany, France, where tides rise and fall up to 13 m (44 ft); the lack of suitable sites for tidal power stations means that it is only ever likely to have local significance. Wave power seems to be more promising. Small experimental plants have been built in several countries such as Britain, Japan, Norway and the United States. These differ in design, but basically work by using air that has

Building for the future (*below*) The NMB Bank building in Amsterdam, Netherlands, is probably the most energy-efficient building in Europe. Solar panels generate heat and power, and modern insulating materials have been used to reduce energy loss.

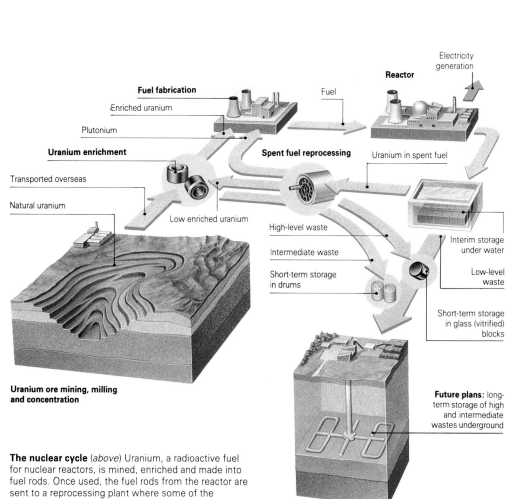

The nuclear cycle (*above*) Uranium, a radioactive fuel for nuclear reactors, is mined, enriched and made into fuel rods. Once used, the fuel rods from the reactor are sent to a reprocessing plant where some of the uranium is recovered.

Smallscale hydroelectric projects (*left*), such as this one in Sri Lanka, exploit fast-flowing mountain streams to generate electricity in rural areas. The only drawback is that the amount of electricity produced depends on the seasonal flow of the stream.

been compressed in chambers by wave action to drive turbines.

Wind power is a clean, relatively cheap and efficient (though noisy) source of energy. However, it is only feasible in certain areas: the turbines need average wind speeds of 22–38 km per hour (14–24 mi per hour). These conditions are found in California and Denmark, which have both become centers for wind power. By 1988 wind turbines provided 15 percent of San Francisco's electricity. Around the world some 60,000 wind-powered turbines generate electricity, and many more are planned for the future.

There are several methods of trapping and using solar energy; one of the most popular is roof-mounted solar panels that collect solar energy as heat and transfer it to a fluid that drives a domestic generator. A promising advance converts sunlight directly into electricity via photovoltaic cells – thin strips of silicon and trace amounts of other substances that produce an electrical current when struck

by sunlight. The principal disadvantage is that many parts of the world do not receive enough sunlight, and the energy supply is not available at night or in cloudy weather.

In areas of volcanic activity, hot water and gases rising to the Earth's surface can be used as a local source of energy. Geothermal energy heats almost all the houses and greenhouses in Iceland's capital, Reykjavik, and a dry-steam well near Larderello, central Italy, has been producing electricity since the early 20th century. In other places, attempts are being made to exploit geothermal energy by drilling deep holes into igneous rocks. Water is then pumped down, and returned hot to the surface.

All these alternative energy sources have the advantages of being renewable, relatively cheap and less damaging to the environment than conventional methods. However, because sunlight, wind and river flows vary, the energy supply is not always constant and may be used only as a supplementary source. Many experts consequently consider that the longterm solution to the energy crisis lies with using all forms of energy more efficiently and cutting down on waste.

Combating Pollution

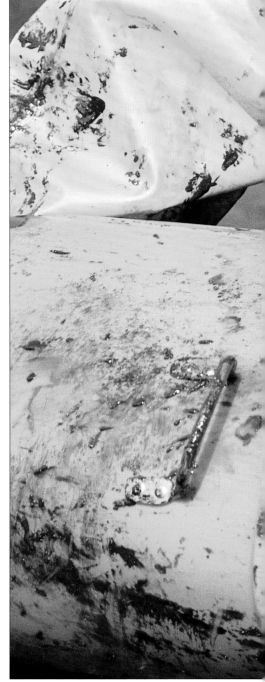

Aır, water and land pollution can be tackled in a number of ways. Regulations can be imposed to reduce industrial emissions. Industry can be required to introduce energy-saving measures and to become more efficient in its manufacturing and transportation processes. Reductions can be made in the use of fertilizers and pesticides. Greater efforts can be made to create plastics that are less chemically complex and can be recycled without potential hazard.

Most industrialized countries have now passed their own laws to curb emissions. Regulations tend to be of two types: ambient quality control, which establishes maximum concentrations of air or water pollution, and technology control. Ambient control is more flexible and can be circumvented. For example, a coal- or gas-burning factory may build very tall smokestacks to reduce the levels of air pollution in the immediate vicinity, but these merely send the pollution higher into the atmosphere and distribute it farther away in downwind areas. Technology controls are more rigid and are easier to police. For example, companies that fail to install scrubbers that remove sulfur dioxide and suspended particulate matter from flue gases can be fined.

Most countries have adopted a combination of these approaches, though the United States and Britain have tended to rely solely on ambient controls, while Germany and Japan favor technology-based controls. Although these impose higher costs on industry, they are generally felt to be more efficient because they make it possible to identify persistent polluters.

The idea that "the polluter pays" is gaining widespread acceptance: in other words, if oil is spilt from a tanker or toxic gases escape from a chemical factory, then it should be the responsibility of the operating company, not the affected community, to repair the damage. After oil from the *Exxon Valdez* tanker contaminated beaches and marine life in Alaska in 1989, the oil company was fined and ordered to pay for the cleanup.

A growing body of opinion maintains that this is the only way to reduce pollution significantly, because the cost of polluting becomes so irksome that producers and consumers are forced to switch to cleaner alternatives. This is already beginning to happen in countries that have introduced cheaper prices for unleaded gasoline and made catalytic converters – devices that chemically reduce emission levels of harmful exhaust gases – compulsory in all new vehicles.

Efficiency measures

The production of one tonne of steel in China and India creates 6 to 8 times the amount of pollution as in Japan, with its more advanced technology. Most experts would argue that the best way to bring down levels of pollution in the longterm is through the introduction of energy-saving measures and the adoption of cleaner fuels. Adopting more environmentally-friendly processes, however, usually involves a trade-off. The new, cleaner and more efficient steel-making technology is expensive; if it is bought then something else must be sacrificed. The costs of installing new technology

Operation cleanup (*above*) A municipal worker scoops up oil from a beach in southwestern England after a tanker collision at sea. Local communities are often left to carry the costs of accidents such as this, but increasingly the "polluter pays" principle puts the burden on the operating company.

Scrubbers (*left*) for cleaning exhaust gases from industrial plant. In the cyclone separator the dust particles are spun from the gases as they spiral downward. The spray tower uses water to wash out dust and dissolve soluble pollutants from the gases. In the electrostatic precipitator, the charged pollutant particles are attracted to collection plates with an opposite charge. All these methods produce hazardous solid waste or sludge that needs to be disposed of safely.

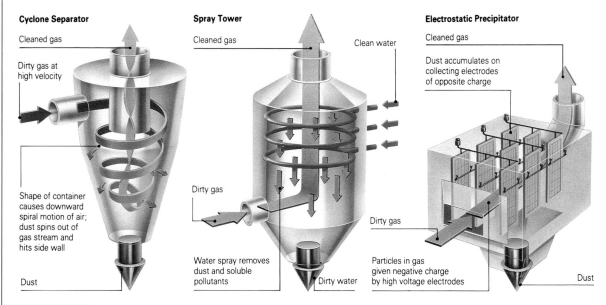

Cyclone Separator
Cleaned gas
Dirty gas at high velocity
Shape of container causes downward spiral motion of air; dust spins out of gas stream and hits side wall
Dust

Spray Tower
Cleaned gas
Clean water
Dirty gas
Water spray removes dust and soluble pollutants
Dirty water

Electrostatic Precipitator
Cleaned gas
Dust accumulates on collecting electrodes of opposite charge
Dirty gas
Particles in gas given negative charge by high voltage electrodes
Dust

CLEANING UP EMISSIONS

Sulfur dioxide, given off when fossil fuels are burned, is a major contributor to acid rain. There are two ways in which levels can be reduced. Sulfur impurities can be removed from coal before it is burned, and desulfurization equipment – or scrubbers – can be installed in factory and power plant smokestacks. These reduce pollution by using liquids or other materials to absorb toxic gases from flue gases to create a solid waste, gypsum, which can then be removed. Scrubbers take out up to 95 percent of sulfur dioxide and over 99 percent of solid particulate matter (though not the most harmful fine particles). They remove only about 30 percent of carbon dioxide.

Electricity production accounts for between 25 and 35 percent of atmospheric pollution worldwide. Germany has installed scrubbers on all its power stations, and if other countries were to follow suit, levels would fall substantially. In the United States, about 200 power stations are responsible for 57 percent of national sulfur dioxide emissions. But scrubbers are extremely expensive. Although countries such as Britain and the United States have justified their resistance to scrubbers on the grounds that the link between sulfur dioxide emissions and acid rain has not been satisfactorily proven, skeptics suggest their resistance is purely financially motivated.

are frequently considered to be too high, especially in developing countries.

The same is true at all levels. If farmers decrease the amounts of fertilizers they use, yields will fall and their livelihood suffer. It is possible to build autos that are far more fuel efficient, and therefore less polluting, than present models, but the trade-off is loss of speed and acceleration. Enormous savings can also be made by reducing the temperatures at which central heating systems are maintained and switching off air conditioning. If pollution is to be seriously reduced, both producers and consumers must reconsider their habits and question whether today's standards of living are being enjoyed at an unacceptably high price to the environment and future generations.

Toward Sustainability

A NEW WORD HAS ENTERED THE ENVIRON-mental debate over recent years: sustainability. It can mean many things according to the point of view of the user, but most would agree that sustainability involves managing all the Earth's resources so that they will continue to be available for present and future generations. This implies that nonrenewable material resources should not be wasted, and that "living" resources such as trees, soil and water should not be exhausted through overexploitation.

It is unrealistic to expect people in less-developed countries not to cut down trees or graze their livestock on marginal land. Another important element in the concept of sustainability is, therefore, the need to tackle the huge disparities in wealth and development that exist between the industrialized and the industrializing world. Only by alleviating hunger and poverty can we hope to achieve a situation in which sustainability could become a reality.

One of the greatest challenges is to find a sustainable agriculture that will continue to allow the world's populations to be fed while reducing pressure on soil and water supplies from overgrazing, deforestation, overirrigation and overuse of chemical fertilizers. Potential solutions involve combining modern technology, which allows more efficient use to be made of resources, with smallscale, traditional farming methods that show an awareness and understanding of local ecological conditions.

Such methods would require massive reinvestment and a fundamental shift in attitudes; farmers would have to switch from largescale monoculture, in which large tracts of land are devoted to growing a single crop, to smaller-scale polyculture – growing mixed crops on much smaller areas of land. Organic fertilizers would replace artificial ones and natural predators would be used to control pests. Sustainability will also only be achieved when an end is put to the practice in many developed countries of giving farmers massive government subsidies that maintain their standard of living but encourage overproduction, create food surpluses and prevent agricultural products from the poorer countries from entering the industrialized market.

Political change needs to go further than this. Many argue that without vastly extended, internationally funded family-planning programs it will be impossible to achieve sustainability, since agricultural productivity will have to continue to keep in step with population growth – at present some 90 million new mouths need feeding every year. In addition, a

Mixing the crops (*right*) In this plantation in Kenya, banana plants are shaded by leucaena trees, which fix nitrogen in their roots, returning fertility to the soil. Their leaves provide fodder for livestock and their branches are harvested for fuelwood.

Hungry mouths to feed (*below*) An overcrowded shanty town in Manila, Philippines. Many experts fear that unless political solutions are found to end Third World poverty and concerted action taken to reduce population growth, sustainability is unachievable.

political solution is desperately needed to the developing world's debt crisis. In 1991, annual interest rates were about $60 billion, though the developing countries only receive about $20 billion in aid. As a result, they are forced to grow nonfood crops such as tea or coffee or overexploit their forests, fisheries or wildlife in order simply to survive.

Can sustainability be achieved?

Individual governments have the power to implement policies such as imposing taxation penalities on companies that fail to implement energy-saving measures or offering economic incentives to "greener" manufacturers or farmers if they so wish. Unfortunately, however, economic or social considerations frequently override goodwill to the environment. Moreover, even if a country does take unilateral action, nature does not respect national boundaries: one country may be affected by another's pollution. So international agreement is vital for sustainability.

In addition, developed countries must broaden their outlook to realize that only by helping to eradicate the problems in developing countries can they help themselves to a healthier environment. Perhaps most importantly of all, the process of education must continue. A better knowledge of the reasons for population control and the potential of alternative energy and new technology in industry and agriculture should help the world to work toward a more sustainable future.

Regreening the Planet

Humans not only have the ability to degrade the environment through inappropriate agricultural and industrial practices that lead to deforestation, desertification, soil loss, salinization and toxification of soil and water. They also have the technology and the knowledge to mitigate or even reverse some of the negative impacts of their activities.

For example, a number of measures can be put in hand to reverse environmental damage on deforested mountain slopes. Check dams built across gullies, terracing on hillsides and the plowing of fields so that the furrows follow the contours of the land rather than run up and down all help to collect rainwater and arrest soil erosion. If high-yielding fodder crops are planted in place of pasture, livestock can be concentrated in areas where they do less damage, and even be stall-fed. Trees planted in areas vulnerable to erosion break the force of rain and bind the soil. These measures not only protect fragile upland watersheds, but also reduce the risk of floods and sediments that clog up watercourses and reservoirs in more populous areas downstream.

In many countries right across the world, including Costa Rica, Kenya, India, Nepal and Thailand, agroforestry (growing trees among crops or on grazing land) is helping to rehabilitate deforested land. By making maximum use of the regional soils and climate, agroforestry brings a number of benefits. The trees help to increase food productivity by restoring nutrients and moisture to the soil and reducing erosion. They yield varied produce such as fuelwood, construction timber, fruit, nuts, organic fertilizers and fodder for livestock. Agroforestry is particularly important in the parts of the world that rely primarily on wood for fuel.

Tree-planting programs have also had some considerable success in rehabilitating land degraded by farming. The soil erosion that created the American "Dust Bowl" of the 1930s was halted through a massive change in agricultural practices

Walls in the desert (*below*) In Burkina, on the southern edge of the Sahara, low walls or *dignettes* are built in intricate patterns. These catch and hold water in the intense downpours of the rainy season, and so prevent runoff from washing away the topsoil.

and the largescale planting of shelter belts. More recently, some 17.6 million ha (43.5 million acres) in the central and northern plains of China have been protected by planting trees as windbreaks. Crop fertility has increased by between 15 and 45 percent as a result.

Regreening programs need not be restricted to rural areas. Old industrial and mining sites despoiled with waste dumps can be transformed through cleansing and replanting. Showpiece examples of what can be achieved through regreening can be found at a number of former iron-smelting sites in western Europe (for example in the Swansea area of south Wales in Britain and in the Ruhr industrial area in western Germany) and in the United States (particularly around Pittsburgh, Pennsylvania). Planting trees and other vegetation in city center parks and gardens is also beneficial, as the plants help to absorb pollutants.

The challenge for the future

Despite these and other initiatives the regreening of the planet is not keeping pace with its degreening. Although reforestation is a well-established practice in the developed world, where over 8 million ha (19.8 million acres) of commercially managed coniferous and deciduous forests are replanted every year in North America, Europe and the former Soviet Union, there is no room for complacency in view of the damage being done to these same forests through acid rain.

The growing realization that forests and other vegetation act as the "lungs" of the planet by absorbing carbon dioxide and releasing oxygen has made the need for renewal and rehabilitation even more pressing. It is estimated that 1 billion ha (2.4 billion acres) of new forests are needed over the next 40 to 50 years to combat the effects of mounting carbon dioxide emissions. The success of community reforestation and agroforestry schemes in many countries suggests that the way forward may be through a large number of small-scale local initiatives. These would help to ease the rate of deforestation in developing countries, encourage wildlife and supplement timber requirements in the developed world.

Replacing the trees A team of local volunteers digs trenches in a hillside in Gujarat, India, preparatory to planting it with tree seedlings. Deforestation has devastated this landscape, and the new trees will help to stabilize the soil.

Global Management

As the world moves toward the 21st century there is increasing awareness of the degree to which the continuing health of the planet depends upon the balanced working of each part of the environment. There is now widespread recognition that the Earth's wildlife and natural resources are not limitless and that the land cannot be exploited with impunity. The challenge for the immediate future is to agree the means and resources to control the pace of environmental change on a global scale.

As a result of the publicizing activities of pressure groups such as Friends of the Earth or Greenpeace, environmental issues are now high on the political agenda of many countries. Most industrialized countries have government departments that regulate pollution controls and have a say in land-use planning. There are also a large number of regional and international organizations that monitor the changes to the environment. For example, the Convention on International Trade in Endangered Species (CITES), the World Wide Fund for Nature (WWF) and the World Conservation Union (IUCN) are concerned with nature and wildlife preservation, while the United Nations Environment Program (UNEP) deals with broader environmental issues. Even some major financial institutions, such as the World Bank, have now set up environmental departments that carefully monitor the effects of their projects.

Almost all the world's governments are now agreed that environmental research is essential to establish more precise links between human activity and environmental change. Analysis of the data collected by satellites is potentially the most important source of information available. In this way scientists can assess the extent of deforestation or desertification, monitor levels of atmospheric pollution, track the movement of oil spills and even detect the onset of drought. At present, however, data is being collected more rapidly than it can be analyzed, and conclusive results are not yet available for environmental trends that could take tens, hundreds or even thousands of years to unfold fully.

Time for action

There is less international consensus when it comes to formulating practical agreements and treaties to control environmental change. Reaching agreement is fraught with difficulties, even within the developed world where the different countries have different environmental goals and economic priorities. The reluctance of both Japan and Norway to halt whaling, despite the ban agreed to by the International Whaling Commission, is just one example.

Reaching international agreements is even harder when the countries involved have very different standards of living. Many developing countries are openly resentful of international pressures to curb development for environmental reasons when the developed countries (which industrialized during an age of scant regard for the environment) are responsible for most of the world's pollution and energy consumption.

Despite the difficulties, an increasing number of major international agreements and protocols on the environment were signed during the 1980s and early 1990s. Although there have been a number of significant dissensions, the international whaling treaty, the Montreal Protocol on the reduction of substances that deplete the ozone layer, and the "Thirty Percent Club" of nations committed to reducing sulfur emissions are examples of what can be achieved through cooperative voluntary action. Similarly, although the first Earth Summit, held in Rio de Janeiro, Brazil, in 1992 failed to live up to the expectations of some, the mere fact that it took place at all was a hopeful sign that the world is beginning to come together to discuss how best we can manage the planet for the future.

Artificial world (*above*) The Biosphere 2 Project is an hermetically sealed enclosure in the Arizona desert containing five self-sufficient ecosystems, planned as a prototype for space colonies. The aim was that volunteers should live there for two years, studying the interaction between humans and the environment in controlled conditions. However, the carbon dioxide level began to rise steadily as soon as the dome's airlocks were closed. Global warming in miniature put the project in jeopardy.

Getting together (*left*) A conference session at the Rio Earth Summit, held under the auspices of the United Nations in 1992. More such meetings between world politicians and environmentalists are needed if internationally agreed targets are to be set for global conservation.

THE RIO EARTH SUMMIT

The United Nations Conference on Environment and Development (the Earth Summit) held in June 1992 was the largest and most ambitious meeting ever held to discuss environmental issues. The Summit had four principal aims: the signing of two treaties, one on global climate change and one on biodiversity; the formulation of a declaration on forest management; and agreeing "Agenda 21", which set out the action necessary for achieving sustainable development.

All these aims were achieved (though the United States refused to sign the biodiversity treaty). Nevertheless, the Summit was accompanied by a widespread sense of frustration that the declaration on forestry and Agenda 21 were not legally binding, and that cautious negotiators had amended many of the most important proposals so much that they lacked real force. Developing nations were particularly disappointed that the developed countries were only prepared to offer an extra $1.8 billion in aid. Despite these shortcomings, the Earth Summit represented a significant step forward. Many observers believed that the proposal to establish a United Nations commission that would monitor how well governments are keeping to the agreed goals would prove to be the Summit's most valuable contribution to future efforts toward global conservation.

REGIONS OF THE WORLD

CANADA AND THE ARCTIC

Canada, Greenland

THE UNITED STATES

United States of America

CENTRAL AMERICA AND THE CARIBBEAN

Antigua and Barbuda, Bahamas, Barbados, Belize, Costa Rica, Cuba, Dominica, Dominican Republic, El Salvador, Grenada, Guatemala, Haiti, Honduras, Jamaica, Mexico, Nicaragua, Panama, St Kitts-Nevis, St Lucia, St Vincent and the Grenadines, Trinidad and Tobago

SOUTH AMERICA

Argentina, Bolivia, Brazil, Chile, Colombia, Ecuador, Guyana, Paraguay, Peru, Uruguay, Surinam, Venezuela

THE NORDIC COUNTRIES

Denmark, Finland, Iceland, Norway, Sweden

THE BRITISH ISLES

Ireland, United Kingdom

FRANCE AND ITS NEIGHBORS

Andorra, France, Monaco

THE LOW COUNTRIES

Belgium, Luxembourg, Netherlands

SPAIN AND PORTUGAL

Portugal, Spain

ITALY AND GREECE

Cyprus, Greece, Italy, Malta, San Marino, Vatican City

CENTRAL EUROPE

Austria, Germany, Liechtenstein, Switzerland

EASTERN EUROPE

Albania, Bosnia and Hercegovina, Bulgaria, Croatia, Czechoslovakia, Hungary, Macedonia, Poland, Romania, Slovenia, Yugoslavia

NORTHERN EURASIA

Armenia, Azerbaijan, Belorussia, Estonia, Georgia, Kazakhstan, Kirghiz, Latvia, Lithuania, Moldavia, Mongolia, Russia, Tadzhikistan, Turkmenistan, Ukraine, Uzbekistan

THE MIDDLE EAST

Afghanistan, Bahrain, Iran, Iraq, Israel, Jordan, Kuwait, Lebanon, Oman, Qatar, Saudi Arabia, Syria, Turkey, United Arab Emirates, Yemen

NORTHERN AFRICA

Algeria, Chad, Djibouti, Egypt, Ethiopia, Libya, Mali, Mauritania, Morocco, Niger, Somalia, Sudan, Tunisia

CENTRAL AFRICA

Benin, Burkina, Burundi, Cameroon, Cape Verde, Central African Republic, Congo, Equatorial Guinea, Gabon, Gambia, Ghana, Guinea, Guinea-Bissau, Ivory Coast, Kenya, Liberia, Nigeria, Rwanda, São Tomé and Príncipe, Senegal, Seychelles, Sierra Leone, Tanzania, Togo, Uganda, Zaire

SOUTHERN AFRICA

Angola, Botswana, Comoros, Lesotho, Madagascar, Malawi, Mauritius, Mozambique, Namibia, South Africa, Swaziland, Zambia, Zimbabwe

THE INDIAN SUBCONTINENT

Bangladesh, Bhutan, India, Maldives, Nepal, Pakistan, Sri Lanka

CHINA AND ITS NEIGHBORS

China, Taiwan

SOUTHEAST ASIA

Brunei, Burma, Cambodia, Indonesia, Laos, Malaysia, Philippines, Singapore, Thailand, Vietnam

JAPAN AND KOREA

Japan, North Korea, South Korea

AUSTRALASIA, OCEANIA AND ANTARCTICA

Antarctica, Australia, Fiji, Kiribati, Nauru, New Zealand, Papua New Guinea, Solomon Islands, Tonga, Tuvalu, Vanuatu, Western Samoa

North America

CANADA AND THE ARCTIC

THE UNITED STATES

CENTRAL AMERICA AND THE CARIBBEAN

SOUTH AMERICA

Central and South America

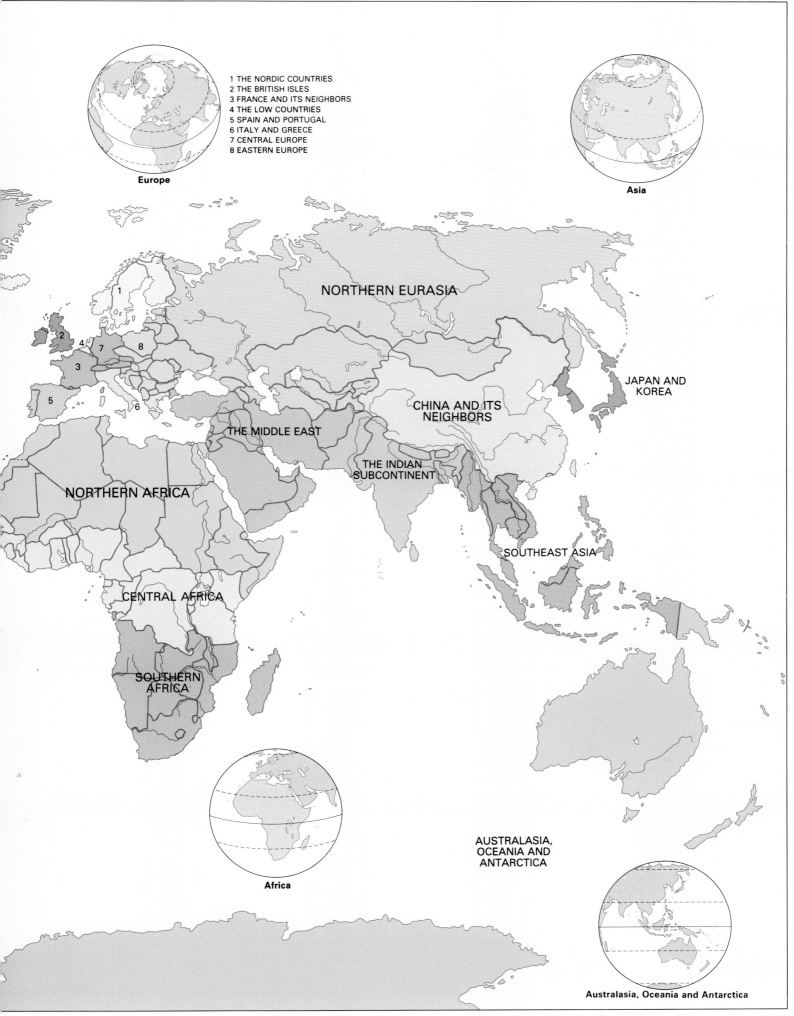

Europe

1 THE NORDIC COUNTRIES
2 THE BRITISH ISLES
3 FRANCE AND ITS NEIGHBORS
4 THE LOW COUNTRIES
5 SPAIN AND PORTUGAL
6 ITALY AND GREECE
7 CENTRAL EUROPE
8 EASTERN EUROPE

Asia

NORTHERN EURASIA

JAPAN AND
KOREA

CHINA AND ITS
NEIGHBORS

THE MIDDLE EAST

THE INDIAN
SUBCONTINENT

NORTHERN AFRICA

SOUTHEAST ASIA

CENTRAL AFRICA

SOUTHERN
AFRICA

Africa

AUSTRALASIA,
OCEANIA AND
ANTARCTICA

Australasia, Oceania and Antarctica

PRESERVING A WILDERNESS

Although Canada's natural resources have only been exploited on a significant scale for about 200 years, the region is not without serious environmental problems. Acid rain, originating from the highly industrialized, urbanized south, has blighted lakes, rivers and forests, while intensive agriculture in the prairie wheat belt has caused severe soil erosion. The vast northern wilderness is vulnerable to soil erosion from logging, and pollution from mining and oil drilling. Wildlife and the lifestyles of the indigenous peoples across the region have been threatened, for example, by huge hydroelectric schemes. However, there is strong local and national environmental awareness. In 1992, the Canadian government announced a 5-year, $3 billion program to clean up and protect the country's air, water and land.

IMPRINTS ON THE LAND

Long before Canada came under the re-shaping hands of the first European settlers in the early 17th century, the region was inhabited by indigenous peoples whose attitudes to the land and its riches were quite different. Native Indians and Inuit lived in harmony with their environment, utilizing its wealth, but not destroying it.

European settlement, located mainly in the south of Canada along what is now the border with the United States, profoundly changed the look of much of the country. It was the railroads that made this possible. In 1885 the 7,000 km (4,350 mi) long Canadian Pacific Railway (CPR) was completed, linking both sides of the continent and opening up the Canadian West to largescale exploitation. By the end of the century, immigrants from Europe were turning the flat prairies of Manitoba, Saskatchewan and Alberta – once home to Native Indians and great herds of buffalo – into vast seas of wheat.

The wheat fields of the prairie provinces have been a huge success in economic terms; in the late 1980s they produced 40 million tonnes of grain annually. But environmentally they have proved a disaster. Soil erosion, loss of soil fertility and the consequent over-reliance on chemical fertilizers and pesticides to maintain or improve productivity – often leading to river and groundwater pollution – are some of the major problems facing Canadian farmers and the environment today.

COUNTRIES IN THE REGION

Canada

POPULATION AND WEALTH

Population (millions)	26.6
Population increase (annual population growth rate, % 1960–90)	1.3
Energy use (gigajoules/person)	291
Real purchasing power (US$/person)	17,680

ENVIRONMENTAL INDICATORS

CO₂ emissions (million tonnes carbon/annum)	120
Municipal waste (kg/person/annum)	630
Nuclear waste (cumulative tonnes heavy metal)	11,000
Artificial fertilizer use (kg/ha/annum)	48
Automobiles (per 1,000 population)	432
Access to safe drinking water (% population)	100

MAJOR ENVIRONMENTAL PROBLEMS AND SOURCES

Air pollution: urban high; acid rain prevalent; high greenhouse gas emissions
River/lake pollution: medium; *sources:* agricultural, sewage, acid deposition
Land pollution: local; *sources:* industrial, urban/household
Waste disposal problems: domestic; industrial; nuclear
Major events: Mississauga (1979), chlorine gas leak during transportation; Saint Basile le Grand (1988), toxic cloud from waste dump fire

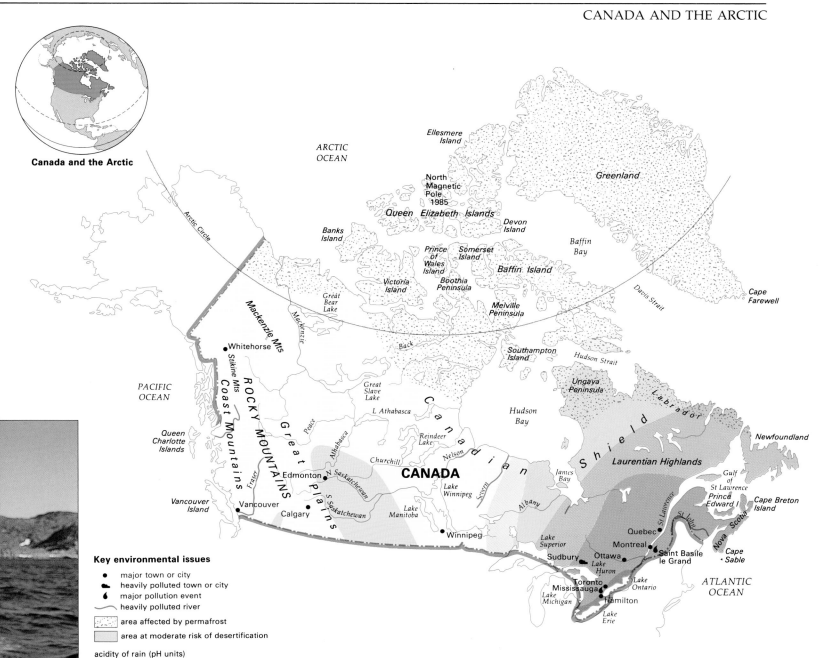

Key environmental issues

- ● major town or city
- ◗ heavily polluted town or city
- ◖ major pollution event
- ∿ heavily polluted river
- ⬚ area affected by permafrost
- ▢ area at moderate risk of desertification

acidity of rain (pH units)

▓	4.2 (most acidic)
▒	4.4
░	4.6
▫	4.8 (least acidic)

Map of environmental problems
(*above*) Global warming threatens the huge permafrost and Arctic areas of the far north, while farther south acid rain is having a major impact on eastern forests, and the western prairie provinces are increasingly at risk of desertification.

Politically or environmentally correct? (*left*) An Inuit fisherman brings in a ringed seal on Cumberland Sound, Northwest Territories. The fight to preserve the Inuits' traditional lifestyle has come into conflict with environmental issues: hunting seals is crucial to the Inuit diet and culture, but the sale of sealskin – a key source of revenue in their marginal economy – has been banned, and the skins must now all go to waste.

Threatened wilderness

Some 60 percent of Canada's 26 million people are concentrated in the corridor of urban development between Windsor, Ontario and Quebec City in the central southeast – only 5 percent of the total land area. This highly industrialized area is the source of considerable pollution of air and water. The country's vast subarctic and arctic areas, by contrast, are still relatively untouched; the isolation, the difficult terrain, mosquitoes and bitterly cold weather have acted as a deterrent to settlement and exploitation.

Nevertheless, though there are few settlements of significant size in the Yukon and Northwest Territories (the largest being Whitehorse in the Yukon, with a population of some 19,000), the area has been increasingly exploited for its mineral resources such as oil, copper, silver, lead, zinc and uranium. Largescale mineral extraction, together with waste disposal, oil spills and noise, are just some of the problems that now pose a serious environmental threat to the unique and fragile habitats of the Canadian Arctic. Global warming, a worldwide problem, is likely to affect the Arctic more seriously than elsewhere.

The original inhabitants of the Arctic, the Inuit, did not have to contend with such problems. As hunters who respected nature they had a strong empathy with their environment. Hunting and fishing provided a subsistence living. In recent decades, however, their traditional way of life has come under increasing pressure. The Inuit population density has always been low – roughly one person to every 1,000 sq km (400 sq mi). According to archaeologists, the number of Inuit living on the south coast of Baffin Island some 4,000 years ago was probably no more than about 100 to 150. By 1984 the figure had risen to just 290.

THREATS TO LAND AND WATER

Although Canada's industry is concentrated in a relatively small area, the country's vast wilderness regions are increasingly vulnerable to wind-blown air pollutants from the industrial south. In the Arctic zone these are increasingly noticeable as a thick haze over the land. In southeastern Canada acid rain or acid snow are common. They form when sulfur dioxide and nitrogen oxides – emitted by ore smelters, power stations and automobiles – react with water vapor in the atmosphere. Among the worst offenders within the region are the smelters of Sudbury, Ontario (the world's largest producer of nickel), which emit huge quantities of sulfur dioxide. Cross-border pollution from the United States exacerbates the situation. Canada now receives some 3.2 million tonnes of pollutants a year from the United States (almost twice as much as the United States receive from Canada).

As a result of this extensive aerial pollution, millions of trees in Canada's eastern forests (including sugar maples, vital to Canada's maple syrup industry) are sick or dying. The leaves or needles of affected trees turn yellow and drop off, while vital plant nutrients are leached from the soil. In addition, more than 14,000 Canadian lakes have now been pronounced "dead" as a result of acid rain; the acidity level of the water is too high to support most forms of aquatic life.

The Great Lakes on Canada's southeastern border have not only been affected by increased acidity; more than 300 toxic chemicals have been recorded in the water. Lake Erie, in particular, has undergone a catastrophic decline in water quality since the 1950s: urban, industrial and agricultural effluents turned parts of it into something approaching the state of an open sewer, and it is only now slowly beginning to recover.

Poisons in the food chain

The Canadian government recognized the grave dangers of waterborne pollution to human health in its "Green Plan", a comprehensive report on the state of the country's environment issued in April 1992: "Fish with tumors and diseases caused by toxins in water, birds with crossed bills and other deformities caused

A GIANT HYDROELECTRIC CONTROVERSY

Hydroelectric power is generally considered to be healthier for the environment than fossil fuel power generated from the burning of coal, oil or gas. However, although HEP produces none of the harmful sulfur emissions of fossil fuels, on a local scale it can be devastating to the environment. The world's largest hydroelectric project at James Bay, in northern Quebec, provides a striking example.

The first stage, completed in 1985, required the building of several dams along the La Grande river, which runs into James Bay. Its effect was to flood 30,000 sq km (11,600 sq mi) of Cree Indian hunting territory. This not only disrupted the Native Indians' tradi-

tional way of life, but also threatened their health: a dramatic rise in mercury levels in the reservoirs (caused by bacteria working on the inundated, mercury-rich vegetation) has poisoned the fish that form a major part of the Indians' diet.

The completed project – which will affect an area the size of France – could increase the salinity of James Bay and threaten its rich wildlife. However, the prospect of further flooding and diversion of rivers led to such strong protests from Native Indians and conservationists alike that in 1991 the second stage of the project was halted by the Canadian government pending an environmental review.

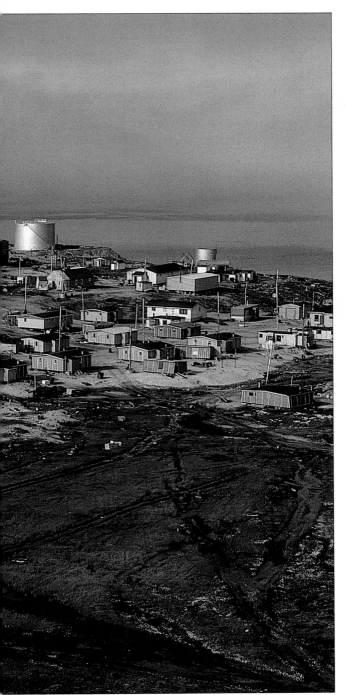

Hungry polar bears (*above*) raid a rubbish dump near an Arctic village. Waste disposal in the cold north is a problem since it cannot be buried and takes a long time to decompose. Dump sites, therefore, provide bears with opportunities to forage for food. The town of Churchill on Hudson Bay even draws tourists to observe the bears raiding the local dump. The bears are often captured and moved away to prevent them from being shot by frightened residents.

Disfiguring the landscape (*left*), a modern village in arctic Quebec surrounded by unsightly tracks in the permafrost. Recovery time from environmental damage in this fragile landscape is painfully slow: vehicle tracks can remain carved into the frozen land for a decade or more after just one vehicle has crossed the winter snow. Arctic tundra plants are extremely sensitive and very slow growing.

by eating contaminated fish, and reproductive failures in mammals feeding on the top predators of the aquatic food web all suggest that human health may be in jeopardy".

Studies of mercury levels in the food chain have highlighted the dangers. Under acidic conditions, bacteria living in the mud at the bottom of lakes can convert naturally occurring mercury into methyl mercury, which is highly toxic. Its noxious effect is magnified as it passes up the food chain. Plankton, for example, may take up only 0.05 parts per million (ppm) in their bodies, but this is concentrated to 0.25 ppm in small fishes and 2.0 ppm in the large fish that feed on them. The uptake of mercury in fish-eating birds and mammals can reach 15.0 ppm.

In humans, high levels can cause liver, kidney or brain damage, and can produce birth defects. In the St Lawrence river, Beluga whales are so contaminated with pollutants that their bodies are regarded as toxic waste.

In the Arctic, pollution in the food chain has reached alarming proportions. High levels of toxic chemicals have been found in all life forms – from zooplankton, through to fish, seals, polar bears and the Inuit themselves, who absorb the chemicals from the animals they eat. The worst contaminants are PCBs (polychlorinated biphenyls) – compounds that are used especially in the manufacture of electrical transformers. Some PCBs are carried by air currents to the polar region from industrialized areas farther south,

while others from marine paint on the hulls of ships pollute the water. In the early 1990s, the breast-milk of Inuit mothers was found to contain five times the level of PCBs found in mothers in southern Canada. The effects include damage to the immune systems of newborn babies, making them vulnerable to disease and infections.

Across northern Canada the land is strewn with hazardous waste from mines and military bases: rusting fuel and chemical drums, dumped into rivers and lakes, stand empty, their contents long since leaked into the water. Rivers downstream of settlements such as Whitehorse are also polluted by untreated sewage.

Perils of development

Massive logging of the great Canadian forests is having a devastating effect on the land. According to the government's Green Plan, the old growth forests of British Columbia in the far west may disappear by 2008. The results will include widespread soil erosion, landslides, the loss of unique animal and plant species, and perhaps a deleterious effect on climate stability.

In the wheatfields of the western prairies, intensive agriculture – including the application of over 1 million tonnes of phosphates and nitrogenous fertilizers on the land annually – has seriously reduced the binding power of the soil. As a result, the prairie soil has lost up to half of its original organic matter, making it vulnerable to erosion by wind and rain. Soil loss costs Canadian farmers between $500 million and $900 million (Canadian) in lost production every year.

ADDRESSING THE PROBLEMS

Since the 1980s Canadian environmental groups have been increasingly vocal in expressing their concern about the country's forests, soil, air and water. The Canadian government's Green Plan took a significant step forward in terms of confronting the issues of environmental deterioration. The government agreed to provide some $3 billion over 5 years in a program to reduce solid waste and eliminate pollutants from the air, land and water; cut back on greenhouse gases and ozone-depleting chemicals; encourage ecologically sound logging, fishing and farming; purchase lands for national parks; and provide for environmental education and research, as well as "outreach" programs to Native Indians and nongovernmental organizations. In addition, in June 1991 the government recognized the importance of preserving Canada's natural heritage when it voted unanimously to protect 12 percent of the country in its pristine state.

Heightened national awareness
A new attitude to the environment can be seen in various areas of development. Typical of the new environmentally sensitive approach to major engineering projects has been the Canadian Pacific Railway route across Rogers Pass in the Rocky Mountains of British Columbia. Completed in 1989, this project was undertaken within the ecologically sensitive Glacier National Park. Before any work was done, there were exhaustive evaluations of the vegetation, fish and other wildlife of the area, the impact of the route on rivers, the possibility of increased erosion and the potential impact of noise and visual disturbance. The effects of construction were monitored throughout, and more than 1 million trees and shrubs were planted after the project was completed.

Replanting is a priority, too, for the British Columbia logging industry. Although the province contains only 2 percent of Canada's forests, it accounts for 25 percent of the country's timber production – much of this from the old growth forests. However, the early period of exploitative clear-cutting without replanting ended over half a century ago. Today British Columbia replants 50 percent of felled sites.

In some areas, progress in environmental protection has only been made very recently and in response to specific threats. In Prince Edward Island, Canada's smallest province, situated in the Gulf of St Lawrence, there is serious concern about chemical contamination of the groundwater, which provides the islanders with their only source of natural drinking water. Most of these chemicals are agricultural pesticides and nitrates, some of them severely toxic, but other contaminants include fuel spilled from underground storage tanks.

To deal with this problem, the island's provincial and local authorities have developed an overall strategy to manage water resources. Future sewage facilities and petroleum storage tanks, for example, may be diverted from groundwater areas likely to be easily contaminated; the storage of hazardous wastes, including materials such as dry-cleaning solvents, is prohibited on the province's territory; and the use of certain agricultural chemicals is restricted. The highly toxic insecticide Aldicarb, widely used on potatoes (Prince Edward Island's main crop), was banned in 1989.

The risks involved in storing toxic waste were belatedly brought home to Canada in August 1988, when a warehouse at Saint Basile le Grand, east of Montreal, in Quebec, caught fire. The warehouse contained 90,900 liters (19,800 gallons) of liquid PCB wastes, as well as other contaminants. The dangerous smoke plume drifted far to the west, and residents living in the area had to be immediately evacuated. Since then, a nationwide inventory of all PCB waste storage sites has been made; stringent and legally enforceable requirements regarding PCB waste storage have been created under the Canadian Environmental Protection Act; and a federal PCB Destruction Program has been established, offering to provide mobile PCB destruction facilities in those areas with large quantities of PCB wastes.

Garbage disposal at source (*left*) An enormous municipal incinerator in Quebec provides an alternative to hauling the city's waste to rural landfill sites for dumping. Correct sorting makes incineration less polluting and more efficient.

Insulated pipes (*right*) in the village of Inuvik, Northwest Territories, ensure that water supplies and sewage disposal pipes do not freeze. The pipes, warmed by radiators, are elevated above the frozen ground to prevent the permafrost from melting.

Protecting the Inuit

Conservation issues are never clear cut, as the debate over the Canadian fur trade has shown. Conservationists from the south regard the sale of sealskins or beaver pelts as an affront to civilized standards, and the slaughter of wildlife as pandering to the luxury market. For Native Indian and Inuit hunters, however, the sale of skins has long been a major source of income. Now, following an international ban on sealskin products, the seals are still being killed for their meat but the skins go to waste. High rates of unemployment exacerbate many of the social problems affecting the Inuit. Such problems are often the result of cultural disruption and displacement.

Critical issues facing Canadian, Greenland and other Arctic Inuit – such as degradation of their land and culture – are the concern of the Inuit Circumpolar Conference, a nongovernmental organization representing the interests of the Inuit peoples. They are also being addressed by the Inuit Regional Conservation Strategy, a project fostering international cooperation on Arctic issues.

KEEPING THE ARCTIC COLD

Major accords such as the Arctic Environmental Protection Strategy cover the development of broad conservation policies for northern Canada, but the day-to-day priorities of those working in the Arctic are more practical. One of the main problems facing construction workers, for example, is not the Arctic cold, as many may think, but the opposite – heat.

The permanently frozen ground (a condition called permafrost) has great strength and resilience when frozen, but becomes weak if it thaws out. Arctic engineers, therefore, concentrate on insulating the ground from any heat generated by the things they build. Roads and runways carry thick gravel pads to prevent their frozen foundations from thawing out and collapsing. Buildings, too, are constructed on gravel foundations or piles. In some cases, heat extractors (giant versions of the domestic refrigerator) are buried on site to keep the permafrost frozen.

Oil pipelines, water supply pipes and sewage disposal pipes, however, have to be kept warm to prevent the contents from freezing solid. They consequently run above the permafrost on stilts, and are fitted with heat radiators and insulated trenching.

Turning on the heat

Scientists working at the tiny settlement of Resolute Bay on Cornwallis Island in the Arctic have found that the ice sheet became thinner by 30 percent between 1976 and 1989. Although local ice-thinning does not reflect global conditions, and computer climate models are imprecise, for more than a decade governments and individuals have been coming to terms with the prospect of "global warming" – a serious rise in the Earth's atmospheric temperature – caused by the build-up of "greenhouse" gases. Methane, CFCs, carbon dioxide (largely emitted as a result of burning fossil fuels for power), and nitrous oxides (from vehicle emissions) warm the atmosphere by trapping the heat that radiates from the Earth's surface.

Forecasts about what will happen as a result of global warming are many and varied, and have been viewed from both scientific and economic perspectives. Some scientists believe that the temperature will rise significantly over the next century, and that this will be greatest in higher latitudes where the increase may be by as much as 6°C or 9°C. This could be accompanied by a rise in sea level over the next century of 50 to 100 cm (20 to 40 in) – though some scientists have forecast a rise of up to 1.8 m (6 ft) – largely due to the melting of glaciers and polar ice.

A rise in the sea level would devastate many low-lying islands and heavily populated delta areas around the world, such as in Bangladesh. The long coast of Canada could also suffer significantly. At the international Earth Summit conference in Rio de Janeiro, Brazil, in 1992, the participating countries agreed to try to reduce greenhouse gas emissions by the year 2000, with the understanding that 1990 levels would be a desirable target.

For better or worse?

Higher temperatures could have some benefits for farmers in the region, enabling them to grow frost-sensitive crops much farther north than at present. However, high daytime temperatures and widespread drought could cancel out many of the gains in cereal-growing provinces, such as Ontario.

In the Arctic itself, scenarios of the future are varied. One view suggests that plants in the treeless tundra may grow quicker and their productivity increase, and that this could even slow down global warming if the mosses, lichens and other tundra plants absorb some of the surplus carbon dioxide. If, however, a lack of nutrients in the soil prevents this growth, then progressive melting of the permafrost might release large amounts of methane – one of the greenhouse gases – which would then reinforce the process of global warming.

The thawing of the frozen ground would have serious consequences for people living in the Arctic. If surface soils become saturated with water and weaken, roads and runways would collapse and many building foundations – previously anchored firmly into the permafrost – would founder. In this scenario, global warming seems set to challenge the ingenuity of polar engineers more than ever before.

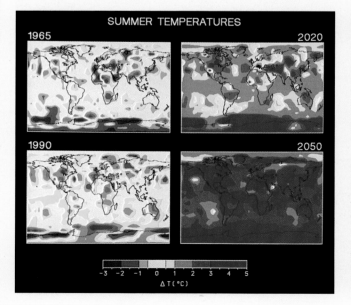

SUMMER TEMPERATURES
1965   2020
1990   2050

$\Delta T (°C)$
-3 -2 -1 0 1 2 3 4 5

Red marks the spot (*left*) of the greatest projected temperature increases due to global warming and the "greenhouse effect" as mapped by the Global Climate Model of NASA's Goddard Institute for Space Studies. Based on current global environmental patterns, a powerful NASA computer program is used to predict likely changes in the Earth's climate up to the year 2050.

Flood in waiting? (*right*) The Greenland ice sheet is the largest area of freshwater ice in the northern hemisphere, up to 3,000 m (10,000 ft) deep in places. Widespread melting in this area due to global warming could have catastrophic effects all over the world. However, some scientists argue that greater snowfall will be associated with the warming and will redress the balance.

THE COSTS OF AFFLUENCE

BEGINNING TO THINK GREEN · AIR POLLUTION · THE RISING TIDE OF WASTE
QUESTIONS OF SUPPLY · THE VULNERABLE WILDERNESS · TACKLING THE PROBLEMS

The United States has probably seen more extensive environmental changes over the past 200 years than any other region. Massive population growth, high standards of living and heavy industrialization have taken their toll on the environment. Some problems, such as air pollution and domestic waste disposal, affect the whole region; others vary from one area to another. In the northwest, forests are disappearing; southwestern water supplies are overloaded; oil spills have polluted the Alaskan coastline; midwestern plains battle with soil erosion, urban parts of California with smog, while the industrial northeast is urgently seeking ways to rid itself of hazardous waste. Since the 1960s the nation has suddenly awakened to its danger, and vigorous programs of environmental rescue are being put into action.

COUNTRIES IN THE REGION

United States of America

POPULATION AND WEALTH

Population (millions)	249.6
Population increase (annual population growth rate, % 1960–90)	1.1
Energy use (gigajoules/person)	280
Real purchasing power (US$/person)	19,850

ENVIRONMENTAL INDICATORS

CO₂ emissions (million tonnes carbon/annum)	1,000
Municipal waste (kg/person/annum)	762
Nuclear waste (cumulative tonnes heavy metal)	17,606
Artificial fertilizer use (kg/ha/annum)	93
Automobiles (per 1,000 population)	550
Access to safe drinking water (% population)	n/a

MAJOR ENVIRONMENTAL PROBLEMS AND SOURCES

Air pollution: locally high, in particular urban; acid rain prevalent; high greenhouse gas emissions
River/lake pollution: medium/high; *sources*: industrial, agricultural, sewage, acid deposition
Marine/coastal pollution: local; *sources*: industrial, agricultural, sewage, oil
Land pollution: locally high; *sources*: industrial, urban/household, nuclear
Land degradation: *types*: desertification, soil erosion, salinization; *causes*: agriculture, industry
Waste disposal problems: domestic; industrial; nuclear
Resource problems: land use competition
Major events: Love Canal (1978) and Times Beach (1986), evacuated due to chemical pollution; Three Mile Island (1979), nuclear power station accident; *Exxon Valdez* (1989), major oil spill from tanker in sea off Alaska; Dunsmuir (1991), pesticide spill during transportation

BEGINNING TO THINK GREEN

The abundant resources of the region's vast wildernesses easily sustained the Native-Americans who were its only occupants until the 16th century. They lived as shifting cultivators in the southwest or as hunter–gatherers in the midwest and along the Pacific coast, while in the southwest the Pueblos carried out irrigated farming. European settlers, however, began to exploit the region on a more dramatic scale from their arrival. Large areas of the virgin forests that covered most of the original 13 states along the North Atlantic seaboard were felled for construction wood or fuel, and the land was cleared for agriculture. Southern woodlands were destroyed to create plantations, while later on the interior's grasslands were plowed for wheat, fenced for ranching and drilled for oil by successive waves of immigrants.

By the 19th century the region was fast on the way to becoming the most heavily industrialized nation in the world. Some people began to question the careless way the land was being plundered, warning that resources should be managed for longterm goals rather than shortterm gains. One result was the creation of the first national parks.

The government steps in

During the 1930s farming expansion on the Great Plains caused serious soil erosion, exacerbated by several years of drought. Thousands of settlers were forced off the land, which had been transformed to near-desert; the whole area became known as the Dust Bowl. The country's economic collapse, or Depression (1929–33), jolted the government into funding massive recovery programs to assist the midwest. In 1935 the Soil Conservation Service was formed to research and prevent erosion. However,

Heavy industry, heavy pollution (*below*) in the highly industrialized area around the Chesapeake Bay estuary near Baltimore, Maryland where marine life has been seriously damaged.

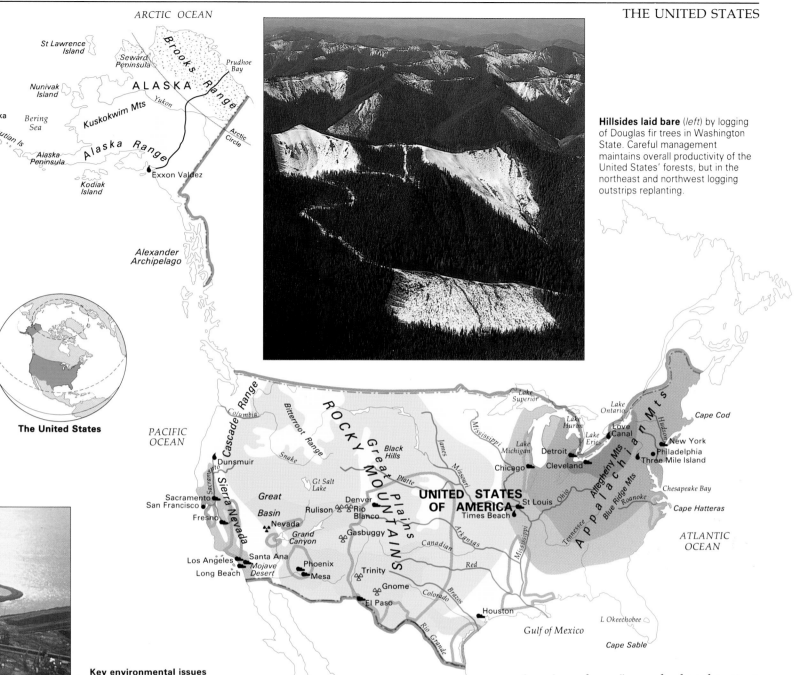

Hillsides laid bare (*left*) by logging of Douglas fir trees in Washington State. Careful management maintains overall productivity of the United States' forests, but in the northeast and northwest logging outstrips replanting.

The United States

Key environmental issues

- ● major town or city
- ◗ heavily polluted town or city
- ◖ major pollution event
- ☢ nuclear test site
- ☣ former nuclear test site
- ∿ heavily polluted river
- — main oil pipeline
- ⬭ area of groundwater depletion
- ⬚ area affected by permafrost

acidity of rain (pH units)

- 4.2 (most acidic)
- 4.4 (least acidic)

areas at risk of desertification

- very high
- high
- moderate
- true desert

Map of environmental problems
(*above*) showing acid rain in the highly industrialized northeast. Groundwater depletion and the risk of desertification are widespread, particularly in the arid southwest.

the problems persist. Between 10 and 12 tonnes per ha (4 and 6 tonnes per acre) of soil are eroded each year in the United States. In the southeast and corn belt states – Iowa, Missouri, Kentucky and Tennessee – rain is the main cause, while wind erosion is greatest in the western Great Plains. Techniques such as crop rotation, contour plowing, terracing, windbreaks and tillage have all been used to conserve soil and prevent erosion.

Other government attempts to assist specific areas included the creation of the Tennessee Valley Authority, a multi-purpose, multiresource agency. Flood control, hydroelectric power and erosion control were all aspects of its programs.

Following World War II, environmental issues were neglected by many Americans, whose aim was to attain "the American dream": a suburban house, a good job and a private automobile. However, after several decades the connection between affluence and pollution became all too apparent, as urban sprawl, land degradation and industrial and vehicle emissions all increased greatly.

Most rivers in the western United States have been heavily dammed to provide water for the expanding urban populations. The dams cause sediment to build up in the reservoirs, while downstream flow is reduced, the banks become eroded, and aquatic habitats are altered. Along the Snake and Columbia rivers in the northwest, for example, salmon spawning areas have been almost wiped out.

Widespread exploitation of energy resources also left its mark on the environment. Strip coal mines have degraded large areas of the west, while gas and oil extraction has caused land subsidence.

AIR POLLUTION

Despite 20 years of pollution control and the world's highest standard of living, cities in the United States are still choked with air pollution. Nearly 75 percent of the population lives in and around some of the largest, most affluent cities in the world, the densest areas of population being in the industrial northeast and in California, where the quality of the air is often hazardous to health.

Smog in the cities
As a result of local and national air pollution laws, emissions of primary air pollutants from industry and automobiles – solid particles such as dust (particulates), sulfur dioxide, nitrogen oxides, carbon monoxide, volatile organic compounds and lead – have been reduced. However, more than half the population still lives in areas that fail to meet the permissible ozone level. Ozone, a gas vital to human survival when high in the

atmosphere but harmful at low altitudes, is the main constituent of smog. It damages leaves, slows plant growth and causes respiratory problems in people. In 1989, 26 of the 64 largest cities in the nation failed to meet the federal health standard for ozone. The highest levels were recorded in Los Angeles, Long Beach and Santa Ana, all in southern California.

Photochemical smog is caused when industrial and vehicle emissions are changed by the action of sunlight into a cloud of harmful gases. The worst affected cities are mainly in the west of the region, often in valleys where the wind patterns and the surrounding mountains create conditions that cause temperature inversions, trapping air pollution close to the ground. Cities on arid plains that lack sufficient moisture to wash away the pollutants are also affected. Denver (Colorado), Phoenix (Arizona), and Los Angeles, for example, are all choking under their own air pollution. In the southwest, air pollution is at its worst in the

Los Angeles' smog (*above*) The Sunbelt cities of the west are particularly badly affected by high concentrations of industrial and automobile emissions that react with sunlight to create hazardous photochemical smog.

Sick home (*right*) The indoor build-up of toxic pollutants from sources such as cigarette smoke, unvented stoves, aerosol sprays or synthetic building materials can cause allergic symptoms such as headaches, eye or skin irritations, and sneezing.

summer when sunlight is more plentiful. In Denver and other western and northwestern cities in the Rockies, however, winter temperature inversions often combine with emissions from wood-burning stoves, widely used for heating, to create high pollution levels.

Carbon monoxide, mainly from automobile emissions, is a serious problem in over one-third of the cities in the United States. In 1989, recorded levels for Los Angeles and Long Beach were 18 parts per million (ppm), twice the accepted health standard of 9 ppm.

Emissions of sulfur dioxide and nitrogen oxides are successfully controlled

INDOOR AIR POLLUTION

Home sweet home is not as safe as it once was. As houses and commercial buildings became more energy efficient during the 1970s and 1980s, concentrations of pollutants built up due to lack of ventilation. Increased cancer deaths from indoor air pollutants focused attention on this issue.

The most widespread cause of indoor pollution is tobacco smoke. Other sources of contaminants include carbon monoxide from gas cookers and fires, asbestos fibers in insulation, and chemical compounds in cleaning solvents.

One of the most insidious contaminants, radon, is a naturally occurring gas that results from the radioactive decay of radium, found in many types of rocks and soils. Radon enters buildings through cracks in the foundations. It was first discovered in the 1960s when houses in western Colorado, registering elevated levels of radon, proved to have been built with materials contaminated by uranium mill waste. In 1984 a nuclear power plant engineer was contaminated not from his work, but from the excessive radon levels inside his Pennsylvania home. This discovery led to a recognition that the radon problem affected many types of geological formations nationwide.

and monitored in all the largest cities of the United States. In 1989 only Los Angeles and Long Beach suffered from levels of nitrogen oxides that were above the legal limit. Monitoring revealed that the most serious particulate pollution occurred in central and eastern industrial cities such as St Louis, Missouri, and in the west as a result of wind-blown dust from fallow agricultural land.

Chemical contamination

More than 58,000 different chemical compounds are used in the United States. At least 400 of these organic and inorganic chemicals are listed as hazardous because they are either toxic, corrosive, reactive or flammable. Many of them are used to manufacture common domestic and industrial products such as plastics, paints and finishes, electrical transformers and fuel additives. These chemicals are routinely emitted into the atmosphere and discharged into water courses. In 1989, 2.7 million tonnes of chemical wastes were released into the air, water, land, groundwater and public sewage systems, or transported to other areas.

Nearly 50 percent of these were emitted into the air alone. Toxic air emissions pose the greatest problems in the 12 states of Washington, Illinois, Indiana, Ohio, Virginia, Tennessee, Arkansas, Louisiana, Alabama, Connecticut, Rhode Island and Massachusetts.

Chemical contamination of the surface water has a long history in the region. The Cuyahoga river, which runs through Cleveland, Ohio, was so heavily polluted by flammable chemicals during the 1950s and 1960s that it periodically caught fire. More recently, however, a train derailment in Dunsmuir, California, in July 1991, spilled 89,000 liters (19,500 gallons) of the pesticide metam sodium into the Sacramento river. Although not classified as a hazardous material on its own, when metam sodium interacts with water it becomes lethal to aquatic life. In the Dunsmuir spill, more than 100,000 trout were killed, aquatic life between the spill and Lake Shasta, 72 km (45 mi) away, was heavily damaged, and public drinking water supplies in Lake Shasta, already seriously compromised through drought, were further threatened.

1 Carbon monoxide from faulty furnaces, stoves, heaters and car exhausts may cause headaches, drowsiness and irregular heartbeat

2 Methylene chloride and other solvents in paint, glues and thinners may cause nerve disorders and diabetes

3 Radon 222 from "radioactive" soils and rock, and some water supplies may cause lung cancer

4 Styrene in carpets and plastic products may cause kidney and liver damage

5 Benzopyrene from tobacco smoke and wood stoves may cause lung cancer

6 Formaldehyde from furniture filling, paneling, particle board in furniture and foam insulation may cause irritation of eyes, throat, skin and lungs, and nausea and dizziness

7 Tetrachloroethylene residue from dry-cleaned clothes may cause nerve disorders, and damage to liver and kidneys

8 Paradichlorobenzene from air fresheners and mothball crystals may cause cancer

9 Chloroform from chlorine-treated water in hot showers may cause cancer

10 1,1,1–trichloroethane in aerosol sprays may cause dizziness and irregular breathing

11 Nitrogen oxides from unvented gas stoves, kerosene heaters and wood stoves may cause lung irritation, cold-like symptoms and headaches

12 Asbestos from pipe insulation, fireproofing, and ceiling and floor tiles may cause lung disease and cancer

13 Tobacco smoke from cigarettes may cause lung cancer, respiratory disease and heart disease

14 Pesticides from fruit and vegetables, and for garden use may cause heart problems and nerve disorders

THE RISING TIDE OF WASTE

The people of the United States generate more than 1.8 kg (4 lbs) per person per day of domestic and commercial (solid) waste, and it all has to be disposed of. Before 1970 most of the nation's solid waste was sent to landfills, where vast heaps of garbage accumulated and were buried. As these became full and so closed, space for new landfills became difficult to find, partly because of the cost and partly because of the expansion of city suburbs onto waste ground. Community opposition to landfills increased so much during the 1970s that disposal options became extremely restricted.

Alternative methods of disposal such as ocean dumping and incineration are also problematic. Dumping in coastal waters has been banned in the United States (as in many other countries) – though until 1992 New York City was allowed to dump sewage into the ocean off the New Jersey shoreline. Incineration, which disposes of 16 percent of solid waste, is a costly and controversial alternative.

Many states have resorted instead to transporting their waste elsewhere; New Jersey, for example, sends some by truck to Pennsylvania, Kentucky and Ohio. This movement of waste became a national scandal in 1987, when a waste-laden barge from Islip, New York, wandered for 164 days round the southern seaboard trying to off-load, only to return still laden to Islip. Its load was eventually incinerated in Brooklyn.

Recycling, another option, is now an integral part of solid waste management

Dangerous freight (*above*) A truck has spilt toxic chemicals over the road after an accident in a residential area of New Jersey. Hazardous chemicals and wastes often have to be transported long distances to suitable disposal sites.

and residents of some states have to separate domestic waste into recyclables by law. One of the nation's fastest-growing industries, recycling generates between $1 and 2 billion a year in profits. One obvious solution to the crisis would be to reduce the amount of waste being generated by, for example, eliminating excessive packaging, making products longer-lasting and recyclable, and reducing production of disposable items such as diapers (nappies).

Hazardous waste

In 1980 Congress passed legislation called Superfund, to identify hazardous waste sites and fund clean-up efforts. By 1991 more than 31,500 sites had been identified. Hazardous waste was at first regarded as a by-product of industry, but municipal landfills containing household products such as pesticides, paint and

Inputs and outputs (*below*) An average city of 1 million people consumes vast quantities of food, fuel and water each day. These are converted into wastes that contaminate the environment and present huge problems of disposal.

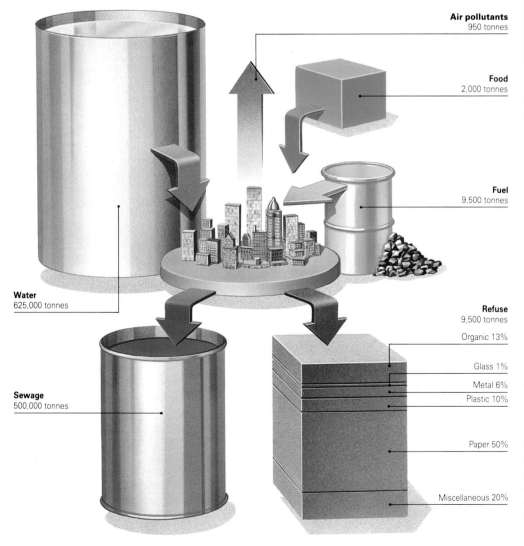

Air pollutants
950 tonnes

Food
2,000 tonnes

Fuel
9,500 tonnes

Water
625,000 tonnes

Sewage
500,000 tonnes

Refuse
9,500 tonnes

Organic 13%

Glass 1%

Metal 6%

Plastic 10%

Paper 50%

Miscellaneous 20%

generate hazardous waste have been changing their manufacturing processes to minimize the amount of waste they produce. For example, one company, Chevron USA Inc., established its Save Money and Rescue Toxics (SMART) program, an initiative that in the first year reduced the company's hazardous waste generation from 135,000 tonnes to 76,000 tonnes, thereby also saving $4 million in waste disposal costs.

Toxic ghost towns

In 1978 Love Canal, a neighborhood of Niagara Falls, New York State, became the first toxic ghost town in the United States. Nearly 1,000 families were evacuated as a result of highly toxic waste leaching from the filled-in canal used by Hooker Chemicals and Plastics Corporation as a dump between 1942 and 1953. Times Beach, Missouri, has also been designated a toxic ghost town. In 1971 a contractor in the town was hired to spray oil on the unsurfaced streets to lay the summer dust – a common procedure in many states. However, a number of people soon began to complain of health disorders, including skin ailments and internal bleeding. Several years later tests on the oil revealed it to be heavily contaminated by several highly toxic compounds including dioxin. The oil had been sold to a chemical company that had failed to clean it up before re-use. In 1983 the government had to buy up all the property and relocate the residents. Today Love Canal and Times Beach are fenced off with warning signs all around.

Another related problem concerns radioactive waste. During the 1980s low-level wastes were accepted at three sites (Hanford, Washington; Barnwell, South Carolina; West Valley, New York), but plans were made for these to close during the 1990s. States are now required to join in regional compacts to handle future low-level waste disposal within their region. High-level waste from spent nuclear fuel is stored in water tanks at the country's 102 commercial reactors, as no permanent repository has yet been constructed. The Waste Isolation Pilot Project in New Mexico is the government's first serious attempt to handle the problem. Highlevel waste from military applications is stored in Department of Energy facilities including those at Hanford and Barnwell – so that these areas too qualify as toxic monuments.

used motor oil are now also classified as hazardous waste sites.

About 1,200 have been placed on the National Priorities List: places so hazardous to public health that they require immediate action. Those on the list are mostly located in California, Florida and the industrialized northeastern states. However, by 1992 only 3 percent of priority sites had been cleaned up. The government made the requirement that those who generated the waste or owned the site must pay for the work, but companies have been reluctant to incur the huge costs of environmental restoration. Some estimates suggest that as much as $1 trillion will be needed before the year 2040 if the sites already on the priority list are to be cleaned up properly.

Since the late 1980s the industries that

THE NIMBY SYNDROME

"Not In My Back Yard!" is the rallying cry for local community opposition to unwanted uses of land – which may include anything from waste landfills to homes for the mentally ill. To site unpopular facilities is virtually impossible in many parts of the region, especially the northeast. The NIMBY syndrome is particularly pervasive in solid waste issues. Polls taken in the early 1990s indicated that nearly 60 percent of the respondents would object to a new landfill in their community, yet these same people are unwilling to offer alternatives to the mounting piles of waste other than to send them somewhere else.

During the 1980s the NIMBY syndrome became more vocal, especially in response to high technology environmental solutions. Even a relatively harmless facility such as a leaf composter provokes opposition from residents who do not want it in their community. While the NIMBY soothsayers may think locally, they certainly do not think globally, preferring to send the problem to someone else's backyard.

The car culture

The United States' love affair with the automobile began in 1908 when the American motor manufacturer Henry Ford (1863–1947) first made the popular Model T Ford widely available through mass production. It was manufactured until 1927. After World War II, economic prosperity enabled many young families to purchase their first automobile and their dream house in the suburbs. The speed, freedom and convenience of the car altered forever the country's landscape. Transcontinental highways provided coast-to-coast links complete with motels, restaurants, gas stations and tourist attractions. Freeways were built to link suburbs to cities. Parking lots arrived to house the automobiles while their occupants worked or shopped. New suburban communities were designed with the car in mind – every house had a two-car garage, with a gas station just down the street.

Instead of high density compact cities, the automobile encouraged low density cities sprawling over vast areas: for example, Los Angeles, Phoenix and Houston. A convenience culture has grown up around the car, with shopping malls, drive-through fast-food restaurants, and highway advertising billboards. Large numbers of people earn a living making, selling, repairing or servicing cars.

By the 1990s there were more than 140 million automobiles in the United States. They travel on more than 5.6 million km (3.5 million mi) of roadways and are driven more than 2.2 trillion km (1.4 trillion mi) a year. Some 7 million new cars are purchased each year. The majority (64 percent) of the population use their cars to commute to work instead of walking or using public transportation. They tend to drive alone instead of organizing ways to share, and a journey to

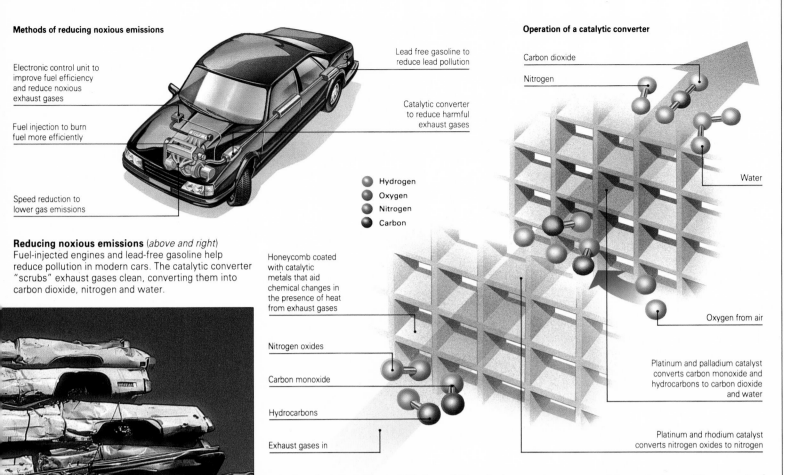

Methods of reducing noxious emissions

Electronic control unit to improve fuel efficiency and reduce noxious exhaust gases

Fuel injection to burn fuel more efficiently

Speed reduction to lower gas emissions

Lead free gasoline to reduce lead pollution

Catalytic converter to reduce harmful exhaust gases

Reducing noxious emissions (*above and right*)
Fuel-injected engines and lead-free gasoline help reduce pollution in modern cars. The catalytic converter "scrubs" exhaust gases clean, converting them into carbon dioxide, nitrogen and water.

Hydrogen
Oxygen
Nitrogen
Carbon

Honeycomb coated with catalytic metals that aid chemical changes in the presence of heat from exhaust gases

Nitrogen oxides

Carbon monoxide

Hydrocarbons

Exhaust gases in

Operation of a catalytic converter

Carbon dioxide

Nitrogen

Water

Oxygen from air

Platinum and palladium catalyst converts carbon monoxide and hydrocarbons to carbon dioxide and water

Platinum and rhodium catalyst converts nitrogen oxides to nitrogen

work of more than 120 km (75 mi) is normal in many parts of the country. California, with 16 million automobiles, has more than twice the number of any other state. About 63 percent of all the petroleum used in the United States is consumed by cars, trucks, planes and trains. Cars consume an average of 1 liter for every 10 km (1 gallon for every 28 mi). In 1987, 142 million passenger vehicles traveled an average of 15,375 km (9,610 mi) per vehicle, consuming a massive 324 billion liters (72 billion gallons) of fuel in just one year.

Noise, smog and junk

The car culture has caused many environmental problems, beginning with traffic congestion and noise pollution. Many cities in the United States are choked with traffic at an apparently permanent standstill. Noise levels are detrimental to the quality of life – and sometimes to health – and noise barriers around highways are commonplace.

The emissions from gasoline-fueled automobiles are the main contributors to urban smog, accounting for 70 percent of

On the scrap heap at a junk yard in California. Recycling automobiles and other appliances now accounts for 43 percent of all steel production in the United States. Recycling scrap to make steel requires 74 percent less energy than refining it from raw ore.

carbon monoxide emissions, 35 percent of hydrocarbons and 40 percent of nitrous oxides. In Los Angeles smog is so bad that the South Coast Air Quality Management District (the government body for air quality in southern California) has produced a plan that will ultimately restructure life in the area. The plan, which runs until 2012, requires residents to change their lifestyle. Gas-powered lawn-mowers and barbecues have been banned. Gasoline is being reformulated to comply with emissions standards. By 1998, 40 percent of cars must be converted to cleaner fuels and by 2007 all gasoline-powered vehicles will be banned. These severe measures are the only way that the people of Los Angeles can reduce the smog now threatening their lives.

Water quality has also been affected by people's reliance on automobiles; in winter, the water draining off roads contains deicing salts, oil, grease and heavy metals that can contaminate freshwater supplies. Leaking underground storage tanks at gas stations also threaten groundwater quality. In addition, used motor oil, antifreeze and tires all require disposal, adding to hazardous waste problems. The landscape of the United States has become littered with broken cars in junkyards, a poignant symbol of the impact of the car culture on the environment.

QUESTIONS OF SUPPLY

Demands made by the United States on its water and energy resources have risen steadily over the century as the population has increased. Unless they are reduced, or supplies increased, the outlook for the region is grave. On average an American uses 1,350 liters (300 gallons) of household water a day. Although overall national supplies are adequate to meet these needs, regionally a different situation has emerged.

Population growth during the 1980s, which was greatest in the arid west and southwest, placed enormous pressure on natural water supplies that were already subject to alternate surplus and drought and strained by the heavy demands of crop irrigation and the coolant systems of thermal power stations. A serious drought began in California in 1986 had showed

no signs of abating by the early 1990s. If the drought continues, it is likely to cause an environmental catastrophe.

Rivers under stress
The western and southwestern states draw most of their water from the rivers of the Rocky Mountains, which have been dammed to create huge reservoirs. The Colorado river and its tributaries provide water for 20 million people, as well as irrigation water for 800,000 ha (2 million acres) of land. In 1922, water withdrawals were divided among seven bordering states, and in 1944 an arrangement was also made to include Mexico. Ten major dams were constructed on the Colorado itself. The border states are free to withdraw water up to their legal limits even when the river is low. (More than 80 percent of the water is used for irrigating cotton and other crops.) However, southern Californian farms and cities have

continually exceeded their allocation, and as other states and Mexico demand their legal share, the strain on the river has become critical.

Another problem in the agricultural areas of California and the Great Plains, especially for the communities that rely primarily on groundwater for drinking water, is nitrate and pesticide pollution. The Ogallala aquifer, which covers 438,600 sq km (170,000) sq mi of the Great Plains, extending from South Dakota to Texas, is the largest source of groundwater in the United States. Extensive withdrawals from the water-bearing rock date back to the 1930s, when efforts were made to stem wheatland erosion by irrigation. Natural replacement has not kept pace, however, and it is estimated that by 2020 the volume of water in the aquifer will have been reduced by 25 percent. In some parts of Texas and Oklahoma, wells have already run dry.

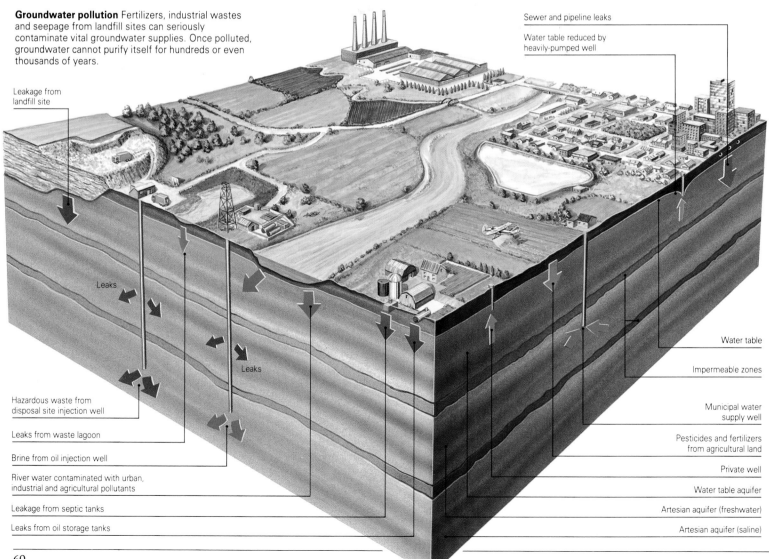

Groundwater pollution Fertilizers, industrial wastes and seepage from landfill sites can seriously contaminate vital groundwater supplies. Once polluted, groundwater cannot purify itself for hundreds or even thousands of years.

Leakage from landfill site

Sewer and pipeline leaks

Water table reduced by heavily-pumped well

Leaks

Leaks

Hazardous waste from disposal site injection well

Leaks from waste lagoon

Brine from oil injection well

River water contaminated with urban, industrial and agricultural pollutants

Leakage from septic tanks

Leaks from oil storage tanks

Water table

Impermeable zones

Municipal water supply well

Pesticides and fertilizers from agricultural land

Private well

Water table aquifer

Artesian aquifer (freshwater)

Artesian aquifer (saline)

Annual rainfall in semiarid California averages 70 cm (28 in) per year, falling mostly during the rainy season from December to March. Since 1986, precipitation has been 75 percent below normal levels, producing the greatest drought in the state's history. Water users, gambling that the drought was temporary, reduced reservoirs to dangerously low levels. Panic increased as the drought set in, and by 1992 water in many parts of the state was rationed for the first time.

Farmers, who use 85 percent of the state's water, had their supplies suspended. Rationing was implemented in urban areas, with stiff fines for lawbreakers. Restrictions were put on washing cars, watering lawns and filling swimming pools. The economy has suffered, too, as business and industry have proved reluctant to relocate into the drought-stricken state. Agriculture – big business in California, generating more than $14 billion in 1989 alone – has also been particularly badly hit, with thirsty crops such as alfalfa, cotton, pasture and rice suffering the most. Increased prices for these and other crops (for example grapes, citrus fruits and almonds) will be felt throughout the nation during the 1990s.

Competition over water became more intense as the drought continued. Urban dwellers in some areas had to pay more for water than farmers and ranchers, while some cities were forced to buy land for the water rights – action reminiscent of the period from 1920 to 1950, when Los Angeles bought 75 percent of the land in Owens Valley, 480 km (300 mi) away, to obtain its water. Although many farmers have been obliged to convert to more efficient irrigation systems and urban residents have learned to conserve supplies, there is no relief in sight for the thirsty Golden State, which in the early 1990s was more brown than golden.

Desiccated landscape (*above*) A farm in California, destroyed by the long drought, is offered up for sale to property developers. There is increasing competition for scarce water supplies between agriculturalists and urban populations.

Where the water goes (*below*) Many domestic users waste water by leaving it running when washing a car, washing the dishes or brushing teeth. Drier states have conducted publicity campaigns urging householders to conserve water.

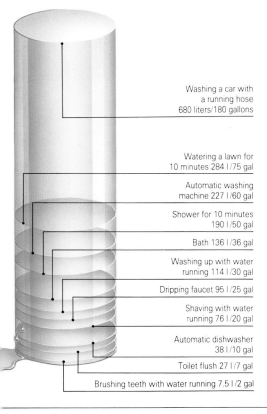

Washing a car with a running hose 680 liters/180 gallons

Watering a lawn for 10 minutes 284 l/75 gal

Automatic washing machine 227 l/60 gal

Shower for 10 minutes 190 l/50 gal

Bath 136 l/36 gal

Washing up with water running 114 l/30 gal

Dripping faucet 95 l/25 gal

Shaving with water running 76 l/20 gal

Automatic dishwasher 38 l/10 gal

Toilet flush 27 l/7 gal

Brushing teeth with water running 7.5 l/2 gal

Shortage of water is not yet a problem in the eastern United States. On the other hand, water quality is. Surface water pollution from both urban and industrial sources, combined with groundwater contamination by chemicals seeping from landfills and hazardous waste sites, has resulted in a considerable deterioration in the quality of water supplies.

Energy and the environment
Despite the widespread concern about pollution from burning fossil fuels contributing to smog, acid rain and the "greenhouse effect", proposals have been put forward to explore new, unspoilt areas for oil, particularly in Alaska. Oil spills such as the *Exxon Valdez* disaster off the coast of Alaska have also caused great public anxiety. Unfortunately the United States as yet has no national policy for examining and balancing the demands of industry for energy against those of the environment.

The accident at the Three Mile Island nuclear generating station near Harrisburg, Pennsylvania, in March 1979 raised public and government awareness about the impact on the environment almost overnight and slowed down the nuclear energy program. As a result of a malfunctioning valve, radioactive steam was able to escape, causing pipes containing radioactive water to burst and spill their contents. No longer covered by coolant, about half the reactor core melted. No significant levels of radioactivity were released and after the accident, new safety regulations were implemented and emergency plans became a legal condition for licensing. However, public mistrust of nuclear power as an energy source remains high, and only 38 new plants have opened since 1979. Despite vigorous attempts by the industry to revive confidence, nuclear power has an uncertain future in the United States.

THE VULNERABLE WILDERNESS

The protection of pristine natural areas has long been a part of American environmental policy. The importance of valuable ecosystems such as wetlands, deserts and forests is recognized through preservation and conservation laws.

Wetlands – tidal marshes, swamp forests, peat bogs and prairie potholes – provide a habitat and breeding grounds for both migratory and nonmigratory birds, and spawning sites for fish and shellfish. They are found throughout the United States, covering more than 25 percent of Alaska, Florida and Louisiana. Wetlands are, however, particularly vulnerable to urban encroachment and land-use changes. Conversion to agricultural

use accounted for 87 percent of wetland loss from 1950 to 1970, while urban development particularly affected wetlands in the states of Florida, Texas, New York, New Jersey and California.

Nearly 56 percent of the original wetlands in the country have been lost, with California (91 percent), Ohio (90 percent) and Iowa (89 percent) suffering most. The rate of loss is also alarming: between 1970 and 1990, wetlands the size of Massachusetts, Connecticut and Rhode Island have all disappeared.

Wetlands serve many functions. For example, those in Florida are home to wading birds, a winter haven for 20 percent of Atlantic migratory waterfowl and a prime habitat for alligators. In addition, they provide an urban water supply and water purification system for

Spoiling the view Passengers on a tour bus stop to admire Mount McKinley, North America's highest peak, in Denali National Park, Alaska. The presence of increasing numbers of tourists is having a detrimental impact on scenic wilderness areas.

communities discharging waste water into estuaries. However, as population pressures have increased throughout Florida, the wetlands have been disappearing at the rate of 28 ha (70 acres) per day. Water has been drained to provide land for housing and agriculture, and the remaining wetlands have been severely contaminated. The Everglades, Great Cypress, Green Swamp, Kissimmee and Lake Okeechobee are all threatened.

In 1990, the federal government established the controversial "no net loss" principle, which states that if developers or farmers wish to fill in wetlands on

Endangered by logging (*above*) Commercial logging in the northwest has progressively diminished the habitat of the Spotted owl. Now a protected species, the Spotted owl has been granted a temporary reprieve by a halt in deforestation.

their property, they must simultaneously create others elsewhere or reconvert former wetlands. However, reclamation is not the only problem to threaten the wetlands of the United States. Wildlife refuges, especially around estuaries, have been polluted by industrial waste, agrochemicals and waste dumps. Leading the fight to clean up wildlife preserves has been a coalition of environmental and special interest groups: the Wilderness Society, Ducks Unlimited and the Izaak Walton League. Among the worst affected has been California's Kesterson National Wildlife Refuge, established to provide waterfowl with nesting and feeding sites. The area also serves as an evaporation pool for water draining off local farmland. However, it has been polluted by pesti-

cides which leached into the area from nearby fields, poisoning wildlife.

Loggers versus the owl

Species loss has been increasing dramatically in recent decades throughout the United States as human activity encroaches ever more heavily on natural habitats. In 1990, 600 species in the region were listed as threatened or endangered, compared with 283 in 1980. Public outrage over endangered species has led to some heated confrontations between environmentalists and commercial companies. In the Pacific northwest, for example, battle lines have been drawn over old growth forests and the Spotted owl.

For decades loggers felled the ancient forests in the Pacific northwest, and by 1992 only 5 percent of the original virgin forest remained. The industry did replant, but logging in these secondary growth forests kept pace until the forests were no longer viable – the trees were too small to harvest. The logging companies then turned their attention to the vast forests under public ownership, many of them virgin tracts of Red cedar and Douglas fir. With federal approval the industry began harvesting the big trees in the national forests – until environmentalists discovered the Spotted owl.

The Spotted owl lives in northwestern old growth forests and its habitat is threatened by the continual logging. Violent confrontations between members of the environmental group Earth First!, the lumber companies and the local residents who were dependent on the timber industry brought the issue to national attention. Now the Spotted owl is a protected species. Logging in these forests has been temporarily halted as environmental groups struggle to preserve the last remaining stands of timber.

Greed and poor management meant that forests were cut at rates far exceeding natural growth, even when there was too much timber on the market. In September 1991 a district judge ruled against a plan to log 26,400 ha (66,000 acres) of public forest. The judge found that the practice of exporting whole logs rather than cut boards contributed more job losses in the area (as sawmills employ local workers) than the halting of the logging for environmental reasons. Perhaps this decision is the harbinger of things to come in the never-ending debate over the environment and economics.

ECOTOURISM HITS THE NATIONAL PARKS

Since the establishment of Yellowstone, (situated in parts of Wyoming, Montana and Idaho) as the first national park in 1872, the United States has actively preserved its natural heritage. By 1992 nearly 32 million ha (80 million acres) of natural landscapes had been preserved within the national park system. In 1905 less than 100,000 people visited these national treasures; by 1990 they were visited by nearly 352 million people in one year alone.

The national parks are now suffering the impact of too many visitors. During the summer tourist season, traffic jams are common in Yosemite Valley in California and around Old Faithful, the famous geyser in Yellowstone National Park. Visitors' demands for accommo-

dation, food, toilets and shopping facilities have placed enormous burdens on the parks. Camping in the interior of many of the national parks is governed by a permit system to keep visitor levels below what the wilderness can accept. Official campsites are always fully booked many months before the start of the season. The popularity of the parks continues to increase, despite the fact that pollution from nearby areas is so great that fine vistas of the Grand Canyon and the Great Smokey Mountains are clouded with the haze of industrial emissions. With visits expected to double by 2000, national parks urgently need protection from the source that first established them: the American public.

TACKLING THE PROBLEMS

Characteristically, the United States responds to environmental crises through technological innovations and increased governmental involvement rather than by proposing fundamental changes in American society or behavior. Although it has one of the best records in the world for legislation and regulatory control over resource use and pollution, this is not enough to tackle all the environmental problems that confront the country.

Ecology and pollution control emerged as foremost concerns of the 1960s. Public alarm culminated in 1970 with the passage of the National Environmental Policy Act (requiring the continual monitoring of the environment and the consideration of the environmental consequences of all policies) and also of a whole series of laws dealing with pollution reduction (such as the Clean Air Act and Clean Water Act) and preservation (the Wilderness Act and the Endangered Species Act).

The costs of environmental clean up have been enormous. In 1990, for example, the United States spent $115 billion on pollution control. Most of these costs are borne by industry. The federal government contributes about 15 percent and the local municipal governments – answerable to local taxpayers – make up the difference.

New incentives

Strategies that involve free market and economic incentives to tackle environ-mental problems, coupled with pollution prevention at source, are increasingly preferred to a narrow program of governmental regulation and enforcement.

Pollution prevention programs have proved attractive options for industry and government, and many different techniques are used. Dramatic changes have been made in industrial processes to reduce the source of pollution (for example, using oxygen bleaching agents in the paper industry instead of chlorine). Certain waste elements are reused (such as in-house waste heat to drive turbines for electricity), and many products have been redesigned so that the different components can be easily separated out for recycling. Creating new products from waste has also increased; for example, playground structures are built from recycled tires, and park benches fabricated from recycled plastic bottles.

Cost-effective care

Environmental audits that consider the costs of waste disposal are now routinely used in financial management decisions, as industries have found that they can actually save money by becoming more environmentally responsible. "Emissions trading", designed to promote cost-effective approaches to air pollution control, is another technique that has gained some popularity. Local firms are awarded pollution "credits" if they reduce their individual smokestack emissions. These credits can then be used by the firm to offset emissions from other smokestacks that do not meet standards, either at the

Calling in the chemicals (*above*) Workers in protective clothing collect hazardous domestic wastes at a temporary stall in Seattle, Washington. These will then be disposed of more safely elsewhere.

Public action (*left*) Californians protest against the toxic waste incinerator at Kettleman City, San Joaquin valley. Growing public concern over environmental issues is beginning to have an influence on government policies in the United States.

Accelerated soil erosion is a problem on any agricultural land, but drought in the mid 1980s provoked visions of the Dust Bowl conditions of the 1930s, prompting federal action to reduce soil erosion.

In 1985 the United States Congress passed the Food Security Act. Among its provisions was the establishment of the Conservation Reserve Program. This paid farmers to remove land at risk from erosion from crop production for at least 10 years and to plant grasses or trees instead.

Farmers had always been aware of the problems of soil erosion and declining soil fertility, but previously had little economic choice but to keep the land in production. This program has been enormously successful in the south: Florida, Georgia and South Carolina have placed more than 75 percent of their eligible land in the conservation program. In western areas, which suffer extensive wind erosion problems, the program has been less successful. Nevada and Colorado, which both average more than 30 tonnes per ha (12 tonnes per acre) of erosion, have no land in the program. With the passage of the 1990 Farm Bill, increased federal support for soil conservation techniques such as conservation tillage and sustainable agriculture is now part of the agricultural community's way of doing business.

same site or other sites, or they can be sold to other firms just like any other commodity, or "banked" by the firm for future use. The result has been a reduction in local air pollution.

Businessmen in the United States have discovered that "green" is profitable. In addition to recycling, the environmental control industry (which includes solid waste companies, hazardous waste treatment firms and consulting companies) is booming, generating over $100 billion a year in revenues. Manufacturers and distributors of environmentally safe products such as canned "dolphin-friendly" tuna, recycled paper, organic food and biodegradable products have found large and lucrative markets among consumers who are willing to pay a little more for these products.

Investor interest in environmental stocks has also increased. Credit cards and checkbooks with the environmental logo of a favored group (such as Greenpeace or the Sierra Club) are regularly available, with a small percentage of proceeds being donated to the group. In many ways the United States has become more aware than ever of the destructiveness of its very affluent, consumer-orientated lifestyles. Consumer activism, grass roots activism encouraged by organized pro-environmental lobby groups such as the Sierra Club, Friends of the Earth and Greenpeace, and a slowing of the economy have all helped to stress the importance of a healthy environment and the need to make environmentally responsive economic decisions.

The *Exxon Valdez* disaster

On March 24, 1989, a supertanker, the *Exxon Valdez*, struck a reef near Bligh Island, 40 km (25 mi) south of Valdez, Alaska. Over 49.5 million liters (11 million gallons) of oil gushed into Prince William Sound, resulting in the worst oil spill in the history of the United States. In addition to the sheer volume of oil, what was so dramatic was the location – one of the most unspoilt areas in the United States. Alaska is the last frontier, the last remaining stretch of vast wilderness in the region.

The American public watched the nightly news broadcasts on television in horror as the oil slick quickly spread, fouling mile after mile of the Alaskan coastline. Two days after the accident, the governor of Alaska declared a disaster emergency. Despite containment efforts, the spill covered nearly 8,000 sq km (3,100 sq mi) within two weeks. The emergency response by the oil company Exxon, the

Coast Guard, the federal government, Alaska, even the local fishing industry was too little and too late. It was clear from the beginning that no one anticipated an accident of this magnitude, nor were they prepared for it.

Cleaning up the spill

Exxon tried to recover by initiating a massive clean-up program in June 1989 that included washing parts of the shoreline with high-powered hoses. The operation cost the firm nearly $1 billion and supposedly cleaned 1,760 km (1,100 mi) of soiled beaches. The state of Alaska, however, was not satisfied with their efforts and began its own cleanup. The salmon industry was hard hit, as were other coastal and offshore fisheries. Tourism may have been damaged as well, since the once pristine beaches have now been affected by pollution.

It was clear that Exxon was liable for the

The cleanup operation (*above*) One of the hundreds of relief workers who battled to remove oil from the beach at Naked Island. This oil was burnt. A variety of methods were tried to treat the oil that did not reach the beaches, such as using chemical dispersants, but none was completely effective.

Black death (*right*) The oil slick near the spillage in Prince William Sound. The disaster is estimated to have killed thousands of marine mammals and over a quarter of a million birds; the true environmental costs may not become apparent for several decades to come.

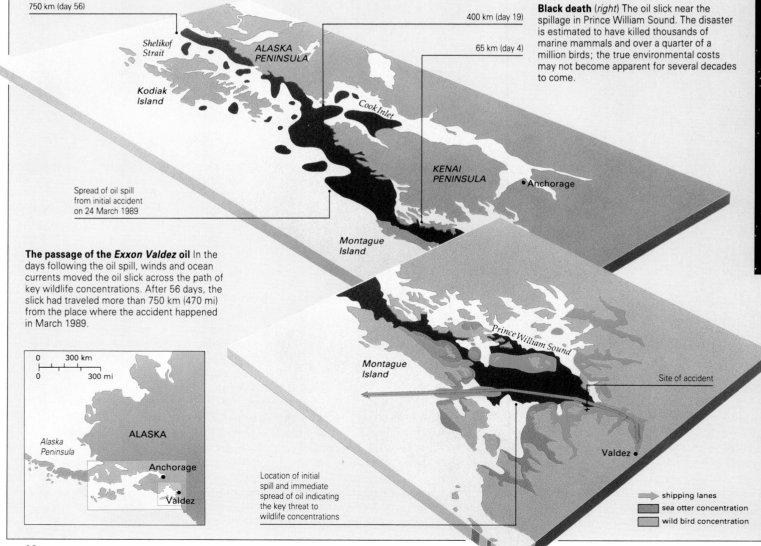

The passage of the *Exxon Valdez* oil In the days following the oil spill, winds and ocean currents moved the oil slick across the path of key wildlife concentrations. After 56 days, the slick had traveled more than 750 km (470 mi) from the place where the accident happened in March 1989.

Spread of oil spill from initial accident on 24 March 1989

750 km (day 56)

400 km (day 19)

65 km (day 4)

Shelikof Strait

ALASKA PENINSULA

Kodiak Island

Cook Inlet

KENAI PENINSULA

• Anchorage

Montague Island

0   300 km
0   300 mi

Alaska Peninsula

ALASKA

Anchorage

Valdez

Location of initial spill and immediate spread of oil indicating the key threat to wildlife concentrations

Montague Island

Prince William Sound

Site of accident

Valdez •

shipping lanes
sea otter concentration
wild bird concentration

spill damages as the captain of the ship was intoxicated at the time of the accident. Nearly 330 lawsuits were therefore filed to recoup financial losses from the effects of the spill. By the spring of 1991 Exxon had settled the criminal and civil cases brought by Alaska and the federal government in exchange for $1.1 billion to restore Prince William Sound. However, lawsuits from fishermen, seafood processors, Alaskan residents and a number of businesses were still pending in 1992. The saga of the *Exxon Valdez* is far from being over.

As a result of this disaster, in 1990 the United States Congress passed the Oil Pollution Act. This required by law contingency planning for catastrophic oil spills and established a wide network of regional oil spill response centers. The Act also required better navigational safety, crew licensing and improved training, and even went so far as to demand that all tankers be differently constructed, with double hulls that would contain some of the spill should the outer hull be punctured. Perhaps the most far-reaching provision of the new law is the increased liability of owners and operators for direct damage when spills occur; not only can fishermen sue for damages, but so can local, state and federal governments. Pollution penalties were also increased.

All this has led to increased environmental concern on the part of the oil industry, which is now imposing more stringent self-regulatory controls as it is in its best interest to avoid accidents in the first place rather than incur the costs later. Public opinion is also on the side of safety; in 1992, 48 percent of people in the United States felt that oil spills posed an extremely serious problem. Only abandoned hazardous waste sites ranked higher on their list of environmental concerns. This is the more positive legacy of the *Exxon Valdez*.

Winds of change

Over the last two centuries prospectors in California have been quick to exploit the region's valuable natural resources: first gold and later oil. But in recent years attention has centered on an unlimited energy source: wind.

Wind has been used to power machinery – for example for grinding corn or pumping water – for thousands of years. There are historical records that show windmills in operation in China over 2,000 years ago. In 1880 they were used for the first time to generate electricity, but low fossil fuel prices in the early 20th century made them uncompetitive in most developed countries. It was not until the oil price shock of the 1970s that governments around the world began seriously to consider wind as a viable alternative and renewable energy source. In the Netherlands, the government gave the go-ahead in 1991 to the biggest windmill-building program since the 17th century. Some 2,000 new windmills will be built by the year 2000.

The mechanics of wind energy are simple: the windmill is located in an area of fairly constant, strong wind, which turns the blades of the windmill. These in turn rotate a turbine, creating electricity. A wind-driven turbine can recover 59 percent of the kinetic (the force moving the air) energy in the wind. Wind farms of multiple wind turbines are the most economical way of converting wind into electricity. However, there are drawbacks to wind energy. Windmills make a very loud, throbbing noise as the blades rotate, and this can scare birds and animals in the area and disturb local people. Some people also consider them to be an eyesore. But despite the drawbacks, wind power appears to be one of the best prospects for an energy-hungry future.

The world's largest windfarm in the California desert in the United States. The Heronemus Scheme, under study in the 1990s, envisages a series of 300,000 windmills spread across the Great Plains.

STOCK-PILING ECOLOGICAL PROBLEMS

COMPREHENSIVE CHANGES · CLEARANCE AND CONTAMINATION · PROSPECTS FOR THE FUTURE

Environmental concerns have only recently been added to the political agenda of Central America and the Caribbean, but are certain to figure more strongly in future. Some success has been achieved in nature conservation, but the task of tackling the region's other pressing issues is only just beginning. Two key concerns are widespread soil degradation and deforestation in the mountains of mainland Central America and some of the high islands. Deforestation, in turn, has caused flooding and siltation. Other dilemmas include severe air pollution in urban areas, the heavy impact of tourism on coastlines and coral reefs, oil contamination in the Caribbean Sea, and disturbance to the marine ecology from overfishing. The main causes are rapid population growth and poverty, which put excess pressure on resources.

COUNTRIES IN THE REGION

Antigua and Barbuda, Bahamas, Barbados, Belize, Costa Rica, Cuba, Dominica, Dominican Republic, El Salvador, Grenada, Guatemala, Haiti, Honduras, Jamaica, Mexico, Nicaragua, Panama, St Kitts-Nevis, St Lucia, St Vincent and the Grenadines, Trinidad and Tobago

POPULATION AND WEALTH

	Highest	Middle	Lowest
Population (millions)	88.6 (Mexico)	5.1 (Honduras)	0.3 (Barbados)
Population increase (annual population growth rate, % 1960–90)	3.3 (Honduras)	2.5 (Panama)	0.3 (Barbados)
Energy use (gigajoules/person)	169 (Trinidad & T)	17 (Panama)	1 (Haiti)
Real purchasing power (US$/person)	6,020 (Barbados)	3,790 (Panama)	970 (Haiti)

ENVIRONMENTAL INDICATORS

	Highest	Middle	Lowest
CO$_2$ emissions (million tonnes carbon/annum)	78 (Mexico)	3.3 (Panama)	0.25 (Barbados)
Deforestation ('000s ha/annum 1980s)	615 (Mexico)	2.3 (Honduras)	0.1 (Cuba)
Artificial fertilizer use (kg/ha/annum)	181 (Costa Rica)	68 (Guatemala)	3 (Haiti)
Automobiles (per 1,000 population)	203 (Trinidad & T)	43 (Panama)	5 (Haiti)
Access to safe drinking water (% population)	100 (Barbados)	71 (Mexico)	39 (El Salvador)

MAJOR ENVIRONMENTAL PROBLEMS AND SOURCES

Air pollution: urban high
Land degradation: *types*: soil erosion, deforestation; *causes*: agriculture, population pressure
Resource problems: inadequate drinking water and sanitation; coastal flooding
Population problems: population explosion; inadequate health facilities; tourism
Major event: Ixtoc 1 (1979), oil rig fire and leak; Guadalajara (1992), series of gas explosions

COMPREHENSIVE CHANGES

Central America and the Caribbean have always been subject to earthquakes and volcanoes, which bring immediate and dramatic changes to the environment. Some areas have been repeatedly buried under lava and volcanic dust and scorched by fire, while others are prone to landslides and inundation by tidal waves. Such disruption, though immediately devastating, can enrich the environment in the long run – volcanic ash, for example, has created some of the region's most fertile soils.

Thousands of years ago, human activities started to alter the region's natural landscapes. Crop cultivation may have begun as early as 7000 BC in Mexico, displacing the natural vegetation. Later, terraces were carved into many of the hillslopes, and channels dug to bring water for irrigation. But until colonial times – from the 16th century onward – change was small scale. The predominant form of agriculture was slash-and-burn cultivation. By this method, small plots of land were cleared of trees and cultivated for only two or three years before being left fallow to allow the forest to regenerate. The system, still in use in many parts of the region, helps preserve soil fertility and conserves land quality – as long as population densities do not become too high and the land overused.

Environmental change accelerated with the introduction of commercial farming methods by European settlers. Spanish, English, French and Dutch colonists created large plantations of single crops for export. Some plantation crops, such as sugar cane, required total deforestation of the land before planting. Although most were established on the fertile, flatter land of the valleys and coastal plains, some spread onto hillslopes, allowing the thin, unprotected soils to be washed quickly away by the heavy rains.

The modernization and commercialization of farming in the course of the 20th century has placed even greater demands on the region's soil and water resources. After 1945 high-yielding hybrid maize, adopted from the United States to boost production levels, became widespread across much of the mainland. The hybrids require far more nutrients from the soil than native varieties, and this has led to an expensive dependency on fertilizers, resulting in river and groundwater pollution. In Costa Rica and elsewhere, widespread cattle ranching – established with American investment to supply the United States' huge market for beef – has transformed vast areas of tropical rainforest to poor grassland.

Pressure of numbers
From 1950 to 1970, the region had the highest rate of population growth in the world, at 3.3 percent per annum. Since then rates have slowed but are still high. Densities of population in relation to the supply of farming land are some of the highest in the world. There are well over

600 people per sq km (1,560 per sq mi) of cultivable land in many of the Caribbean islands, and roughly half this on the Central American mainland. Yet the amount of cultivable land – usually 15 to 30 percent of the total land area – is diminishing because of degradation.

The growing number of landless cultivators, who have no option but to squat on and cultivate infertile mountain land, are the main cause of land degradation in the 1990s. Although they practice traditional slash-and-burn cultivation, the population pressure is such that they are forced to overuse the land in order to feed their families. In southern parts of the mainland especially, forests are being

Buried in the jungle (*right*) Temple ruins at Tikal in northern Guatemala. These monuments to the Mayan civilization were deserted in the 9th century and remained undiscovered until the mid 19th century. The Mayans built many such cities, but the forest was able to recover after the decline of their early civilizations. Without proper management, modern civilization will cause permanent damage to the forests.

Map of environmental problems (*below*) At present rates of loss only 5 percent of the region's original rainforests will remain by 2010. Where forests have been cleared, erosion and desertification are serious problems. Volcanoes, earthquakes and hurricanes are common natural disasters.

Central America and the Caribbean

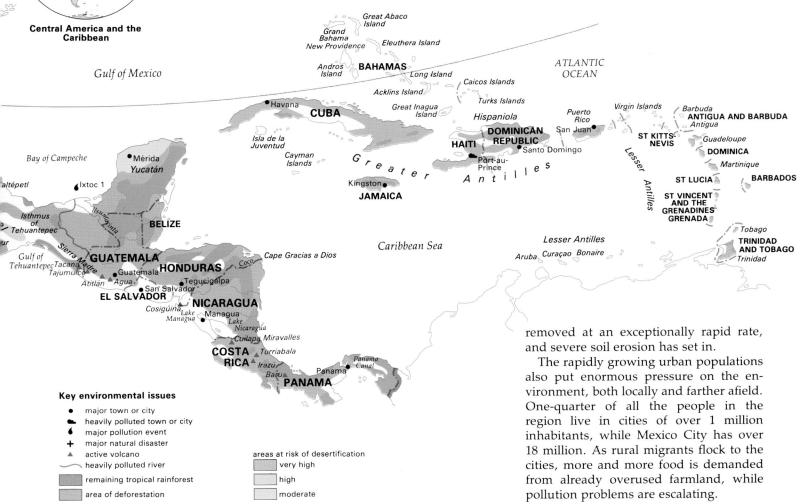

Gulf of Mexico

Great Abaco Island

Grand Bahama
New Providence

Eleuthera Island

ATLANTIC OCEAN

Andros Island

BAHAMAS

Long Island

Caicos Islands

Havana

CUBA

Acklins Island

Great Inagua Island

Turks Islands

Hispaniola

Puerto Rico

Virgin Islands

Barbuda
ANTIGUA AND BARBUDA
Antigua

Bay of Campeche

Mérida
Yucatán

Isla de la Juventud

Cayman Islands

DOMINICAN REPUBLIC

San Juan

Santo Domingo

ST KITTS-NEVIS

Guadeloupe

DOMINICA

Martinique

altépetl

Ixtoc 1

HAITI

Port-au-Prince

G r e a t e r

A n t i l l e s

BARBADOS

ST LUCIA

Isthmus of Tehuantepec

Usumacinta

BELIZE

Kingston

JAMAICA

L e s s e r

Antilles

ST VINCENT AND THE GRENADINES
GRENADA

Gulf of Tehuantepec

Sierra Madre

GUATEMALA

Tacaná
Tajumulco

Guatemala

Coco

Cape Gracias a Dios

Caribbean Sea

Lesser Antilles

Aruba Curaçao Bonaire

Tobago

TRINIDAD AND TOBAGO
Trinidad

Atitlán

Agua

HONDURAS

Tegucigalpa

San Salvador

EL SALVADOR

Cosigüina

NICARAGUA

Managua

Lake Managua

Lake Nicaragua

Miravalles

Cuilapa

COSTA RICA

Turriabala

Irazú

Baru

Panama Canal

Panama

PANAMA

Key environmental issues

- ● major town or city
- ◗ heavily polluted town or city
- ◖ major pollution event
- ＋ major natural disaster
- ▲ active volcano
- 〜 heavily polluted river
- ▩ remaining tropical rainforest
- ▩ area of deforestation

areas at risk of desertification
- ▩ very high
- ▩ high
- ▩ moderate

removed at an exceptionally rapid rate, and severe soil erosion has set in.

The rapidly growing urban populations also put enormous pressure on the environment, both locally and farther afield. One-quarter of all the people in the region live in cities of over 1 million inhabitants, while Mexico City has over 18 million. As rural migrants flock to the cities, more and more food is demanded from already overused farmland, while pollution problems are escalating.

Seasonal rainforest

Small (0.5 ha/1.2 acre) plots for Amerindian agriculture

Plots revert to seasonal rainforest after the removal of Amerindian groups by Spanish colonizers

Forest burnt

Last forest removed for small-scale peasant agriculture

Irrigation canals

Irrigation canals destroyed

Sugar cane cultivation

Sugar cane cultivation abandoned

Livestock grazing on abandoned land

1492: Amerindian period; no soil loss

1600: Hispanic period; soil loss through gullying along trackways of domesticated animals

1700: Sugar cane agriculture; increase in soil loss from cultivated land and from gullying

1800: Revolution and post-revolution period; severe soil loss and gullying

1900–1980: Modern period; major loss of top and subsoil with extreme gullying and sheet erosion

Deforestation in Haiti: the gradual process of deforestation and subsequent erosion from precolonial to modern times. After the arrival of Europeans, each new period of exploitation prepared the way for the devastation that now exists.

CLEARANCE AND CONTAMINATION

Five centuries of increasing environmental deterioration in Central America and the Caribbean have left the region extremely vulnerable to damage from human activities. The consequences of forest loss present some of the region's most intractable problems. Less than 40 percent of the region's natural forests remain, and many of those that do are being stretched beyond their capacity for survival. This is especially the case along the central spine of the mainland, where illegal logging for valued timber and squatting by landless peasant cultivators is widespread.

Mexico, Nicaragua, Guatemala, Honduras, Panama and Costa Rica all have exceptionally high rates of deforestation. Costa Rica has one of the highest rates in the world at 3.9 percent per annum: of the 1.5 million ha (3.7 million acres) of forest that remain, 60,000 ha (148,000 acres) disappear every year. In other parts of the

region – especially in Cuba, the Dominican Republic and northern Mexico – timber removal for fuelwood (for cooking and heating) outstrips natural regeneration in the forests.

Disappearing soils

Forest loss quickly leads to land degradation. This involves first a decline in soil fertility and then soil erosion, especially on steep slopes. Tropical soils are heavy in laterites (iron and other oxides), which in normal circumstances are leached downward through the soil. Removal of the tree cover erodes the topsoil, exposing the laterites and making the land unworkable. Guatemala has lost 40 percent of the productive capacity of the land through erosion. In the Caribbean, sugar-cane plantations and factory sugar estates have been especially destructive. Sugar cane does not tolerate shade and requires total

HAITI'S POVERTY-STRICKEN ENVIRONMENT

The tourist image of a Caribbean island paradise could not be further from the truth for Haitians. Extreme poverty and overpopulation have greatly exacerbated its environmental dilemma.

Ruined by years of neglect, economic mismanagement and corruption, Haiti is the poorest country in the western world, with a per capita income of only $320 per year. People attempt to scratch a living from the most marginal of resources. Less than 2 percent of the country remains forested and the rest is under severe threat from people cutting trees for fuelwood and charcoal, desperate for some form of income.

Much of the country is mountainous and severely eroded. The topsoil disappeared long ago, and now the country is losing even its subsoil to erosion – 14 million tonnes of it are swept by rain down to the sea every year. Farmers are forced ever higher to find usable land for their crops, but the combination of overcultivation and thin soils means that the new plots are quickly eroded.

Although groundwater is plentiful, safe drinking water is scarce. Most of the population of 6 million live without piped water, and inadequate sewerage makes the sources they have vulnerable to contamination and disease. In the "cardboard cities" erected by rural emigrants on landfill in the capital, Port-au-Prince, thousands wash in (and drink) a water-sewage mix.

Defiling an underwater paradise A plastic bag on elkhorn coral in the Bahamas is a visible reminder of the environmental threats to coral reefs, which are highly vulnerable to pollution, fishing and physical damage, all related to urbanization, agriculture and tourism.

clearance of the natural vegetation for it to flourish. The soil between individual crop plants is kept as weed-free as possible during cultivation, leaving the exposed soil highly vulnerable to erosion from rainstorms. Barbados, the first "sugar island", had been cleared of all its tree cover, and suffered consequent severe soil erosion, well before 1700. Haiti, once the richest of all the sugar nations, no longer has any topsoil.

In 1952 most of the forest within the Lake Gatún watershed in central Panama was intact. Today much has gone, leaving bare hills on which soil erosion is severe. The Panama Canal route runs through Lake Gatún and is now threatened by heavy siltation as the stripped soil accumulates in the water. It is estimated that more earth has been removed from the canal since it opened than during its initial construction. The canal is in danger of becoming clogged, and some larger vessels already have to be diverted around Cape Horn. Moreover, because there are now so few trees to intercept rainwater and stabilize run-off, the discharge of water into the canal system has become irregular. Often there is too little water to refill the high-level locks.

Mining activities also cause forest loss. Former silver mining in Zacatecas, Mexico, denuded the forest for kilometers around, while in Jamaica, huge open-cast bauxite mines have left deep-red gashes against the green forested landscape.

Pollution problems

Pollution in the region is becoming ever more widespread. Levels of air pollution in Mexico City are among the worst in the world, with serious implications for the health of its citizens. In Guadalajara, Mexico's second largest city, at least 191 people were killed in April 1992 when industrial gases were discharged into underground sewerage pipes and caused a series of explosions. Raw sewage, discharged from holiday homes around Lake Atitlán, Guatemala, has caused such excessive algal growth in the lake that the oxygen supply has plummeted and 80 percent of the wildlife has died.

At sea, oil spills and oil exhaust products discharged by ships can be spread rapidly by the ocean currents. Increased offshore oil production within the Bay of Campeche in the Gulf of Mexico resulted in a severe oil spill in 1979. In 1986 spilled oil from a refinery in Panama devastated coral, sea grasses and mangroves, killing more than half the coral down to a depth of 3 m (10 ft).

Coral reefs are especially sensitive to pollution. Sewage discharged from hotels clouds seawater, inhibiting the growth of corals by depriving them of sunlight. It also increases nutrient levels in the water, encouraging the growth of aquatic plants, which smother and kill the coral. Overfishing, and sudden declines in algal predators (such as sea-urchins), have also encouraged algae to proliferate. Thus, the ecosystem of the coral reef is destroyed. A new coral reef may take hundreds of years to develop.

PROSPECTS FOR THE FUTURE

Urgent action is required to combat environmental problems in Central America and the Caribbean, but political difficulties and lack of money make effective action hard to achieve. Intergovernmental agreements on the environment have always proved difficult to negotiate, monitor and uphold. Many countries have huge foreign debts: Mexico owes over $96 billion. The scale of such debts often constrains nations in their efforts to direct money toward conservation, reclamation or preservation projects.

Land reform over most of the region would go far to relieving the growing problem of peasant cultivators exhausting marginal farming plots. In Guatemala, for example, 80 percent of the land is owned by only 2 percent of the population, leaving the bulk of the rural population to make a living from the remaining land, which is of inferior quality. Many other social problems hinder the resolution of environmental problems, including population growth, poverty, and lack of education (particularly in how to deal with environmental problems).

The return of trees
Some positive action, however, has started in most countries of the region, notably in the designation of protected parks and reserves. Often tourism has provided the spur for conservation. The demands of wildlife tourism, for example, have led to a new forest preservation plan in the Petén area of northern Guatemala.

In Costa Rica, action on the deforestation issue has been taken a stage further, from protection of the remaining forests to the actual reforestation of previously cleared and often degraded land. Efforts to recreate the dry forests of the Guanacaste National Park in the Pacific lowlands are being paralleled upslope in the Monteverde cloud forest of the central mountains. Here, a project has been launched to buy bare, abandoned farming plots and replant them with forest trees propagated in special nurseries. Farmers have also been paid to give over parts of their land for replanting, which will not only provide windbreaks and renewable sources of timber and fire-

COSTA RICA'S PATH TO SURVIVAL

Since the 1970s Costa Rica has led the way in forest conservation in Central America. The government has worked closely with farmers, conservation bodies and foreign banks to overcome the country's devastating record of deforestation by incorporating surviving forests into protected areas. The latest scheme involves creating a green corridor between the Monteverde cloud forest and another preserve in order to save fast-disappearing species of plants and animals.

The privately owned Monteverde reserve is a remarkably complex ecosystem. It straddles the mountains that divide the Caribbean and Pacific parts of the country, and can be divided up the steep slopes into clearly defined zones of climate and vegetation, each forming a distinct habitat for wildlife. However, land outside the protected area has been extensively deforested to make way for cattle ranching, and conservationists have seen their efforts thwarted as forested areas have been reduced to pockets or islands of refuge amid bare grazing land. Several bird and mammal species that breed in the protected highlands have been deprived of their lowland wintering grounds, and have disappeared from the area.

Local biologists, galvanized into action by threats to the habitat of the rare Umbrella bird, formed the Monteverde Conservation League, which aimed to expand the area of protected forests by linking individual forest stands along a network of forested "paths". These would allow migratory access through woodland. The League negotiated with farmers to buy and reforest 4,000 ha (9,900 acres) of land between the Monteverde forest and another reserve, creating a continuous refuge of over 21,000 ha (51,900 acres).

Symbol of conservation (*left*) A brilliantly colored male quetzal, the national bird of Costa Rica, is just one of thousands of species benefiting from the region's conservation programs.

A self-appointed conservationist (*below*), tending his patch in the rainforest of Panama. The Kuna Indians' way of life, based on harvesting the many resources of the forest without overexploiting them, is in harmony with nature: the Kuna are natural conservationists of their home in the Kuna Yala Reserve.

wood, but will also help to prevent soil erosion on the slopes.

The Panamanian government has also encouraged tree planting in deforested upland areas, and has discouraged – through more vigilant policing – illegal logging and settlement in pristine forests of the Chocó, in the far east of the country. Communities in the interior of Haiti have been encouraged to look after reforestation schemes set up by environmental groups; 10 million tree seedlings have been planted there since 1981. Meanwhile, in Jamaica the government has insisted that bauxite mining companies restore land that has been worked out. The companies must first replace the topsoil and then either reforest the areas or plant them with pangola grass, which provides good feed for cattle. Badly eroded land both in Jamaica's Yallah's valley and Barbados' Scotland district has been remolded and terraced with the help of large subsidies from the government.

Marine protection

Many countries have established marine reserves to protect coral reefs, and while some have set up underwater trails for tourists, others are closed to visitors for full conservation. Overfishing is increasingly being controlled by military patrols in the Caribbean Sea, sanctioned by the international Law of the Sea convention, signed in Jamaica in 1982. Along both Caribbean and Pacific inshore waters, the threat of oil contamination has been reduced by the introduction of new types of oil booms for holding back slicks and new dispersant technology used for breaking up floating oil.

Financial shortages, however, constantly restrict environmental schemes in the region. One recent innovative measure has been the decision of several American banks to write off their foreign debts, and donate the money to conservation schemes in debtor nations. Costa Rica has been one major beneficiary since 1988. Many hope that these debt-for-conservation schemes – very positive contributions to environmental protection – will spread more widely in the region in the immediate future.

Crisis in Mexico City

Few cities in the world so graphically illustrate the environmental hazards of urban overcrowding as Mexico City, where the population is expected to rise to more than 25 million by the year 2000. The city already has severe problems of air pollution, access to clean water and the disposal of sewage and waste. Its crisis reflects the widespread problems of high rural–urban migration, rapid urban growth and poor infrastructure that beset much of the developing world.

Shaky foundations

Mexico City's environmental plight is not helped by its geography. The city lies in a dry mountain basin on lake beds deposited in the old Lake Texcoco. The beds are spongy and vulnerable to subsidence when water is extracted from underneath them. They also accentuate the effects of earthquake tremors, as in 1985, when the Guerrero-Michoacan quake claimed 7,000 lives, demolished at least 1,000 buildings in the center and caused over $4 billion damage. Except for the rainy season from July to September, when floods can occur, the climate is arid and surface water resources are limited.

Subsidence of the lake beds caused by past water extraction (up to 8 m or 26 ft in places) is now so great that both water and sewage pipes have been dislocated. Moreover, most local sources of groundwater have now either dried up or become contaminated by sewage seepage. Today, water supplies are pumped in over the mountain rim from adjacent high valleys at a cost of $47 million per year.

Sewage is also expensive to manage. A new deep system of sewers is being laid 40 m (130 ft) below ground level so as to avoid subsidence effects. Much of the raw sewage from this is piped north for many kilometers along the city's Gran Canal, and then tunneled through the mountain rim and out of the basin. However, 40 percent of homes in Mexico City have not been provided with piped water or sewerage. Even if the city authorities could supply these utilities, they would still have to cope with the influx of 400,000 new immigrants every year, most of whom squat on public land, with no access to clean water or waste facilities.

Noxious air

Mexico City's most intractable environmental hazard, however, is its air pollution, which is probably the worst in the

Mexico City site

- Lake area in 1519
- Present-day lakes
- Urban areas

Sierra las Cruces

Lake Zumpango

Lake Xaltocan

Lake Texcoco

Mexico City

Sierra del Ajusco

Iztaccihuatl (5,286 m/17,343 ft)

Popocatepetl (5,452 m/17,887 ft)

0 — 20 km
0 — 10 mi

A city threatened by its location (*above*) Mexico City is built on gradually subsiding lake beds that exacerbate the effects of earthquake tremors. The surrounding mountains trap air pollution and also make water supply a major problem.

Smoking rubble (*left*) of a collapsed hotel is extinguished by firefighters after the Mexican earthquake of 1985. Mexico City is located at the junction of two great mountain chains – the area of greatest seismic activity in the country.

Traffic in Mexico City (*right*) accounts for 60 percent of solid-particle air pollution. During rush hour, visibility can drop to less than three city blocks. Pollution-related illnesses kill 70,000 city dwellers annually, and others have excessive levels of lead in their blood from vehicle exhaust.

world. The surrounding high mountain rim traps air contaminants in the city basin, preventing their dispersal by air currents. Contaminants, including carbon monoxide, hydrocarbons, sulfur dioxide and nitrogen oxides, are emitted by the city's 3 million motor vehicles and 30,000 factories, most of the latter burning highly polluting low-grade, sulfur-rich fuel. Ozone levels have been recorded at 10 times the normal atmospheric levels.

The Mexican government has launched a 5-year pollution reduction program costing $4.6 billion. It has closed a large oil refinery within the city and set exhaust limits for vehicles and factories. But these are rarely observed and pollution continues to worsen. In March 1992 ozone levels climbed so high that the government temporarily closed schools so that children could stay at home: learning disabilities have been linked to the amount of exhaust that children breathe. It also banned 40 percent of vehicles from the streets, and ordered more than 200 factories to reduce production.

BURNING ENVIRONMENTAL ISSUES

UNEVEN EXPLOITATION · PRESSURES OF DEVELOPMENT · DEMANDS AND ACTION

South America currently faces enormous challenges of environmental management. Outside the region, one issue above all dominates the discussion of the continent's environmental problems – the destruction of the Amazon rainforest. As the largest remaining area of rainforest it is often seen as a symbol of the destruction already wrought elsewhere in the world by humankind. Within the region, economic development and improvements to living conditions have been made a high priority, and the impact on the environment has been considerable. Soil erosion is severe in many areas, especially the Andes, while heavy industrial and urban pollution and poor conditions of sanitation exist in and around some of the biggest cities. Massive hydroelectric and mining schemes have proved environmentally highly controversial.

COUNTRIES IN THE REGION

Argentina, Bolivia, Brazil, Chile, Colombia, Ecuador, Guyana, Paraguay, Peru, Surinam, Uruguay, Venezuela

POPULATION AND WEALTH

	Highest	Middle	Lowest
Population (millions)	150.4 (Brazil)	13.2 (Chile)	0.4 (Surinam)
Population increase (annual population growth rate, % 1960–90)	3.3 (Venezuela)	2.5 (Colombia)	0.7 (Uruguay)
Energy use (gigajoules/person)	88 (Venezuela)	22 (Brazil)	8 (Paraguay)
Real purchasing power (US$/person)	5,790 (Uruguay)	3,810 (Colombia)	1,480 (Bolivia)

ENVIRONMENTAL INDICATORS

	Highest	Middle	Lowest
CO₂ emissions (million tonnes carbon/annum)	610 (Brazil)	69 (Colombia)	0.3 (Guyana)
Deforestation ('000s ha/annum 1980s)	9,050 (Brazil)	270 (Peru)	3 (Surinam)
Artificial fertilizer use (kg/ha/annum)	162 (Surinam)	43 (Uruguay)	3 (Bolivia)
Automobiles (per 1,000 population)	126 (Argentina)	33 (Guyana)	2 (Ecuador)
Access to safe drinking water (% population)	96 (Brazil)	61 (Peru)	35 (Paraguay)

MAJOR ENVIRONMENTAL PROBLEMS AND SOURCES

Air pollution: locally high, in particular urban; high greenhouse gas emissions
River pollution: medium; *sources*: agricultural, sewage
Land degradation: *types*: soil erosion, deforestation, habitat destruction; *causes*: agriculture, industry, population pressure
Resource problems: fuelwood shortage; inadequate drinking water and sanitation; land use competition
Population problems: population explosion; urban overcrowding; inadequate health facilities
Major events: Cubatão, Brazil (1984), accident in natural gas/oil refining facility

UNEVEN EXPLOITATION

Throughout history the intensity of the human impact on South America's natural environment has been markedly uneven. Substantial areas of the continent have been marginal for human activity – too high, too dry, too cold or too steep. Episodic hazards such as drought, flood, frost and earthquakes affect some areas. Over much of the continent soils are poor, and some areas are also susceptible to parasitic diseases such as malaria and schistosomiasis (or bilharzia).

It is not surprising, therefore, that overall population densities are low (about 17 people per sq km or 44 per sq mi), and that large areas of unmodified natural vegetation remain – almost half the continent is still forested. Until recent decades the greatest human impact was confined to southeastern Brazil, the pampas of Argentina and Uruguay, central Chile, the Andes and the coastlands from Peru to Colombia. However, the pattern has been changing recently as develop-

ment has spread into the often vulnerable environments of the hinterland.

Until the 16th century, human use of the land and its resources was mostly limited to hunter–gathering and simple farming. The Incas, who practiced intensive farming in the Andes, were an exception. They terraced the mountain slopes and channeled seasonal rainfall along stone aqueducts to irrigate the plots. They even diverted water from some 50 streams to develop irrigated oases in the coastal desert of Peru, growing maize and other crops.

The scramble for resources

With the arrival of Spanish and Portuguese settlers in the 16th century, however, exploitation of the continent's resources intensified. The European settlers established an extractive economy based on precious metals. Silver was mined at Potosí in Bolivia and gold and diamonds in Minas Gerais, Brazil, with lesser mining centers developing in Chile, Colombia and Ecuador. Commercial agriculture and ranching developed to supply the mines and towns, and to provide exports. Natural vegetation was cleared and replaced by plantations of a single crop such as sugar cane and by vast cattle ranches. In northeastern Brazil, central Chile, the Peruvian coast and the Andean valleys, farms took the form of large landholdings (*latifundia*).

In the 19th century the temperate grasslands of the pampas became grazing lands for cattle and sheep or were used to grow cereals. From the 1880s even the cold, dry plateau of Patagonia in southern Argentina was turned into sheep pasture. In tropical regions, more forested land was cleared as the demand for coffee and other plantation crops increased. Opencast mining also developed, the surface of the land being stripped away so that ores such as copper could be extracted.

In the 20th century agricultural change came about through intensification and expansion. Farming, particularly for cash crops, has involved large units of production, mechanization and increased use of fertilizers and pesticides. Although yields and output have improved, pollution has

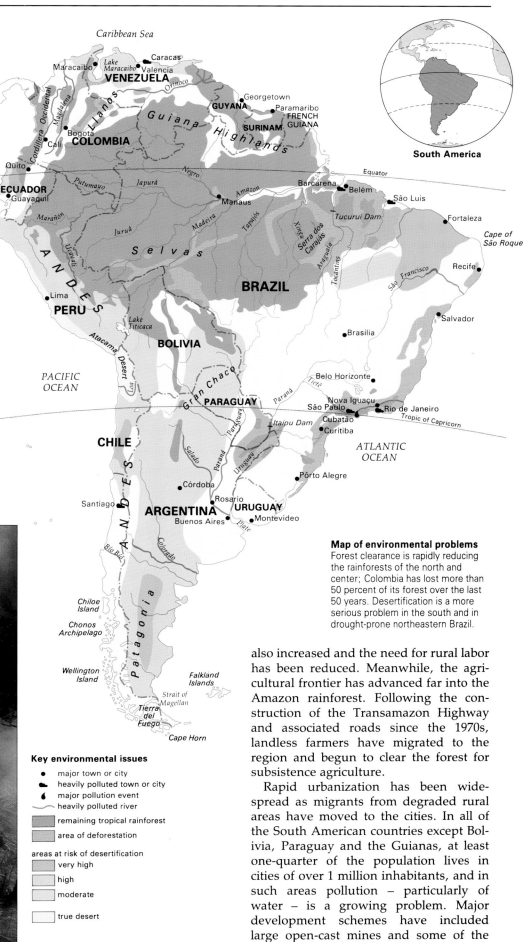

South America

Map of environmental problems
Forest clearance is rapidly reducing the rainforests of the north and center; Colombia has lost more than 50 percent of its forest over the last 50 years. Desertification is a more serious problem in the south and in drought-prone northeastern Brazil.

also increased and the need for rural labor has been reduced. Meanwhile, the agricultural frontier has advanced far into the Amazon rainforest. Following the construction of the Transamazon Highway and associated roads since the 1970s, landless farmers have migrated to the region and begun to clear the forest for subsistence agriculture.

Rapid urbanization has been widespread as migrants from degraded rural areas have moved to the cities. In all of the South American countries except Bolivia, Paraguay and the Guianas, at least one-quarter of the population lives in cities of over 1 million inhabitants, and in such areas pollution – particularly of water – is a growing problem. Major development schemes have included large open-cast mines and some of the biggest hydroelectric dams in the world. These schemes are seen as essential for economic advance, but they also have a major impact on the environment.

Key environmental issues

- • major town or city
- heavily polluted town or city
- major pollution event
- heavily polluted river
- remaining tropical rainforest
- area of deforestation

areas at risk of desertification
- very high
- high
- moderate
- true desert

Rainforests at risk (*left*) South America contains the world's greatest area of rainforest, but it is being cut down, burnt or bulldozed up to 20 times more quickly than it is being replaced.

PRESSURES OF DEVELOPMENT

Most of the major environmental issues facing South America stem from the continent's struggle for economic development. The goals of providing food, jobs, basic services and export revenue have often clashed with the well-being of the landscape and its resources.

As is the case elsewhere in the world, intensive cash crop production usually entails an increase in the application of fertilizers and pesticides, which seep into rivers causing pollution problems for the wildlife and people who use the water. Organic waste from Brazil's sugar-cane crop, when dumped into watercourses, has also contaminated water supplies. The area planted with sugar cane more than doubled to 4 million ha (10 million acres) in the 1980s through the country's drive to produce cane alcohol as a substitute for imported petrol.

The increasing demands made on existing farmland and new marginal plots by subsistence and commercial farmers has increased soil erosion. Lack of either the knowledge or the means to conserve soil particularly affects the smallholders who work the thin soils and steep slopes of the high Andes, as well as the slash-and-burn cultivators who clear plots for farming in the forested foothills and lowlands. The clearance of trees from remote upland areas in the Colombian Andes so that illicit drug crops can be grown has also created severe erosion problems. In Patagonia, overgrazing of sheep pasture has led to soil deterioration across 5 million ha (12.5 million acres).

Into the forest
The greatest advance of agriculture into virgin land has been in the Brazilian rainforest. Both ranchers and squatters have contributed to the potent image of the burning forest by clearing land for livestock and crops. Brazil's environment agency claimed forest losses of 24,000 sq km (9,500 sq mi) in 1988–89, 14,000 sq km (5,500 sq mi) in 1989–90 and below 14,000 sq km (5,500 sq mi) in 1990–91. Aside from the conservation and soil erosion issues, forest burning is seen as one of the contributors to worldwide global warming; carbon dioxide, one of the so-called "greenhouse gases" is soaked up when the trees are alive, but released when they are cut down and burned.

For the continent as a whole, the estimated forest area in 1990 was about 8.6 million sq km (3.3 million sq mi). This represented a reduction of 11 percent since 1980 (against losses of 5 percent in Africa and 9 percent in Asia). Agriculture, urbanization and commercial lumbering are largely responsible. Wood also remains a significant source of fuel. In Brazil, it provided one-fifth of primary energy during the 1980s.

Inundation and pollution
Environmental controversy has accompanied many of the enormous hydroelectric developments and mineral extraction projects launched in South America. Dam schemes have caused great loss of land and displacement of local people by flooding, and they create breeding sites for disease carriers such as mosquitoes. The Itaipu project on the river Paraná between Paraguay and Brazil inundated 1,460 sq km (560 sq mi) of land and expunged the spectacular Guaira Falls.

The search for oil in eastern Ecuador has caused extensive river pollution and spurred deforestation. A single drilling site may occupy only 5 ha (12.5 acres) but access roads open up the forest to settlers. In Brazilian Amazonia, thousands of people extract gold from surface deposits and rivers. The mercury they use to settle the gold during panning washes into streams and has accumulated in fish, a major source of food in the region. The effects in humans can include liver, kidney and brain damage, as well as birth defects.

Industrialization of the landscape (*above*) A huge copper mine and processing plant in northern Chile. The drive for economic development is being conducted at high cost to the environment: habitats are disturbed, people and animals displaced, and pollution damages human and environmental health.

POVERTY, DIRT AND DISEASE

The ravages of disease are intimately linked with the quality of the environment for millions of poor South American people who lack adequate sanitation and clean water. This link became evident in January 1991, when there was an outbreak of cholera in coastal Peru. Cholera is carried by bacteria in water, and in areas of poor sanitation quickly spreads through contaminated drinking supplies and through fish, shellfish and vegetables washed in infected water. The coastal fishing communities of Peru provided an ideal breeding ground for cholera. Within one month of the first case, 32,000 more were recorded, including 139 fatalities. The epidemic quickly spread along the coast, and by April there were cases in Ecuador, Colombia, Chile and Brazil. By mid 1992 the number of deaths throughout South and Central America had reached 4,000.

Dramatic though the rapid spread of cholera has been, another illness, Chagas disease, is responsible for even more deaths throughout South America every year. Less well known than cholera, this disease is carried by a blood-sucking insect that is common in poor, rural homes, especially those with mud-brick walls and dirt floors; each occupant may suffer up to 25 bites per night. An estimated 500,000 people become infected with Chagas disease every year in Brazil, Bolivia, Paraguay, Uruguay, Argentina, Chile and Peru. Tens of thousands die from the associated fevers and longterm effects such as heart disease.

Combating disease Women from the slums of Lima, Peru, learning about the transmission of cholera and other diseases. Most urban residents do not have access to clean water or adequate sanitation facilities.

Industrial growth has been a high priority in South America, but it has brought pollution from fumes, noise and waste. One of the largest industrial concentrations is Cubatao, near São Paulo in southeastern Brazil – a site hemmed in by mountains that slow the dispersal of pollutants in the atmosphere. Its factories include steel mills, oil refineries, and fertilizer and cement works, which together emit thousands of tonnes of sulfur dioxide, nitrous oxides and hydrocarbons. The severe pollution causes health problems, including respiratory illness in local children, and damages rivers and vegetation. Despite a 6-year program of environmental improvement, states of emergency are still called in the area when pollution levels are very high, and production has to be shut down.

Rapid urban expansion has also generated severe problems – uncontrolled urban sprawl, shortages of clean water and drainage, and congestion. In Santiago, Chile, pollution from vehicles, factories and domestic fires masks the view of the Andes some 60 km (37 mi) away.

The shanty towns that have grown up in many of the region's major cities face added threats. Commonly located on hillsides, ravines and tidal flats, they are especially vulnerable to flooding and landslides. In Rio de Janeiro, southeastern Brazil, torrential rainstorms occasionally flood the hillside shanties, spreading effluent and rubbish, causing disease and even washing dwellings away. In 1988 300 people died and at least 10,000 lost their homes following heavy rains.

DEMANDS AND ACTION

South America faces crucial decisions regarding the management of its lands. From both inside and outside the continent, there is increasing pressure for environmental factors to be taken into account during the planning of future development, and for action to be taken to redress existing problems.

The "Great Drought" of 1877, in which an estimated 500,000 people in northeastern Brazil lost their lives, perhaps provides the first example in the region of an environmental hazard being met by direct action: water storage reservoirs were built and irrigated agriculture extended to avert further disaster. Periodic drought remains a scourge in this area. In 1970 a particularly severe drought prompted the government to plan the Transamazon Highway and its agricultural colonies, to encourage people to migrate from the northeast.

South America has a good record in creating national parks. Chile was a pioneer in 1912, followed by Brazil, Venezuela and Argentina in the 1930s, and others more recently. However, the mere designation of a national park is not enough to ensure the conservation of the habitats within it. Funds are needed to maintain parks and other natural reserves, and to protect them against incursions by squatters and miners. Protected areas in South America are generally given low priority in the allocation of

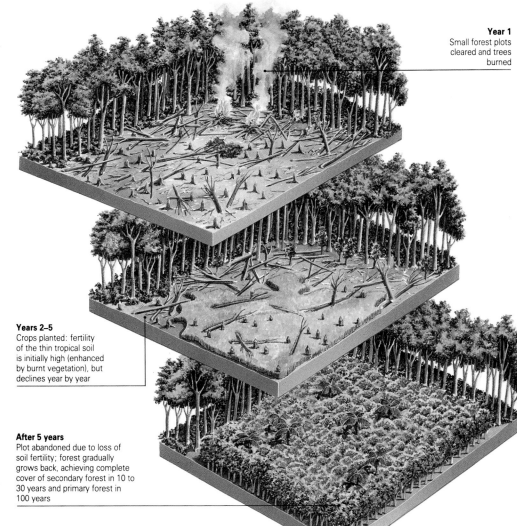

Year 1
Small forest plots cleared and trees burned

Years 2–5
Crops planted: fertility of the thin tropical soil is initially high (enhanced by burnt vegetation), but declines year by year

After 5 years
Plot abandoned due to loss of soil fertility; forest gradually grows back, achieving complete cover of secondary forest in 10 to 30 years and primary forest in 100 years

Sustainable forest farming The traditional method of smallscale slash-and-burn farming allows the forest to regenerate. Land is only cultivated for a few years before soil fertility diminishes, the farmers move on, and the forest grows back.

TIME FOR ACTION

With an overall population density of fewer than 3 people per sq km (8 per sq mi), the Guianas – the three territories of Guyana, Surinam and French Guiana – have yet to suffer environmental damage on a scale seen in other countries. Most of their land is still covered in dense virgin tropical rainforest. But population growth, road-building and mineral exploration now make them highly vulnerable to change. Many conservationists argue that now is the time to safeguard the environment, before disruption becomes widespread. There are already plans to create a 4,000 sq km (1,600 sq mi) reserve of remote and pristine rainforest in Guyana for conservation, genetic research and some sustainable forestry. However, this is threatened by the construction of a road

between Manaus in northern Brazil and Georgetown on the Guyanese coast that will open the area to migrant farmers and miners.

At present, French Guiana, an overseas department of France, has no protected zones whatsoever. Although its population is small (about 100,000, mostly concentrated along the coast), incursions by hunters, often in areas opened up by forestry, have already depleted local wildlife. The World Wide Fund for Nature is pressing for France to start protecting the rainforests in French Guiana. It is campaigning for the creation of nature reserves and for strict control over logging, hunting and road building, and is promoting the idea of a "European Tropical Park" run by the European Commission.

scarce capital, and conservation programs may be expendable in areas that are found to contain mineral reserves or oil.

Programs to conserve soil and timber have been slower to get under way, partly because of a lack of knowledge regarding the consequences of exploitation. Most countries now have basic soil surveys that provide information on soil quality, potential use and management. An important consequence of Brazil's program to develop the Amazon was an aerial survey that provided information for detailed maps, and resulted in the compilation of reports on soil, vegetation, land use and land-use potential.

In northern Argentina attempts are being made to regenerate the semiarid Chaco thorn forest and savanna, which were devastated by several decades of tree felling, charcoal burning and cattle

A mixed audience (*above*) Leaders of indigenous forest peoples and foreign visitors listen intently to proposals for global environmental management from world leaders at the United Nations' Earth Summit held in Rio de Janeiro in 1992.

herding. Controlled grazing has improved the quality of the grassland and permitted the regeneration of the natural woodland. The trees are used for their hardwood timber and tannins.

Some countries have embarked on extensive reforestation programs, mainly for commercial purposes. In Brazil *Eucalyptus* trees have been planted in the southeastern state of Minas Gerais to supply charcoal for blast furnaces, and *Araucaria* pines planted for timber in Paraná, in the south. In southern Chile tax incentives have encouraged commercial planting in excess of 60,000 ha (150,000 acres) a year.

Industrial and urban clean-up

In recent years there have also been moves to counter some of the environmental problems of industry and urban

growth. In 1992 the United States-owned Southern Peru Copper Corporation (responsible for 60 percent of Peru's copper output) proposed to invest heavily in a program to improve the handling of mine waste and smelter gases at its two mines, and sewage disposal from one of its mining settlements. In eastern Ecuador, an oil leak that polluted lakes and rivers and killed fish in the Cuyabeno National Park prompted the Ecuadorian government and the oil companies concerned to sign an agreement to improve environmental management.

In 1991, flooding of the Brazilian city of São Paulo by the river Tietê, heavily polluted by major industries, sewage and shanty town garbage, prompted a 3-year plan to regulate polluters and provide new sewerage and water treatment facilities. Many South American cities are trying to improve other basic services. "Self help" improvement schemes operate in many cities, whereby efforts are being made to provide settlement areas with water, drainage and electricity.

A global concern

On an international level, there have been proposals for "debt-for-nature" swaps, in which part of the considerable foreign debts owed by South American countries are cancelled by the indebted countries on condition that the money be spent to finance environmental projects. However, such schemes have been criticized for making only a limited contribution to reducing debt, while reawakening the vexed issue of foreign control over national territory and resources.

In February 1992 the eight Amazonian Pact countries (Venezuela, Guyana, Surinam, Brazil, Bolivia, Peru, Ecuador and Colombia) signed the Declaration of Manaus, which requested money and technical help from the developed countries to protect the rainforest. It claimed that such environmental problems required new forms of assistance from the "super-rich" countries such as the United States, Japan and Germany, including freer trade, increased financial aid and better access to technology.

The Grande Carajás project

Environmental controversy has surrounded one of South America's biggest ever development schemes. The Grande Carajás project, proposed by the Brazilian government in 1980 as a major regional plan for eastern Amazonia, affects 895,000 sq km (350,000 sq mi) or one-tenth of Brazil's total area, and it involves the development of mining, forestry, agriculture, industry, hydroelectricity and transportation. The key elements are four "mega-projects" – the Carajás iron ore mine, the Tucuruí hydroelectric plant, and aluminum smelting at Barcarena and São Luis. The huge mining concession – covering 4,290 sq km (1,710 sq mi), of which 500 sq km (200 sq mi) was used for the mine site and planned town – preceded the Grande Carajás project.

A watery end (*above*) Rainforest flooded as part of the work on the massive Tucuruí hydroelectric dam. When completed, the dam is expected to supply a significant proportion of Brazil's electricity, but it has already destroyed huge areas of forest and increased the risk of malaria in the country.

Open-cast mining has a poor image in environmental terms because of its direct damage to the land surface. However, World Bank assistance to the Grande Carajás project incorporated environmental conditions, and since 1972 the mining company responsible for the scheme has fostered a research program and management strategy for the mine area as well as the region surrounding its 900 km (560 mi) railroad link to the coast. This research involved surveys of the area's terrain, its rivers and drainage, wildlife,

Making inroads A new road (*left*) and unfinished rail link (*right*) slice through virgin rainforest to the Carajás mines. Such construction work opens the way to destruction of large tracts of the Amazonian rainforest as settlers clear plots beside the road, gradually moving away from it as plot fertility diminishes.

archaeology and its indigenous peoples. Strategies were then formulated for conservation, the acquisition of forest reserves, landscaping, pollution control and the protection of Amerindian lands.

Hidden dangers

Other elements of the Grande Carajás project have been subject to more intense criticism. The giant Tucuruí dam, which when fully complete will house the world's fourth largest hydroelectric plant, has flooded 2,400 sq km (925 sq mi) of rainforest and displaced 20,000 people. Since the reservoir was created, it has become a breeding ground for huge numbers of mosquitoes, greatly increasing the threat of malaria in the region.

Lesser industrial schemes included 25 blast furnaces using Carajás ore and charcoal. Plans to plant eucalypts for charcoal production have foundered because of doubts about their ecological viability in the lowland rainforest region. Only four furnaces had come into operation by 1990 and they have continued to use charcoal from forest trees. Although much of the timber is surplus from cleared land and sawmills, there are fears that the demand for charcoal will generate wider deforestation in the future. The heaviest deforestation would result from proposals to develop agriculture and ranching over 70,000 sq km (27,000 sq mi) of eastern Amazonia, but to date these proposals have not been implemented.

In spite of careful planning, additional environmental pressures have arisen in the Grande Carajás region from the unplanned arrival of thousands of migrants at the project sites. Settlements were created at the Carajás mine and Tucuruí site for the workers required and their families, but no provision was made for the huge number of people who arrived spontaneously seeking work. The population of the Tucuruí site, for example, grew from 800 in 1974 to 85,000 in 1985. The railroad and new highways required by the project have also opened up land to migrant peasants, who have cleared forested areas for farming, caused river pollution, and come into conflict with the Amerindian population.

THREATENED LAKES AND FORESTS

THE CHANGING LANDSCAPE · POLLUTION IN THE SEAS AND WINDS · LEADING THE WAY

Most environmental problems in the Nordic Countries are related to air and water pollution, much of which originates outside the region. Acid rain, carried by the prevailing winds from the industrial areas of Great Britain and continental Europe, is poisoning lakes, rivers and forests. Many rivers in the region, as well as others rising in Germany, Poland and the former Soviet Union, drain into the shallow Baltic sea, which is becoming increasingly polluted by industrial waste and pesticides. Oil spills here and in the North Sea endanger marine life. Concern for the preservation of their remaining wildlife habitats and for the health of their fishing and forestry industries is leading the people of the region to tackle these problems with an energy and willingness that is not always apparent in other countries.

THE CHANGING LANDSCAPE

The Nordic Countries are more sparsely populated than other parts of Europe, and human impact on the environment has on the whole been less: sizeable areas of natural wilderness are still to be found, especially in the far north of the region. In the past, the biggest changes to the natural landscape resulted from farming. The region's fertile soils are restricted to coastal areas of southern Norway and Sweden, parts of western Norway, and to the whole of Denmark (except west Jutland), where most of the native mixed deciduous forests were cleared to make way for agriculture several centuries ago.

In Denmark, reclamation of land began in the mid 19th century when large areas of heathland and bog were improved to create additional farmland. Conifer plantations have been established on the North Sea coast to prevent the coastal sand dunes encroaching onto agricultural land. Iceland's native woodlands were cleared about a thousand years ago by the original Viking settlers, and today the windswept island is sparsely vegetated.

With fishing, timber has always been the region's greatest natural resource, exploited for building, shipbuilding and fuel. During the 19th century, with the rising export demand for lumber and forest products, the mixed coniferous and deciduous forests in southern and central

Drastic measures for a drastic situation (*below*) A helicopter releases a cloud of limestone into a Swedish lake to counteract the effects of acid rain. The Nordic Countries are pioneers in ways of tackling pollution.

COUNTRIES IN THE REGION

Denmark, Finland, Iceland, Norway, Sweden

POPULATION AND WEALTH

	Highest	Middle	Lowest
Population (millions)	8.4 (Sweden)	5.0 (Finland)	0.3 (Iceland)
Population increase (annual population growth rate, % 1960–90)	0.8 (Iceland)	0.4 (Sweden)	0.4 (Finland)
Energy use (gigajoules/person)	199 (Norway)	157 (Iceland)	147 (Sweden)
Real purchasing power (US$/person)	16,820 (Iceland)	13,980 (Finland)	13,610 (Denmark)

ENVIRONMENTAL INDICATORS

CO₂ emissions (million tonnes carbon/annum)	15 (Denmark)	8.7 (Norway)	0.4 (Iceland)
Municipal waste (kg/person/annum)	474 (Norway)	408 (Finland)	317 (Sweden)
Nuclear waste (cumulative tonnes heavy metal)	1,900 (Sweden)	400 (Finland)	0 (Norway)
Artificial fertilizer use (kg/ha/annum)	2,917 (Iceland)	234 (Denmark)	136 (Sweden)
Automobiles (per 1,000 population)	406 (Sweden)	377 (Iceland)	311 (Denmark)
Access to safe drinking water (% population)	100 (Sweden)	100 (Iceland)	97 (Finland)

MAJOR ENVIRONMENTAL PROBLEMS AND SOURCES

Air pollution: acid rain prevalent
River/lake pollution: high; *sources*: acid deposition
Marine/coastal pollution: medium; *sources*: industrial, agricultural
Land pollution: local; *sources*: industrial; acid deposition
Waste disposal problems: domestic; industrial
Major events: Aker river, Oslo (1980), acid leak from factory; Ålesund (1992), oil spill from tanker *Arisan*

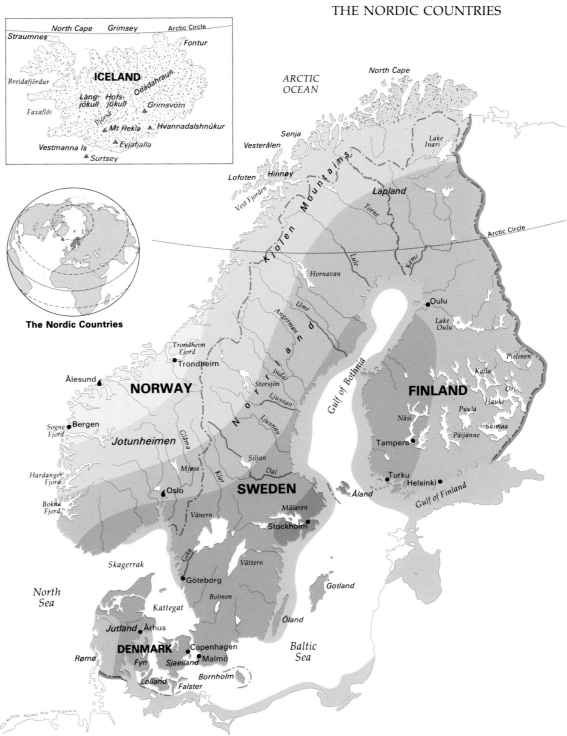

Scandinavia and the boreal coniferous forest that covered most of Finland, northern Sweden and parts of Norway began to be replaced by conifer plantations. Today they have almost entirely supplanted the native forests in these areas, and swamps and marshes have been drained to create new areas for forestry. Conifer plantations provide a relatively sterile environment for wildlife, and numbers of forest-dwelling and wetland species have declined alarmingly throughout the region.

The environmental impact of industry

The lack of indigenous coal reserves delayed the industrialization and urbanization of the region until the late 19th century, considerably later than most other parts of Europe. Change came about

Key environmental issues

- ● major town or city
- ♦ major pollution event
- ▲ active volcano
- ∿ heavily polluted river
- area affected by permafrost
- severe sea pollution

acidity of rain (pH units)

- 4.2 (most acidic)
- 4.4
- 4.6
- 4.8 (least acidic)

Map of environmental problems (*above*) Acid rain, particularly devastating to Sweden and Finland, affects trees, soil and thousands of lakes and rivers. Pollution in the Baltic Sea is worst around the coast, but also affects deep water. The delicate permafrost area of Iceland could be at risk from global warming.

with the development of hydroelectricity. The fast-flowing rivers of the region were well suited to provide this, and many valleys were flooded in the course of constructing dams and reservoirs for the new hydroelectric plants (11 were built in Norway alone between 1896 and 1900). As roads were built to service them, the mountainous interior of the Nordic peninsula was opened up for the first time.

Because of the difficulty in transporting electricity across long distances, industrial development was scattered across the region, initially at waterfall sites and later where metal ores were extracted or on the coast where lumber, floated down the rivers, was processed for export. This

limited the spread of heavy, polluting industry. Large urban agglomerations, comparable to those of the European coalfield areas, developed only around the capital cities and Sweden's main port of Göteborg.

After World War II, however, the expansion of the metal smelting, wood pulp, heavy engineering, food processing and oil related industries created new problems of air and water pollution. Although industries are monitored and pollution limits have been established, regulations are not always observed: six out of ten Norwegian industries still discharge more than their official quotas permit. The growing affluence of the region's population has also increased levels of air pollution as the demand for energy to heat houses and run appliances has risen and car ownership has soared.

POLLUTION IN THE SEAS AND WINDS

The geographical position of the Nordic Countries makes them particularly vulnerable to pollution imported from outside the region. The North Sea has long been used as a dumping ground for western Europe's industrial and domestic wastes; industrial and sewage effluents discharge into it from all the countries around its shores. Currents carry this pollution toward the coastline of Denmark and Norway, where it contaminates beaches and also threatens fish stocks already made vulnerable by overfishing in the North Sea and the North Atlantic.

The exploitation of offshore oil reserves, with the risk of oil discharges, pose an additional environmental threat

Effluent pours out (*above*) from a paper mill at Grums on the northern edge of Lake Vänern, Sweden, where it is treated. High levels of mercury from effluent were found in local fish as early as the 1960s, and their consumption was banned.

to marine life – one that is likely to grow as exploration extends farther into the North Atlantic and the southern Barents Sea. The construction of large terminals to store and process the oil have affected the Norwegian coastline. Spills from the heavy volume of international shipping that passes through coastal waters also endanger marine life.

In 1980, more than 50,000 seabirds were killed in the Skagerrak strait, between Denmark and Norway, when the Greek tanker *Stylis* discharged oil-contaminated ballast water into the sea. In January 1992 the bulk carrier *Arisan* ran aground near Ålesund on the west coast of Norway: the

THE ENERGY DILEMMA

Environmental awareness in the region discourages the use of fossil fuels, primarily oil, which give off high levels of carbon dioxide when burned and contribute to global warming and other environmental ills. However, alternative sources of energy often have only limited potential or bring with them new environmental problems of their own. There is no easy solution. Nowhere is the dilemma more acute than in Sweden.

Hydroelectric power provides nearly half of Sweden's electricity production. However, most of the untapped potential is in the north of the country, far away from the main markets in the south, and grid links are expensive to build. Furthermore, the creation of reservoirs for hydroelectricity floods large areas of land, threatening the way of life of the seminomadic reindeer-herding Sami (Lapps) and destroying wildlife habitats. Opposition from environmental groups and the Sami themselves halted major hydroelectric development on four rivers – the Torne,

Kalix, Pite and Vinde – in northern Sweden in the 1970s.

Nuclear power accounts for the other half of Sweden's electricity production, and is also opposed by environmentalists. The industry was developed in Sweden and Finland from the 1960s onward to reduce the heavy reliance on imported oil, used to fuel power stations and industry. However, public opposition increased in the wake of the Chernobyl nuclear disaster. Concerns had already been raised about safety and the disposal of nuclear waste, and in 1980 a referendum in Sweden resolved that no further reactors should be commissioned and that all existing reactors should be phased out by 2010. This will leave the country with a mounting energy crisis. Although it is developing both biofuel – produced from organic waste – and wind power programs, this will only go a small way to making good the shortfall in supply, and Sweden will have to rely on importing Danish or Norwegian natural gas to generate electricity.

---

fuel oil that spilled from the wreck ruined beaches and threatened the nearby seabird sanctuary at Runde. Seepages from the rusting tanks of the many ships (more than 2,109 since 1914) that have been wrecked along the Norwegian coast are also a source of contamination.

A growing environmental problem in the enclosed waters of the Skagerrak and Kattegat straits and the Baltic Sea is the seasonal appearance of algal "blooms" caused by eutrophication. Fertilizers draining into the sea from agricultural land cause high levels of nitrogen and phospherous to build up in the water. This results in the excessive growth of microscopic algae, and in sheltered coastal waters they may cover the surface for many kilometers. As this blanket is rotted down by bacteria the water is starved of oxygen, killing off aquatic life.

In 1988, a toxic algal bloom 10 km (6 mi) wide and 30 m (100 ft) deep moved through the Skagerrak strait, between Denmark and the Scandinavian peninsula, damaging marine life along more than 200 km (125 mi) of the coastline of western Sweden and southern Norway. Fish farms along the Norwegian coast were badly affected by the algal bloom. Fish farming, an increasingly important economic activity in the fjords of the long Norwegian coastline, itself pollutes coastal waters with waste feed and great quantities of fish feces, while the antiparasitic chemicals used to treat the fish harm other marine life.

A threatened sea

The shallow waters of the Baltic Sea, which is virtually tideless and almost landlocked, are heavily polluted with toxic waste and are especially vulnerable to eutrophication. Heavily polluted rivers drain into the Baltic from the industrial and agricultural heartlands of Poland and the former Soviet republics. Although intensive farming and forestry in Sweden and Finland contribute to the problem of fertilizer runoff, 60 percent of the nitrogen enrichment in the Baltic comes from the former Eastern bloc countries.

The effects have been quite devastating: 100,000 sq km (39,000 sq mi) of the Baltic – nearly half of its deep waters – are virtually "dead".

A particular danger comes from the high levels of mercury discharged into the Baltic from the pulp and paper factories around its shores. Organic mercury compounds enter the food chain and affect in turn shellfish, fish, seals and seabirds. Levels of mercury in some fish are high enough to harm human health, and Denmark and Sweden have now banned the consumption of cod livers from the Baltic.

Pollution from the air

The prevailing winds carry air pollution to the region from other parts of Europe. It falls as acid rain, poisoning lakes and rivers, killing fish, and damaging trees and buildings. Most of the acid rain that occurs in Denmark, Norway and Sweden comes from Germany and Britain; the former Soviet Union and Germany are the major sources of acid pollution in Finland. In addition, soils and vegetation over large parts of the region were contaminated by radioactive fallout from the nuclear accident at Chernobyl, Ukraine, in 1986, affecting the major grazing areas for livestock and reindeer.

LEADING THE WAY

Politicians and the public in all of the Nordic Countries are quick to identify local and global environmental problems and to take action to solve them. For example, Sweden has set itself the target of reducing sulfur emissions from the burning of fossil fuels to 35 percent of their 1980 level by 1995; Norway's goal is a 50 percent reduction by 1993, and Denmark and Finland are aiming for 50 percent by 1995. In contrast, most European and North American levels are set to fall to only 70 percent by 1993.

Steps have also been taken to develop renewable and nonpolluting sources of energy. Because Denmark is low-lying it is able to make considerable use of wind power. Clusters of three-sail windmills have been installed, mainly in Jutland, and many more are planned. There is potential for tidal power along the coasts of Denmark and Norway. Geothermal energy from volcanic springs is widely used in Iceland. Methane from farm waste and biofuels such as wood chips or pelleted domestic refuse are used to heat factories, greenhouses and homes in Denmark, and to a lesser extent Sweden and Finland. The burning of peat, traditionally used as a domestic fuel in Finland and Sweden, is actively discouraged. Conventional power station cooling systems provide hot water to heat apartment blocks and private houses in Denmark.

Curbing the car
The Nordic Countries were among the first to take effective measures to remedy the environmental problems posed by mass automobile ownership. Catalytic converters, which reduce by some 75 percent emissions of nitrogen oxides from gasoline-fueled engines – a major contributor to global warming – have been compulsory in all new automobile engines since 1989. They may only run on unleaded gasoline.

Since unleaded gasoline was first introduced into the Nordic Countries in the mid 1980s there has been a sharp fall in the amount of lead in the air. Samples taken from mosses in southern Norway, for example, show that lead levels in 1990 were about a third of those in 1977; cadmium levels also fell. A special tax on gasoline, with higher rates for leaded than unleaded, was introduced by the

A wind turbine in Sweden (*above*) is used to generate power as part of the country's attempt to become less dependent on nuclear energy and imported fossil fuels. Neighboring Denmark also has 3,000 wind turbines, mostly smallscale.

Hot water on tap (*right*) at a geothermal spring in Iceland. Bathers frolic in the naturally heated pool in close proximity to the "clean" power station in the background. Geothermal energy heats more than three-quarters of Iceland's households.

WASTE RECYCLING

The very high level of concern about environmental issues in the Nordic Countries is reflected in the efforts and time spent by ordinary individuals and the municipal authorities in recycling waste products, thereby reducing the amount of incineration and landfill required to dispose of domestic waste, and conserving resources. Glass, paper and aluminum are commonly recycled, and measures such as charging a small returnable sum of money on all glass bottles encourage reuse.

It is now a common practice in major cities such as Helsinki and Stockholm to sort household waste into different categories at home. The waste is collected as usual by the municipal authorities, but the presorting makes subsequent recycling a much easier process. Denmark has pioneered schemes to incinerate suitable domestic waste locally to fuel district heating systems. Items that present particular problems – batteries, for example, can leak heavy metals if not disposed of properly – can be returned to the store they were bought from for recycling or safe disposal.

To deter the dumping of old automobiles, a returnable deposit on vehicles in Norway is repaid when they are disposed of at a breaker's yard, while since 1975 the price of new Swedish vehicles has included a repayable scrapping fee. Manufacturers of automobiles and other products are increasingly using types of plastics that are suitable for recycling.

Norwegian government in 1991 to further reduce emissions and discourage car use.

A number of other measures have been put into effect to limit car use. These include the pedestrianization of many inner city shopping streets. In Copenhagen and Stockholm, severe city center parking restrictions have been imposed, and in some Norwegian cities, such as Bergen and Oslo, tolls have to be paid by vehicles using the city center streets. Modern, efficient public transportation systems are provided to encourage people to leave their automobiles at home.

Cleaning up the seas
With so much pollution coming to their shores from outside, all the countries of the region have played an active role in promoting international agreements to clean up the seas, often running ahead of their European neighbors. Three conferences on pollution in the North Sea, held between 1984 and 1990 by all the countries bordering its shores, set a target for halving the 1985 level of discharges of most toxic chemicals and nutrients by 1995, and for ending the dumping

and incineration of hazardous industrial waste by 1994. However, the United Kingdom agreed only to halt dumping sewage sludge by 1998, and was alone in persisting with the dumping of radioactive waste. The 1988 Nordic Countries Action Plan to end marine pollution set earlier target dates and was much more wide-ranging.

Finland, Sweden and Denmark, with East and West Germany, Poland and the Soviet Union, were signatories in 1980 of the Helsinki Convention on the Protec-

tion of the Marine Environment of the Baltic Sea. The discharge of all hazardous chemicals and all dumping in the Baltic was banned, and limits were placed on other discharges. The pesticide DDT was banned, and the use of polychlorinated biphenyls (PCBs) – toxic organic compounds widely used as coolants and insulators – was strictly limited.

Action to reduce the pollution of rivers and seas is expensive, but governments and people are prepared to pay the costs. Norway, for example, targeted $5.6 bil-

lion in 1990 to improving methods of sewage and waste disposal and treating industrial and agricultural effluents. The large amounts to be spent on installing and monitoring industrial pollution controls may lower profits, reduce market share and even bankrupt some firms unless international competitors apply the same measures. Nevertheless, in 1988 Sweden ordered its pulp and paper industry – vital to the country's economy – to halve the chlorine pollutants used in the bleaching process.

A hidden poison

When sulfur dioxide (released when oil and coal are burned) or nitrogen oxides (from motor vehicle exhausts) react with oxygen and water vapor in the atmosphere they form acid rain – a sulfuric or nitric acid solution that can be carried over great distances before falling to earth as rain, snow, fog or mist, or as dry depositions or acid particles. As a result, soils, lakes, rivers and groundwater become acidified and poisoned.

Most parts of the Nordic Countries are covered by thin, naturally acidic soils. By adding to the acid levels already present, acid rain leaches out nutrients and releases heavy metals in the soil, which drain into lakes and rivers. Soil in the worst affected parts of southern Sweden is up to 50 times more acid than it was 60, or even 30, years ago, and the store of nutrients has declined by up to a half since 1950. In urban areas acid rain damages buildings and corrodes drinking water pipes.

Dead lakes and dying forests

As the acid levels of lakes and rivers rise, increasing harm is done to fish and other aquatic life. Acid water disrupts the mechanisms by which fish maintain their balance of fluids and they lose body salts; aluminum damages the gills of fish, causing them to suffocate. Young fish are particularly vulnerable, and as acid levels are highest during the breeding season in the spring thaw – when the melting snow dissolves the winter's accumulation of acidic depositions – fish populations have been decimated.

Clear but toxic (*above*) Lake Hästevatten, an acidified lake on Sweden's west coast. Most of the region's lakes are underlain by granite, and so are naturally acidic. Others with limestone or sandstone beds are more resistant to acid fallout, due to alkaline bicarbonate being released from the rock. Most rock in Sweden, however, produces too little bicarbonate to prevent increased acidity from acid rain. Fish and most aquatic plants die. Water lilies, shown here, are a hardy exception, while some algae flourish.

What goes up must come down (*right*) Industrial aerial pollutants are distributed by the wind and converted into even more dangerous substances in the form of acid rain. Areas affected can be great distances from the sources of the pollutants.

Spread of acid rain in Europe
acidity of rain (pH units)

4.3 (most acidic)
4.5
4.7
4.9 (least acidic)

prevailing winds

→ summer

→ winter

Average acidity of rain and snow throughout Europe Prevailing winds from the southwest carry pollutants to the Nordic Countries from Europe, particularly Britain. Less frequently, but more damagingly, light winds (not shown) from industrial Germany can bring severe locally generated air pollution. Only about 10 percent of acid rain in Norway and Sweden is generated by their own industries.

In southern Norway as early as 1921 some lakes were already found to be too acid for fish to survive; now, all the lakes and rivers in an 18,000 sq km (7,000 sq mi) area of southernmost Norway have lost nearly all their fish stocks and the problem is spreading up the west coast and inland. One-fifth of Sweden's 90,000 lakes and 10,000 km (6,200 mi) of rivers and streams are affected: 4,500 lakes have no fish life. Where fish do survive, they are often unfit to eat as they are contaminated by poisonous metals. Finland has no fish-dead lakes, but 500 lakes have become acid in recent decades.

Acid rain causes serious damage to forests by lowering soil fertility and releasing aluminum in soluble form, which harms tree roots. The crowns of the trees lose their density, there is needle loss, and eventually the trees die. In southwestern Sweden in the 1980s a fifth of middle-aged and mature spruce trees were found to have lost at least 20 percent of their needles, and similar damage was found in northern inland forests where soils are thin and poor. Pine trees in parts of Finland had suffered 10 percent loss. Only half of Norway's trees had full density crowns by 1988. Acidification also harms other forms of plant life, including food and fodder crops.

Attempts may be made to counteract water acidity by adding limestone to lakes and rivers. Sweden applies 200,000 tonnes of finely ground limestone annually. However, the cost is high and the process must be repeated every 2–5 years; moreover, liming does not stop the acidification of soils. The only long-term solution is to reduce levels of acid rain. The Nordic Countries have led the way in cutting sulfur and nitrogen emissions, but as most of the acid rain they receive originates outside the region, their actions will be of limited effect unless they can persuade countries responsible for the pollution to pursue the same policies.

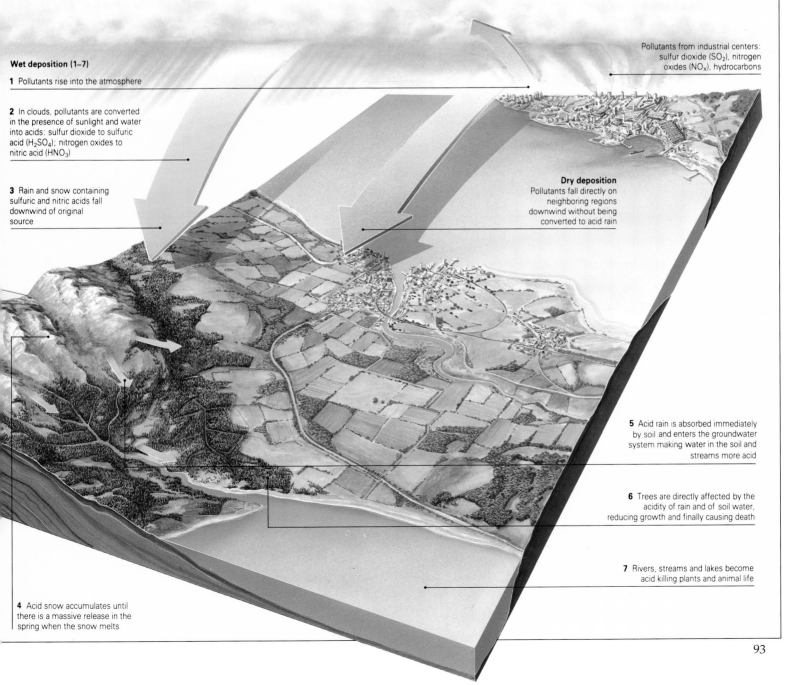

Wet deposition (1–7)

1 Pollutants rise into the atmosphere

2 In clouds, pollutants are converted in the presence of sunlight and water into acids: sulfur dioxide to sulfuric acid (H_2SO_4); nitrogen oxides to nitric acid (HNO_3)

3 Rain and snow containing sulfuric and nitric acids fall downwind of original source

4 Acid snow accumulates until there is a massive release in the spring when the snow melts

Pollutants from industrial centers: sulfur dioxide (SO_2), nitrogen oxides (NO_x), hydrocarbons

Dry deposition
Pollutants fall directly on neighboring regions downwind without being converted to acid rain

5 Acid rain is absorbed immediately by soil and enters the groundwater system making water in the soil and streams more acid

6 Trees are directly affected by the acidity of rain and of soil water, reducing growth and finally causing death

7 Rivers, streams and lakes become acid killing plants and animal life

TACKLING INDUSTRY'S AFTERMATH

THE CHANGING LANDSCAPE · ISLANDS UNDER SIEGE · A GREENER, PLEASANTER LAND

The British Isles changed over 200 years ago from an agricultural to an industrial society, and in doing so laid the foundations of many of the region's current environmental problems. These include heavy use of fossil fuels and pressure on rural areas from the demands of housing, industry and leisure. In the second half of the 20th century an accelerated rate of change has created new hazards, among them radioactivity in the Irish sea from nuclear waste dumping; acidification of fresh water and damage to trees from acid rain; loss of wetlands from excessive peat extraction; and soil erosion brought about by new methods of farming. Although initially slow to take action against environmentally harmful practices, the region has some notable success stories to its credit.

COUNTRIES IN THE REGION

Ireland, United Kingdom

POPULATION AND WEALTH

	Ireland	UK
Population (millions)	3.7	57.2
Population increase (annual population growth rate, % 1960–90)	0.9	0.3
Energy use (gigajoules/person)	101	150
Real purchasing power (US$/person)	7,020	13,060

ENVIRONMENTAL INDICATORS

	Ireland	UK
CO₂ emissions (million tonnes carbon/annum)	0.4	150
Municipal waste (kg/person/annum)	309	313
Nuclear waste (cumulative tonnes heavy metal)	0	30,900
Artificial fertilizer use (kg/ha/annum)	682	356
Automobiles (per 1,000 population)	199	353
Access to safe drinking water (% population)	100	100

MAJOR ENVIRONMENTAL PROBLEMS AND SOURCES

Air pollution: locally high, in particular urban; acid rain prevalent; high greenhouse gas emissions
River/lake pollution: local; *sources*: agricultural, sewage, acid deposition
Marine/coastal pollution: medium; *sources*: industrial, agricultural, sewage, oil
Land pollution: local; *sources*: industrial, agricultural, urban/household
Waste disposal problems: domestic; industrial; nuclear
Major events: *Torrey Canyon* (1967), oil tanker accident; Camelford (1989), chemical accident; river Mersey (1989), oil spill

Environmental battleground Heathland is a fast diminishing habitat, often at the center of conflict over land use, as at this new housing estate in Dorset, southwestern England. Areas of open countryside are increasingly at risk from urban encroachment.

THE CHANGING LANDSCAPE

Change is a perpetual feature of the British landscape. Thousands of years of intensive human activity have so altered it that very little is left in its original form. Across southern, eastern and central England, in the lowlands of Scotland and Wales and in eastern Ireland, over 5,000 years of cultivation and settlement have left only a few remnants of the original habitat of dense, deciduous forest. Even in the uplands and mountainous areas of Wales, Scotland and central northern

England, past enclosure of common land for grazing and previous – but now abandoned – cultivation have left little that is truly natural.

In the past 200 years, moorland, heath and woodland have been further diminished by spreading conurbations, the construction of a road and rail network and the expansion of small towns and villages: land-uses that destroy farmland, scenically beautiful landscapes and wildlife habitats. Since 1945 some 45 percent of Britain's surviving ancient woods have disappeared.

The arrival of modern farming methods after 1945 increased pressure on the environment. Intensive agricultural activity has changed the landscape of eastern England, where over half the hedgerows have been destroyed since 1965, turning a varied pattern of small fields into featureless tracts of agricultural land. Such "factory farming", supported by subsidies from the European Community (EC), has resulted in heavy use of nitrate fertilizers and pesticides, and also in soil erosion as unsheltered topsoil is blown off by the wind.

Modern leisure activities have also changed the face of the countryside. Golf courses, country parks and tracks for walkers proliferate in unspoilt areas and historical buildings have become a focus for visits, as urban dwellers with increasing affluence, time and mobility seek opportunities to leave the cities. On any fine Sunday, some 18 million urban Britons are to be found having a day in the country.

Finding homes for urban dwellers who prefer the space of the suburbs to the inner cities has put even greater pressure on land resources. "Green belts" – areas of land set aside never to be built on – were established round London and other major cities in the late 1940s, but they have since come under constant threat from housing developments, light industry and shopping superstores, and some have been lost as a result.

The industrial legacy

The taking of land for industrial development affected large areas of Britain from the early 19th century onward. Subsequent decline in industry in many parts of the midlands and the north of England, the central valley of Scotland and the Welsh valleys has left a legacy of disused mines, crumbling factories and slag heaps. However, these are progressively being cleaned up by the removal of unsightly tips and industrial ruins. In 1968 the government established "country parks" close to large towns, many made by flooding former quarries to make lakes, or grassing over heaps of industrial waste. In northwestern England miles of pleasant walking track, the Wirral Way Footpath, have been created from an abandoned railroad skirting heavily built-up areas.

Government-appointed bodies such as the Countryside Commission (founded in 1968) and the former Nature Conservancy Council (now divided into four separate agencies, one for each country of the United Kingdom), together with the independent National Trust, currently monitor the rural landscape of Britain. Their constant vigilence, together with deepening public concern, offer hope that future changes to Britain's landscape will be closely scrutinized before they are allowed to happen. The ecology lobby, for example, put great pressure on the government not to go ahead with a planned bypass for the southern city of Winchester that would destroy part of Twyford Down, a highly valued local habitat. Despite the pressure, the government gave the go-ahead in 1992.

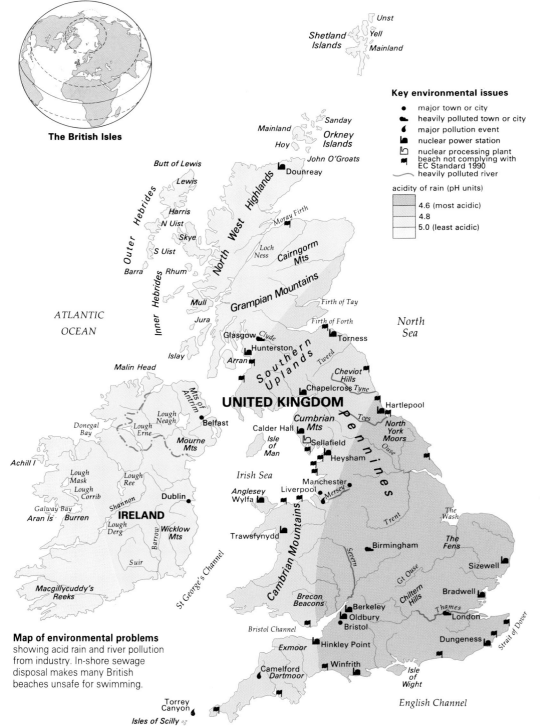

The British Isles

Map of environmental problems showing acid rain and river pollution from industry. In-shore sewage disposal makes many British beaches unsafe for swimming.

ISLANDS UNDER SIEGE

The key environmental issues in the British Isles arise from the fact that the large population is urbanized, industrialized and has a high standard of living. Major problems include the disposal of waste and potential pollution of land, water and air.

About 20 million tonnes of domestic waste is produced annually in the region. Industrial waste amounts to 30 million tonnes, usually far more toxic. Most is disposed of in landfill sites, but disused quarries and gravel pits are running out and during the last few decades the surrounding seas have been used as an alternative. Britain annually dumps over 5 million tonnes of sewage sludge and about 1.5 million tonnes of power station wastes, together with toxic chemicals, into the North Sea and the Irish Sea. About 17 percent of all sewage, over 1,350 million liters (300 million gallons) a day, is discharged directly into the sea, most of it untreated. As a result about one-quarter of all British beaches are, according to EC directives, unfit for bathing.

Dangerous oil

Approximately 550 ships pass through the English Channel daily, more than through any other sea lane in the world. The potential for pollution through collision or the discharge of foul liquids from ships' bilges is therefore high. In the North Sea, 150 oil and gas rigs discharge oil-contaminated drill cores and pump out over 150 million tonnes of oil and 1.1 billion cu m (40 billion cu ft) of gas a year, adding to possible hazards.

Spectacular disasters underline the precarious nature of extraction and sea transportation of oil. They include the fire on the *Piper Alpha* oil rig in July 1988, which claimed 188 lives; an oil spill of 100,000 tonnes from the wreck of the *Torrey Canyon* off the southwest coast in 1967; and the fracture of an oil supply pipe under the river Mersey in northwestern England in 1989.

Nuclear power stations generate about one-fifth of the region's electric power. Much of the spent uranium fuel is reprocessed for re-use at Sellafield on the northwest coast of England, where radioactive waste is stored. In reprocessing, radioactive materials including caesium-137 and plutonium are discharged into

the Irish Sea, where they can contaminate shellfish and thence humans. Public concern has resulted in the reduction of discharges to 3 percent of their 1970s level, but the need for any reprocessing is increasingly being queried, given the world's plentiful supplies of uranium. Debate was fueled by the discovery that the rate of child leukemia in one village near the Sellafield plant was 10 times the national average, though direct links to the plant have not been proved.

British rivers, canals and lakes are generally in good condition, but figures conflict on whether pollution levels are improving or deteriorating. The government claims that 95 percent of Britain's waterways contain good to fair quality water; the National Rivers Authority, the government's water pollution watchdog body, suggests that there has been an overall deterioration since 1985. Nearly all the pollution comes from intensive farming: pesticides, nitrates and unadulterated farm slurry make their way into surface and groundwater, affecting the quality of drinking water and encouraging the growth of bacteria, which in turn starves fishes and plant life of the oxygen they need for survival.

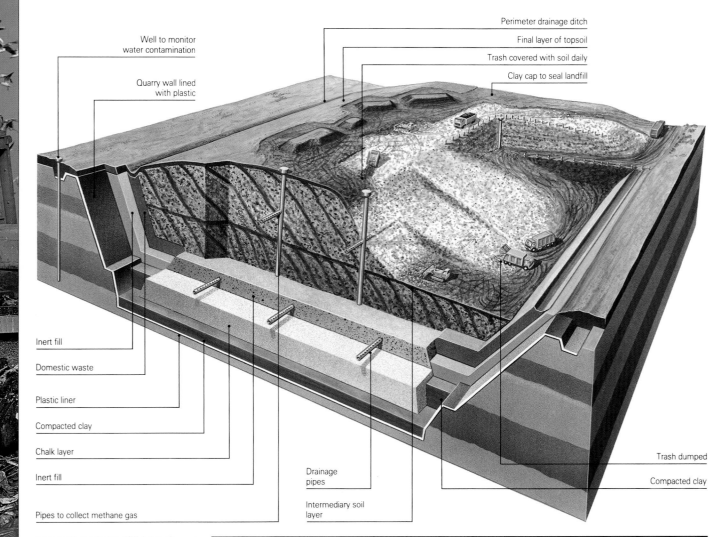

Well to monitor
water contamination

Quarry wall lined
with plastic

Perimeter drainage ditch

Final layer of topsoil

Trash covered with soil daily

Clay cap to seal landfill

Inert fill

Domestic waste

Plastic liner

Compacted clay

Chalk layer

Inert fill

Pipes to collect methane gas

Drainage
pipes

Intermediary soil
layer

Trash dumped

Compacted clay

The well-made landfill (*above*)
About 90 percent of Britain's solid
waste is buried in landfill sites such
as old quarries. A well-run site is a
complex engineering system
designed to contain contaminated
water and any explosive gases that
may build up in underground pockets
as the waste rots.

A scavengers' paradise (*left*) A
landfill site in northern England. The
concrete pipe below the bulldozer is
used to test for and collect methane
– an explosive gas that has been
recognized as a potential alternative
energy source.

Farewell to London fog

One of the most famous health hazards in
19th and early 20th century Britain were
the "peasouper" fogs that blanketed large
cities in winter, caused by emissions
from thousands of coal-burning fires and
exacerbated by industrial and traffic
fumes. The "Great London Smog" of 1952
killed 4,000 sufferers from heart and respiratory diseases, and led directly to the
passage of the Clean Air Act (1956), a
model for similar legislation elsewhere.
The Clean Air Act created "smokeless
zones" where coal fires were banned and
industrial air pollution was strictly con-

GETTING RID OF WASTE

A new approach to the problem of
waste management in Britain was taken
in the 1992 Environmental Protection
Act. Under this legislation, the operator
of a waste disposal site that is causing
pollution must rectify the situation,
however much that may cost. In other
words, the polluter must pay. (The
legislation does not apply to closed
landfills, and in 1992 at least 1,000 of
these were in need of remedial action.)

Over 90 percent of the region's waste
is disposed of in 4,000 licensed and
controlled landfill sites. Most are well
managed and have few instances of
toxic leakages. The Act's powers, however, were put to the test in 1992 over a
case where 450,000 liters (100,000 gal-

lons) of pesticides were dumped annually on two sites north of the city of
Peterborough, in the English midlands,
contaminating the city's underground
water supply. Water treatment has
since removed the residues.

Lack of suitable landfills near some
urban areas encouraged the building of
municipal incinerators. At least four of
these run at temperatures sufficiently
high to cope safely with toxic waste.
However, when such chemicals were
sent from Canada to Britain in 1989 for
incineration, direct action by seamen
stopped their entry. Since then EC
legislation has required all countries to
become self-sufficient in disposal facilities for toxic waste.

trolled, resulting in a dramatic fall in
urban sulfur dioxide levels.

Ground level pollution has been improved, however, at the expense of high
level pollution. Tall smoke-stacks on coal-
burning power stations carry fumes and
particles far upwind, including to Scandinavia. By the early 1980s Britain was the
world's fourth biggest producer of sulfur

dioxide and was suffering badly from
acid rain. The increased water acidity
kills fishes and other freshwater life;
damages plants, crops and trees; eats into
the stonework of historic buildings and
destroys wildlife habitats such as upland
peat bogs. Acid rain has replaced smog as
the major air pollution problem in late
20th century Britain.

A GREENER, PLEASANTER LAND

Public concern in Britain over environmental issues has mounted to stop the worst excesses of pollution. In 1990 the government published a comprehensive review of environmental policy, which pointed to many successes and firsts in the world: improved air quality in London and other cities; the clean-up of the river Thames; some of the world's most stringent planning legislation to control urban sprawl; and tax incentives to encourage the use of lead-free gasoline. The North Sea is also being cleaned up; pressure from European neighbors and protest groups such as Greenpeace has ensured that dumping of chemical industrial waste will end by 1993, coal wastes by 1995 and sewage sludge by 1998. Incineration of toxic chemicals at sea has already ceased.

By 2000, all British drinking water should meet EC standards. A new National Rivers Authority has independent powers of inspection to monitor water quality and filtration standards. Massive government grants have been given to improve all sewage sludge disposal.

Facing up to the problems

Transnational problems of excess radioactivity and acid rain have not yet been satisfactorily solved. Radioactive fallout from the Chernobyl reactor disaster in the former Soviet Union in 1986 still affects areas of high rainfall such as upland north Wales, where sheep cannot be sold for public consumption. About 20 percent of British electricity is obtained from nuclear power. Although the industry argues that it provides the most environmentally clean way to generate electricity and that storage and filtration have revolutionized nuclear waste disposal techniques since the 1970s, public opinion is not yet convinced. Extensive public inquiries have delayed the opening of some planned power stations such as Sizewell B in southeastern England.

The British electricity industry – mainly run off fossil fuels such as coal and gas – fought hard against accepting responsibility for producing acid rain, which in the 1970s caused serious environmental damage in Scandinavia, and only in 1988 did Britain agree to substantial national controls of sulfur dioxide emissions.

Under EC law, these must be reduced by 60 percent of 1980 levels by 2003. The industry faces huge costs; it will have to install "scrubbers" to strip out pollutants from emitted gases, and must also import coal with a lower sulfur content than British coal. The effect on the British coal industry could be catastrophic, but the environment will benefit.

Alternative sources of power for electricity generation have been sought: wind power, geothermal heating, tidal power. One long-discussed scheme is the construction of a barrage across the Severn estuary between England and Wales to trap an average head of 6.6 m (22 ft) of water on every tide, producing 7 percent of the region's electricity needs. The scheme has foundered on massive capital costs and on the damage that will be caused to habitats by the flooding of upstream wetlands.

Waste as a source of fuel is also being explored, but the endproduct does not yet meet EC standards for heavy metal emissions. Waste is, however, being successfully reprocessed to make compost at one plant at Byker, in northeastern England; if this industry can be developed, not only will the volume of waste be reduced, but the compost may prove a substitute for diminishing peat reserves in Ireland and southwestern England.

Recycling is another way to reduce waste, but the region's record is poor by western European standards; only 1.6 percent of British waste is currently recycled. However, pilot government schemes are now operating in several cities including Sheffield (England), Cardiff (Wales) and Dundee (Scotland).

Restoring the forests

Tree-planting in Britain is increasing, but while nearly half of existing deciduous woodland has been felled since 1900, replanting has been mainly of conifers for timber. The Forestry Commission has planted nearly 2 million ha (5 million acres) since 1900, and is currently planting 33,000 ha (82,500 acres) each year. However, these commercial plantations are not always welcomed. In northern England and Scotland, for example, peatland habitats have been destroyed as a result. Large blocks of conifers not only disfigure the landscape and discourage wildlife; they also tend to acidify soils.

Bigger grants are now available for planting deciduous trees, especially to

Shades of green (*left*) Planted areas of coniferous (dark green) and deciduous (light green) trees in the New Forest in southern England. Deciduous plantations are preferable to coniferous as they support more wildlife, but they are less commercially valuable.

Roadside sanctuaries (*above*) Major new road systems in the British Isles have created pockets of natural vegetation on their verges. Areas once kept mowed and sprayed to kill so-called weeds now support diverse flowering plants and birds of prey.

screen conifer plantations. Some 12,000 ha (30,000 acres) of deciduous woods are being planted on farmland that has been taken out of production (set-aside land). One scheme under debate is the planting of a vast "amenity" forest between cities in the English midlands; another is the creation of a 364 sq km (140 sq mi) forest in Scotland between the cities of Edinburgh and Glasgow.

Although the region was at first slow to respond to threats to its environment, the situation is now changing; environmental issues are assuming greater priority, and Britain is playing a larger part in international environmental debate. However, the British Isles still lags behind other European countries, such as Germany and the Netherlands, in terms of taking environmental action.

KEEN GREEN CONSUMERS

It is a measure of the rise of environmental awareness that many consumers demand goods that are "environmentally friendly" or "green". Such goods take many forms: paper recycled from waste; hair and deodorant sprays that do not contain chlorofluorocarbons (CFCs) that harm the ozone layer; hardwood articles from sustainable forests; washing machines that use less water and energy; unleaded gasoline. In the food market, shoppers are more conscious of the effect that the residues of sprays and fertilizers may have on their health, and are critical of the methods used in intensive animal farming. In the 1980s vegetarianism rapidly increased in popularity and shops and restaurants quickly adapted to accommodate this change. Some people are also willing to pay more for organically grown vegetables and fruit. Although often less attractive in appearance, because of

their irregular shape, they are grown without the use of pesticides and are therefore better for health.

Just how big this market has become is difficult to determine, but surveys in the early 1990s suggested that 20 percent of consumers actively sought out "green" goods; another 30 percent thought of environmental considerations when shopping; while 30 percent said that they were concerned about the environment but did not consider it when shopping. It seemed that about 5 percent were willing to pay more for "green" goods than for their ordinary alternatives.

"Green consumerism" is, however, more complex than just selling green products. Companies now have to regard issues such as environmental performance, trading policies and community involvement as important, and businesses that neglect these may face increasing sales resistance in the future.

Rebirth of the Thames

From Roman times onward the river Thames was the lifeblood of the city of London: the means of communication to a busy port, the source of drinking water and the main sewer. Its banks were flat and muddy, and although the tide scoured the river twice daily, its waters carried pollution and disease. As early as 1357 there were complaints about the stench and filth of the waterway. By the 1840s, when London was the largest city in the world, the situation had reached crisis point, and there were regular epidemics of cholera, a disease caused by water-borne bacteria.

By 1850 the dissolved oxygen in the water, on which aquatic life depends, was reduced to nil and there were no

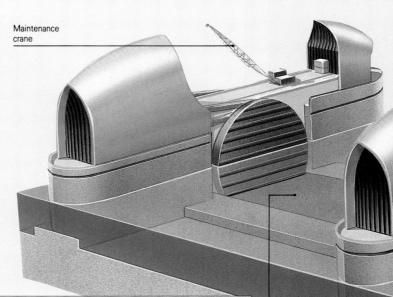

Maintenance crane

Gate arm (1,100 tonnes)

Gate in lowered position to allow boats to pass

Pier

Thames flood barrier and defenses (*above*) The risk of the Thames flooding because of massive surges of storm tides increases each year as the land in this corner of England sinks. To protect London, a huge flood barrier was built east of the City at Woolwich, with gates that can be raised to hold back rare but potentially devastating tidal surges from the North Sea. Upstream defenses (riverside banks) were increased in height and a small barrage built to protect Barking Creek, and downstream defenses were raised by 2 m (6.6 ft). All these defenses should be effective until 2100. The map shows the areas liable to flood before the barrier was built. While only parts of central London would have been directly affected, services such as power, water and sewerage would all have been disrupted for some distance around.

Harvesting the returns (*left*) of the campaign to clean up the Thames, two fishermen by Tower Bridge have caught a basketful of eels. Pollution in the Thames was so severe by 1840 that salmon disappeared from the river and stayed away for 140 years.

longer any fish in the river, the smell from which was so bad in summer that politicians were evacuated from the newly built Houses of Parliament on the riverbank. During the 1850s, however, the situation slowly began to improve. After the cholera epidemic of 1854, which killed thousands, a connection was made between the disease and the water supply.

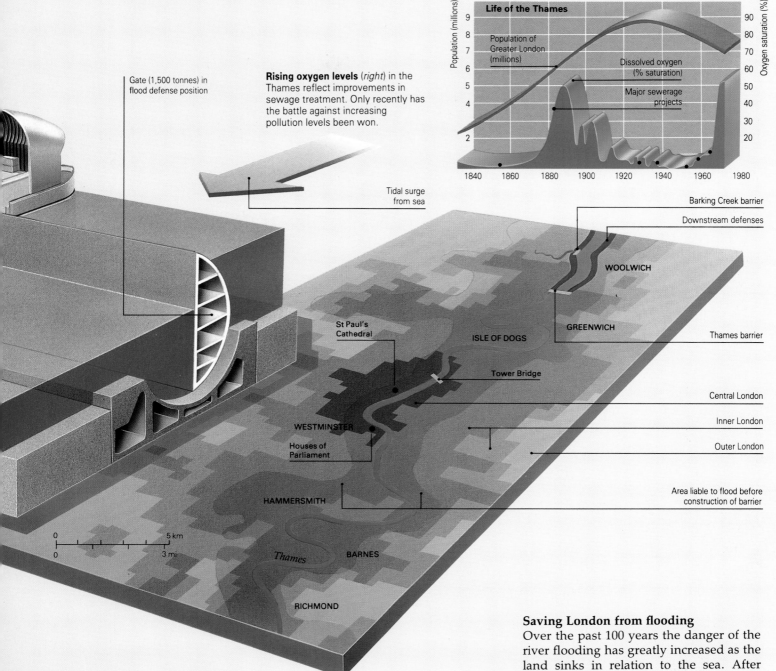

Gate (1,500 tonnes) in flood defense position

Rising oxygen levels (*right*) in the Thames reflect improvements in sewage treatment. Only recently has the battle against increasing pollution levels been won.

Tidal surge from sea

Life of the Thames

Population of Greater London (millions)

Dissolved oxygen (% saturation)

Major sewerage projects

Population (millions)

Oxygen saturation (%)

1840   1860   1880   1900   1920   1940   1960   1980

Barking Creek barrier

Downstream defenses

WOOLWICH

GREENWICH

Thames barrier

ISLE OF DOGS

St Paul's Cathedral

Tower Bridge

Central London

WESTMINSTER

Inner London

Houses of Parliament

Outer London

HAMMERSMITH

Area liable to flood before construction of barrier

0   5 km

0   3 mi

Thames   BARNES

RICHMOND

This led to a massive public works program of connecting sewers and treatment works, with two systems emptying below the urban area on either side of the river. Stone embankments were built, speeding up the flow of the river through the center of the city. For a time the quality of the water improved.

London's population continued to grow rapidly, however. Expanding industry emptied large quantities of toxic phenols and ammonia chemicals into the water, which became more polluted. Flush toilets became mandatory in new housing and cesspools had to be linked up to the sewers, increasing overall pollution. By 1915 the 60 km (37 mi) stretch from Tilbury on the coast to Fulham in central London was declared "dead".

After World War II, when London's population had reached over 8 million, the river's condition again deteriorated as power stations emptied in their warm waste water and a new hazard of non-biodegradable detergents appeared, decreasing oxygen and building up foam on the waterway. Major improvements to the sewage treatment works began in 1950 and over the next 20 years a huge sum was spent on upgrading the system. Industrial pollution was strictly monitored, and in 1964 three species of fish were spotted in the Thames. By 1975 salmon had returned for the first time in 140 years, and in 1992 over 100 species of fish were recorded in the river.

Saving London from flooding

Over the past 100 years the danger of the river flooding has greatly increased as the land sinks in relation to the sea. After some alarming "near misses", it became obvious that the lives of thousands of people living near the river in London were seriously at risk and a flood defense scheme was essential. Construction of the Thames Flood Barrier began in 1974 and was completed in 1982. Spanning the river to the south, its 10 separate movable steel gates can be hydraulically raised to prevent surging tides backed by stormy winds from flowing over the top of the embankments.

When the cleaned-up river Thames with its flood barrier in place is put alongside the Clean Air Act of 1956, which eliminated the London smogs, the city's record of environmental action is seen to be notable both for its pioneering quality and its success.

The great plastic debate

Lightweight, versatile plastic is the preferred packaging of the industrial world. By 1989, 40 percent of Europe's plastic was manufactured for packaging – some 9.2 million tonnes annually. In Britain the figure is 1.3 million tonnes annually, one-third of all plastics produced.

Following the international rise of the "green" movement in the 1980s, plastic gained a reputation as environmentally unfriendly: manufacturing it required too much energy and produced ozone-damaging chlorofluorocarbons (CFCs), as well as creating chemical waste. In addition, the finished product ended up as nonbiodegradable waste, taking up landfill space and killing marine life when dumped at sea. Paper, a biodegradable "natural" product, was claimed to be a better alternative.

However, many experts now say otherwise. An American archaeologist found that plastics make up less than 5 percent by weight and 12 percent by volume of dumped garbage. "Biodegradable" paper, which accounts for a much higher proportion, decomposes very little in airless landfills. Furthermore, the production of paper is less efficient and clean than the production of plastic: one tonne of paper cups (100,000 cups) requires 10,000 kg of steam and 6,400 kilowatt-hours of electricity, compared with 5,000 kg and 180 kilowatt-hours for one tonne of plastic cups (650,000 cups). Paper bag production creates twice as much sulfur dioxide, three times as much carbon monoxide, six times as much dust and fifty times as much waste water as plastic bag production.

Bags of waste polythene are sorted prior to recycling. The waste is turned into pellets that are reprocessed by an extrusion method into polythene sheeting, used in the building industry.

DILEMMAS OF A MODERN AGE

France only began to develop a wide-spread interest in environmental matters after World War II, when the region changed from being predominantly agricultural to a major industrialized power. From the late 1960s the problems of air and water pollution were addressed, though with mixed results, and since then a growing emphasis has been placed on the disposal of domestic and industrial solid waste. The region is relatively little affected by problems such as acid rain, found in most other parts of Europe. More pressing concerns relate to the potential environmental hazards associated with nuclear power, which provides about three-quarters of France's electricity, and also to the growing pressures on land use and resources from the rapid development of tourism and recreational activities.

COUNTRIES IN THE REGION

Andorra*, France, Monaco*

POPULATION AND WEALTH

Population (millions)	56.2
Population increase (annual population growth rate, % 1960–90)	0.7
Energy use (gigajoules/person)	109
Real purchasing power (US$/person)	13,590

ENVIRONMENTAL INDICATORS

CO$_2$ emissions (million tonnes carbon/annum)	120
Municipal waste (kg/person/annum)	272
Nuclear waste (cumulative tonnes heavy metal)	12,700
Artificial fertilizer use (kg/ha/annum)	299
Automobiles (per 1,000 population)	391
Access to safe drinking water (% population)	100

MAJOR ENVIRONMENTAL PROBLEMS AND SOURCES

Air pollution: locally high, in particular urban
Marine/coastal pollution: medium; *sources:* industrial, sewage, oil
Land pollution: medium; *sources:* industrial, nuclear
Waste disposal problems: domestic; industrial; nuclear
Population problems: tourism
Major events: Val d'Isere (1970), major avalanche; *Amoco Cadiz* (1978), oil tanker accident; Les Arcs (1981), landslide; le Grand Bornand (1987), major flood; Nîmes (1988), major flood; Protex plant, Tours (1988), fire at chemical plant

* *Not included in figures*

TWENTIETH-CENTURY TRANSFORMATION

Until the mid 20th century, changes to the French landscape were characteristically slow and steady, rather than sudden and dramatic. Woodland clearance began as early as 2000 BC, and by the early Middle Ages probably over half the country's temperate and evergreen forests had been cleared for cultivation. In parts of the south of France, subsequent overgrazing of deforested land by sheep and goats removed much of the protective ground-cover, leaving the soil highly vulnerable to erosion: extensive areas were transformed into scrub vegetation, known as maquis, or a scattered shrub habitat called garigue. In the west, the coastal marshes were drained and reclaimed – either to allow settlement or to extend the agricultural area – and hillsides in the south were terraced for farming.

Subsistence farming, centered either around the village or the family plot, persisted as the main form of land use up until the mid 20th century. In contrast to many other western European regions (notably Britain) the environment was relatively little changed. (One exception was the extensive planting of pine forests during the 19th century across the Les Landes region of southwestern France, creating the largest pine forest in western Europe.) Widespread urbanization had not yet taken place, and there was little largescale industrialization outside the capital and the traditional mining areas of north and northeastern France.

Progress brings sudden change

Extensive and radical transformation of the environment came from the mid 1950s onward. By this time France was experiencing rapid population growth, together with profound changes to its economy. The increase in population,

France and its neighbors

Industrial landscape (*above*) The port of Le Havre in northern France is one of the most polluted areas in the country – the result of rapid industrial growth, especially in the oil refining and petrochemical industries, after 1945.

Map of environmental problems (*above*) Nuclear power has reduced air pollution from fossil fuels, but tourism and industry have polluted the coasts, and in the south avalanches and forest fires are common.

Key environmental issues

- major town or city
- heavily polluted town or city
- major pollution event
- + major natural disaster
- nuclear power station
- nuclear processing plant
- beach not complying with EC standard 1989
- heavily polluted river
- main area of coastal tourism
- main skiing area
- area of fire risk

remaining forest
- coniferous
- mixed
- broadleaf
- sclerophyllous

allied to accelerating migration away from rural areas, led to the rapid expansion of towns and cities. Huge, hastily built apartment complexes (*grands ensembles*) spread onto the rural areas on the outskirts of towns, defacing the landscape. While the population of the rural Massif Central or southern Alps rapidly dwindled, in many urban centers the pressure on land and resources intensified.

Industrial development gave rise to large estates of heavy and frequently polluting industries (including oil refining, chemicals and petrochemicals, and heavy metallurgy) around major ports such as Marseille in the south and Le Havre and Dunkirk on the north coast, and in the northeastern areas of the Nord, Lorraine and the lower Seine valley. Water and

power supplies, and road and rail transportation were all necessary to support these new developments. But by the 1990s some of the traditional mining and steel-making areas of the north had already become derelict industrial land.

Countryside under pressure

The rural landscape also changed significantly after the 1950s as modern farming practices were introduced, including increased use of chemical fertilizers and pesticides. Hedgerows were bulldozed to create larger fields, disrupting local ecosystems and modifying micro-climates. Half the wetlands of the Camargue on the Mediterranean coast were turned into rice fields, and much of the Marais Poitevin, the main winter home of wildfowl on the western Atlantic coast, has been drained to grow wheat and sunflowers.

Added pressure on rural areas has come from the increased demand for recreational facilities. As incomes have risen,

the leisure time of city dwellers has increased and more and more people have turned to the country – particularly around Paris – for leisure pursuits such as golfing, sailing and walking. Demands from the highly profitable tourist industry have also increased, particularly along the Mediterranean seaboard, including Monaco, and to a lesser extent in Andorra. Residential overcrowding and congested roads have become serious problems.

LIBERTY, ECOLOGY, CATASTROPHE?

Questions of water supply and pollution are major environmental issues in the region, despite improvements since the early 1960s. Industrial effluents are still discharged, inadequately treated, into the rivers, lakes and the sea, particularly around large port–industrial zones such as Le Havre, where toxic substances including phosphates, sulphates, zinc, chrome and lead are pumped into the Seine estuary. Near Marseille the Etang de Berre, a shallow saltwater lake, is heavily polluted, while the dumping into the Rhine of waste salts – extracted with potash in southern Alsace – has long been a source of controversy with other countries bordering the river.

Methods of disposing of domestic sewage are also unsatisfactory and fall below the standards set by some neighboring European countries. Along the densely urbanized stretch of the Mediterranean coast to the east of the Rhône delta, beaches have had to be closed because of the high levels of effluent pollution, exacerbated during the summer by the massive influx of tourists.

In agricultural areas such as the Paris basin, increased use of fertilizers and pesticides on cereal crops has caused excessive levels of nitrates to build up in rivers and streams. The same problem is common in Brittany, where effluent run-off from intensive pig and poultry rearing has also raised nitrate levels in watercourses. This can stimulate the rapid growth of aquatic plants and algae. As these die and are broken down by bacteria, dissolved oxygen in the water is reduced, and fish and other forms of aquatic life are unable to survive.

Increased irrigation of crops has also brought problems. Some 1.2 million ha (3 million acres) were being irrigated in the early 1990s, more than twice the total in 1970. As a result, there have been severe water shortages in southwestern France, an area already hit by droughts in 1976 and 1989. The construction of a series of small dams and reservoirs could help solve the problem, but only at high cost and the risk of ecological damage.

Air and ground pollution
The main contributors to air pollution in the region are the chemical and oil

Water shortage (*left*) A thin trickle in a riverbed in the Alpes de Provence in southern France is the result of a prolonged period of drought. The problems of erratic water supply due to irregular rainfall have been exacerbated by the heavy demands made by the expansion of irrigated agriculture, which has boosted crop yields.

Unfit for fish (*right*) Coastal overdevelopment along the Mediterranean is hazardous to aquatic life and human health. Huge fleets of boats – from small craft to oceangoing yachts – dump fuel into the harbors, while industry and tourist hotels contribute their own waste. Many beaches in the region do not meet EC standards for environmental safety.

refining industries (and a few thermal power stations) that emit huge quantities of sulfur dioxide into the atmosphere. In some of the larger conurbations – such as Paris, Lyon and Strasbourg – air pollution is exacerbated by vehicle exhaust fumes. Local climatic conditions and temperature inversions can prevent the dispersal of pollutants, causing smog to build up to a critical level – in which case local factories are required to reduce emissions (and therefore productivity).

The effective disposal of solid industrial and domestic wastes has become an ever-increasing problem as their volume continues to grow. Less than half of such wastes are processed before being dumped in landfills or, increasingly, incinerated. Toxic materials and radioactive waste from France's massive nuclear power program pose particular difficulties: the strength of local opposition has frequently hampered the search for suitable storage sites.

Environmental hazards

Potentially hazardous industrial activities have also aroused concern. In 1966 public opinion was alerted to the dangers when a fire engulfed the Elf oil refinery south of Lyon, causing 18 deaths; the blaze started when escaping gas was ignited by a car that had caught on fire in an accident on the adjacent expressway. Another serious fire among oil storage tanks in the same area in 1987 emphasized the dangers of this 10 km corridor beside the river Rhône, where a heavy concentration of petrochemical and related industries is sited next to the main transregional road and rail links.

Oil was also responsible for a catastrophe affecting the coast of Brittany in 1978, when the wreck of the tanker *Amoco Cadiz* spilled over 200,000 tonnes of crude oil off the shore. In 1990 the region was awarded damages of $120 million against the American oil company; these reflected not only cleanup costs, but also compensation paid to local fishermen, oyster-farmers and hotel-keepers, whose livelihoods were affected for several years.

Natural disasters in mountainous areas – avalanches, landslides and flash floods – have been exacerbated by the deforestation of mountain slopes to provide ski runs and resorts. By removing the protective tree cover, the land is left vulnerable to erosion, and plant and animal habitats are disrupted. In 1970, for example, an avalanche at the ski resort of Val d'Isère killed 35 people, and in 1981 part of the mountainside was washed away at the winter sports resort of Les Arcs. In 1987 a flash flood in the mountains to the south of Lake Geneva devastated a camp-site at le Grand Bornand, killing 37 people and causing widespread destruction. Such events have not been confined to mountainous areas: in 1988 Nîmes, in the southeast, was left in chaos after flash floods swept through the town killing eight people and damaging property.

FEAR ENGENDERS PROTEST

Despite widespread international opposition from environmentalists to the building of nuclear power stations, France has pressed ahead since the early 1970s with a major nuclear energy program. Fifty-five power stations were in operation by 1992, with a further six under construction, producing 75 percent of the region's electricity. No other country in the world produces so much of its electricity from nuclear energy. France's nuclear industry also includes the world's largest reprocessing plant for radioactive waste at La Hague, near the port of Cherbourg on the north coast, as well as the uranium enrichment plant at Tricastin in the lower Rhône valley, jointly owned by a consortium of countries.

Development of these activities has not gone unchallenged, especially in the wake of the accidents at Three Mile Island in the United States (1979) and Chernobyl in the Ukraine (1986). Despite the economic benefits of the La Hague processing plant, and the French government's willingness to process other countries' waste there, such activity has proved highly controversial. At the same time, strong local opposition has blocked the establishment of storage sites for radioactive waste, particularly since plutonium contamination was found in 1990 at a "decontaminated" nuclear waste site near Paris.

There was also strong opposition to the construction of the Superphénix fast-feeder reactor at Creys-Malville, east of Lyon in the Rhône valley, and in 1987 further outcry greeted the discovery of leaking sodium from this plant. The leak itself was small and not dangerous, affecting a storage tank in which fuel rods were kept rather than the reactor itself, but the threat of danger (sodium ignites on contact with air or water) was enough to stir up strong protest.

MOVES TOWARD PROTECTION

The French government's interest in protecting the environment is a recent development. Although two important measures were taken in the 1960s with the creation of a series of national parks and six regional water authorities, it was not until the 1970s that a more comprehensive official approach emerged. In 1971 the Ministry of the Environment was established, and since then government initiatives in this field have multiplied.

During the 1980s the government came under increasing pressure to give greater priority to the environment. Belatedly – compared with Germany – a Green Party was established, while at regional level numerous independent bodies protested the need to protect the environment: FRAPNA, the Rhône-Alps Federation for the Protection of Nature, is a typical example. Pressure groups sprang up to oppose potentially damaging development schemes: for example, there was strong local opposition to plans to extend the highspeed (TGV) rail link southward to the Mediterranean, which would have entailed cutting straight through the picturesque countryside of Provence.

Another emotive issue concerned plans to build a series of dams along the upper Loire and its tributaries in order to regularize the river's erratic flow. The local community was made aware of the dangers, following the completion on the river Rhône of 20 dams (for hydroelectric power) that had left the river sterile. Opponents stressed the damage that would be done by the permanent flooding of meadows in the upper valleys of the Loire – a haven for wildlife.

A ten-year plan
In 1990, reaffirming its commitment to the environment, the government published a 10–year national plan outlining future policy. The plan aimed to reduce atmospheric pollution by up to 30 percent, notably with respect to chlorofluorocarbons (CFCs), which damage the ozone layer, and carbon dioxide – one of the "greenhouse gases" that contributes to global warming. It also proposed far greater control of water pollution, a commitment to recycling and improvements to over 200,000 homes exposed to excessive noise. Government legislation, alongside an increasing range of Euro-

pean Community directives, now cover problems relating to toxic waste disposal, hazardous industries and the protection of animal and plant life.

Despite such moves, France has been less successful in tackling water pollution and waste disposal than some of its European neighbors: Germany and Britain both spend more on the environment. There is no national research institution in France specializing in environmental matters, and the work of the Ministry of the Environment has been hampered by lack of cooperation with other ministries, such as Industry and Agriculture, on whose decisions it depends.

Government efforts to protect the environment have also been opposed by

Harnessing the tide (*above*) A tidal barrage at La Rance in Brittany, on France's northwest coast, takes advantage of unique conditions. The 13 m (44 ft) fall of the tide is used to turn turbines to generate electricity – proving a clean and relatively cheap energy source.

local communities with conflicting interests. Such was the case with France's six national parks. Each was planned as two zones: a central nature reserve in which building was forbidden and the number of visitors controlled, and a less restricted outer zone. Local communities within the parks – dependent on tourism for their living – though conscious of the need to protect rare or endangered species and landscapes of outstanding natural beauty, argued for greater development opportunities. Such controversy

has not affected France's many regional parks, the first of which was created in 1967. These are spread throughout the country and are the responsibility of regional authorities, who have successfully combined the preservation of rural life with recreational provision.

Industry plays its part

In recent years, industry in France has shown a greater interest in environmental matters. This is partly because stricter legislation has forced it to control pollution, and partly because of a change in public opinion and consumer demand in favor of "environmentally friendly" products. Air pollution has been reduced by largescale investment to produce cleaner emissions, while the massive nuclear power program – though controversial – has led to the closure of many power stations burning fossil fuels.

Industries renowned for being heavily polluting – such as chemicals and oil refining – have invested heavily to reduce the toxic nature of their waste products. By 1992, Rhône-Poulenc, for example, one of the region's largest chemical firms, was spending nearly $0.5 billion annually on measures to reduce pollution in the manufacture of products ranging from sulfuric acids to pesticides. The two national automobile manufacturing "giants", Renault and Peugeot, now produce cars with cleaner exhaust emissions, and have set up specialized centers for recycling components from used automobiles. In addition, it is estimated that some 1,500 companies either produce materials to combat pollution, or are involved in treating and eliminating waste products.

Emptying the bottle bank (*above*) Glass from a bottle bank in Paris is collected and taken away for recycling, helping to reduce pressure on rapidly diminishing landfill space.

FROM DUST TO ASHES

Each year France produces over 18 million tonnes of household refuse, representing an average of over 300 kg (660 lb) per person; in Paris the figure is more than twice this amount. However, the traditional method of disposing of domestic waste, by tipping it onto dumps or into disused quarries that are eventually filled with earth (landfills), has increasingly been rejected, partly because of the lack of suitable sites and partly because of associated environmental hazards, such as groundwater pollution. Consequently new methods of disposal have had to be found.

One solution has been the development of highly sophisticated incinera-tion plants located in major urban centers. Typical of these is the ultra-modern incinerator opened in 1990 at St Ouen in the northern suburbs of Paris. The plant cost some $200 million to build, and is capable of dealing with 630,000 tonnes of waste each year. Noxious fumes have been reduced to a minimum, and the energy from waste gases is converted to electricity, which is used not only to run the incinerator, but also to provide heat for 70,000 apartments in Paris, and heat and light for 1,500 nearby homes. By recycling the waste gases from this one incin-erator, France saves the equivalent of 112,000 tonnes of fuel oil every year.

Fighting the flames

Skies clouded by smoke, charred tree-stumps, blackened and reeking hillsides and vacationers evacuated from their campsites in the face of advancing flames: all have become familiar features of the Mediterranean coasts of France during the summer months. Every year an average of between 30,000 and 40,000 ha (72,000 and 96,000 acres) of French forests are destroyed by fire.

In 1989 over 5,000 ha (12,000 acres) of pinewoods in the southwestern area of Les Landes went up in flames; it was also there that France's worst ever forest fire broke out in 1949, leaving 24 people dead and destroying over 130,000 ha (312,000 acres) of woodland. But the areas most frequently affected are the heavily wooded mountain slopes of Corsica, and the extensive pine and cork oak forests of the south, particularly in the mountainous area of Provence between Marseille and Nice on the Mediterranean coast. Since 1933, nearly 500,000 ha (1.2 million acres) of forest in this area have been destroyed.

The price of such devastation is high. Spending on prevention and fire-fighting amounts to over $200 million each year, and in addition, wildlife and habitats are destroyed, homes are burnt out and livelihoods lost; in 1986 the local mimosa crop, which supplies the cut flower market, was almost completely wiped out. Worse still, firefighters may be injured or even killed. Devastated slopes need to be cleared and replanted, otherwise soil erosion quickly sets in, provoked by the heavy rains of spring and fall.

Lack of management of the Mediterranean forests arguably contributes to the spread of the fires. Much of the woodland is privately owned rather than grown for commercial purposes, and the traditional practices of coppicing and grazing have largely been abandoned as agricultural activity in the area has declined. Consequently the woods are thick with dry undergrowth, particularly in July and August when temperatures are high and there is little rain. The forest fires generally start among the leaf and dead-wood covering on the forest floor. Flames spread rapidly to the undergrowth, then to the resinous and highly inflammable Maritime and Aleppo pine trees.

The risk of fires, already great, has increased dramatically as picnic areas for tourists have multiplied: one carelessly

dropped cigarette end or the sparks from a barbecue are all that is needed to set the landscape ablaze. Once started – whether accidentally or ever more frequently by arsonists – fires may be fanned by strong winds such as the mistral, which funnels down the Rhône valley with great force and can cause a fire to advance at up to 6 km (nearly 4 mi) an hour.

Tackling the blaze
A range of fire-prevention practices and fire-fighting techniques has been developed in recent years to combat this ever-increasing hazard. Forest owners are now required by law to clear the forest undergrowth regularly in order to help prevent fires, while new access roads and cleared areas are designed to act as fire

Airborne attack (*above*) An aircraft with a hull full of water swoops low over burning forest to dowse the flames. Swift action is vital to halt a major fire, which can spread at 6 km (4 mi) per hour.

breaks. Efforts have also been made to encourage forest owners to keep goats and sheep, which consume the scrub vegetation, keeping the ground clear of potential tinder. A system of fire watchers in risk areas has also been set up, and the number of light aircraft and helicopters used for bombarding the fires with water has been increased. Sophisticated vehicles, capable of negotiating steep slopes, have also been developed to fight fires from the ground.

Research into the manner in which these disastrous fires are started and how they spread has been intensified, and will hopefully suggest other methods of prevention for the future. Already introduced have been the planting of deciduous trees that ignite less easily, such as holm oaks, and successful experimentation with a technique of shrouding vegetation threatened by an advancing fire in a cloud of minute water droplets.

Long ignored, now exploited

White horses, black bulls, pink flamingoes, duck hunters and rice fields are among the contenders competing for the precious resources of the vast wetlands at the delta of the Rhône river on France's southeast coast. Nearly half of these wetlands – known as the Camargue – has been damaged or has now disappeared.

The Camargue is a large, flat, water-logged plain with no large towns and very few villages. For thousands of years the area was virtually uninhabitable. Since the 1920s, however, most of the northern half of the Camargue has been drained, desalinated, irrigated and turned into vast rice fields.

The southern half of the wetlands is much the same as it has always been, except that some land has been drained and elevated roads have been built. An extensive series of dikes has also been built to prevent flooding, and a number of major canal systems (*far right*) run through the marshes. Apart from the 16,200 ha (40,000 acre) national reserve at Vaccares, most of the southern Camargue is privately owned and divided into about 30 ranches. Each ranch raises a herd of the famous black bulls used in local festivals. The equally famous white horses of the Camargue still roam freely across the marshes.

A growing industry for the ranches is duck hunting. Millions of ducks, geese and other waterfowl pass through the Camargue ever year. Ranchers have begun to clear some areas of the wetlands to make lakes for duck hunting. Increasing demands could spell the end of the remaining unspoiled environment.

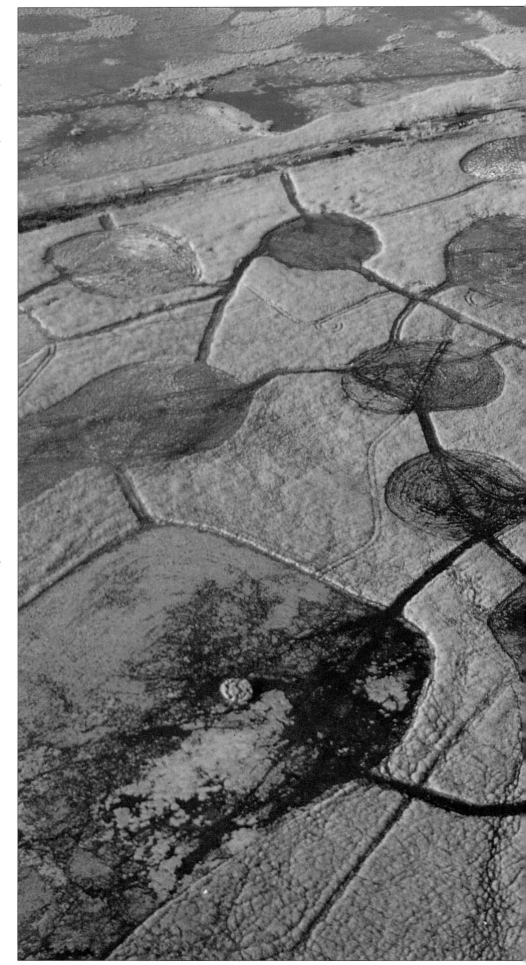

Duck hunting in the Camargue is increasingly popular. Small lakes are cut into the reedbeds to attract the waterfowl, which are shot by hunters who remain concealed in the reeds. The hunters move about the wetlands in small boats, using the canal systems to reach the hunting areas. Once a lake has been abandoned, the reeds quickly grow back (*paler circles*).

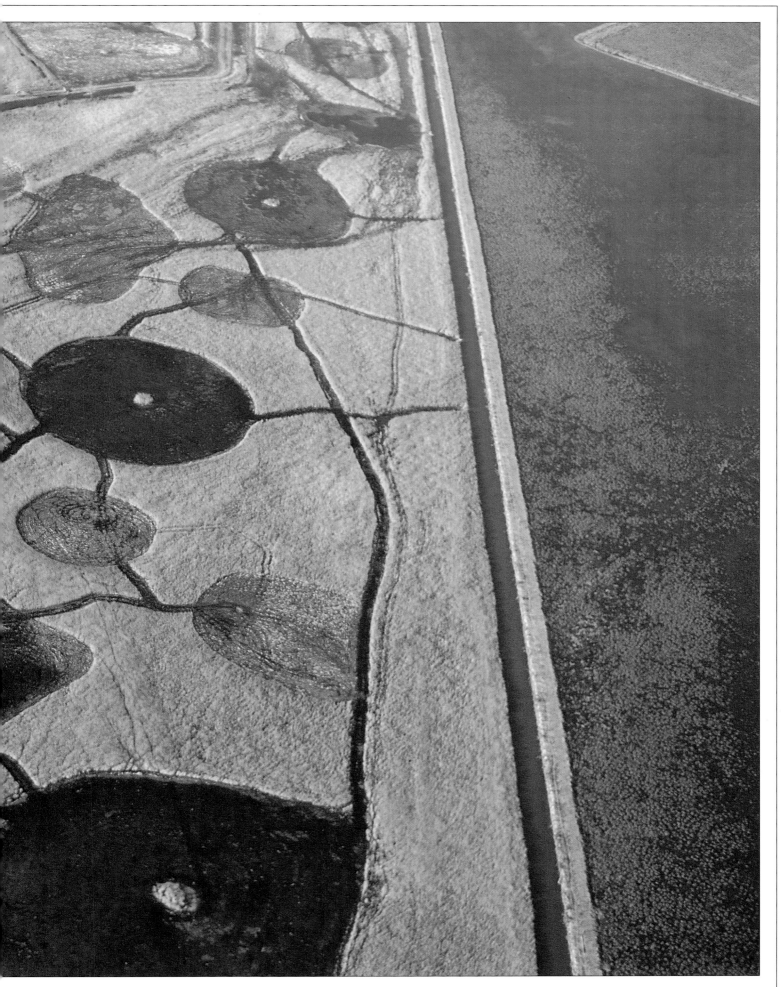

In the Low Countries – the most densely populated region of Europe – the human impact on the environment is extremely intense. The spread of settlements, farms, communications and industry has transformed almost all the landscape, leaving little room for wilderness areas. Low-lying coastal areas have been drained and the land reclaimed for agriculture. With so much industry and agriculture crowded together, the region suffers persistent pollution problems. Industrial chemicals and livestock effluent contaminate its freshwater supplies, while vehicle fumes pollute the air. But persistent problems have prompted concerted action. The Low Countries have put considerable effort into seeking ways to cure their environmental ills, through new technology, management, legislation and public will.

COUNTRIES IN THE REGION

Belgium, Luxembourg, Netherlands

POPULATION AND WEALTH

	Belgium	Luxembourg	Netherlands
Population (millions)	9.9	0.4	15
Population increase (annual population growth rate, % 1960–90)	0.2	0.6	0.9
Energy use (gigajoules/person)	163	326	199
Real purchasing power (US$/person)	13,010	14.290	12,680

ENVIRONMENTAL INDICATORS

	Belgium	Luxembourg	Netherlands
CO_2 emissions (million tonnes carbon/annum)	25	1.6	43
Municipal waste (kg/person/annum)	313	357	449
Nuclear waste (cumulative tonnes heavy metal)	700	0	200
Artificial fertilizer use (kg/ha/annum)	510	n/a	688
Automobiles (per 1,000 population)	349	405	346
Access to safe drinking water (% population)	100	100	100

MAJOR ENVIRONMENTAL PROBLEMS AND SOURCES

Air pollution: locally high
River/lake pollution: high; *sources*: agricultural, sewage
Marine/coastal pollution: high; *sources*: industrial, agricultural, sewage, oil
Land pollution: high; *sources*: industrial, agricultural, urban/household
Waste disposal problems: domestic; industrial; nuclear
Resource problems: land use competition; coastal flooding; water level control and flooding
Major event: Lekkerkerk (1980), toxic waste dump discovered

LANDS TRANSFORMED

It is hardly surprising that many serious environmental issues confront the Low Countries. The Netherlands is Europe's most densely populated country, with Belgium coming second; only Luxembourg lags behind. The concentrated mass of people requires as much space as possible for living, working, travel and recreation, with natural landscape as the loser. In the Netherlands, the area of woodland and other natural landscapes has fallen from over 6,000 sq km (2,300 sq mi) in 1900 to just 1,500 sq km (580 sq mi) in the early 1990s.

The greatest transformation of the landscape has come about through land drainage and flood control. About two-thirds of the Netherlands and large parts of Belgium are so low-lying that without the protection of dunes and dikes they would be subject to regular flooding, either by the sea or by rivers. From the beginning of the 15th century, windmills were used to pump water from the lowest inland stretches. Later steam, diesel and finally electric pumps took their place, draining the once extensive wetlands to create "polders" – areas in which the water level can be controlled, making farming possible. The efforts of the past can be seen today in the network of canals, ditches, dikes and embankments that still spread across the land.

Along the coast, dikes and drainage systems have created large polders from land close to or even below sea level. Since the early 13th century, as much as

627,500 ha (1.5 million acres) in the Netherlands have been reclaimed from the sea in this way. One area – now a residential quarter of Rotterdam – lies as low as 6.7 m (22 ft) below sea level.

Intense land use

Along with the transformation of the landscape, there has been an enormous increase in the intensity of land use. Agricultural yields in the Low Countries, per hectare or per animal, are among the highest in the world, and the same can be said for the levels of energy consumption, the density of traffic networks and vehicle numbers. All this intense activity generates severe problems of air, water, soil and noise pollution.

The Netherlands took little part in the 19th century Industrial Revolution and

The Low Countries

Key environmental issues

- major town or city
- heavily polluted town or city
- major pollution event
- heavily polluted river
- eastern limit of salt penetration
- area of intensive animal husbandry

acidity of rain (pH units)

- 4.6 (most acidic)
- 4.8 (least acidic)

Map of environmental problems (*above*) About half of the pollution in the Netherlands – including acid rain – originates in nearby countries. The rest is locally generated: the Netherlands and Luxembourg burn more fossil fuel per person than anywhere else in Europe. Local problems can be traced to salt water penetrating low-lying lands, and to heavy industry and dense populations of people and animals.

Old world, new world (*left*) An old windmill stands in the shadow of a nuclear power plant at Doel, Belgium. Energy from polluting fossil fuels has declined, and almost 10 percent of Belgium's energy now comes from nuclear power. The Netherlands plans to build 2,000 windmills to harness wind power – a clean, safe source of energy.

never gained the concentrations of heavy industry – with their attendant problems of land degradation – that exist in both Belgium and Luxembourg. After World War II, however, oil imports into the Rotterdam-Europoort area have helped the Netherlands to develop Europe's largest petrochemical industry – now a notorious source of air pollution.

The geographical position of the region also contributes to its pollution problems. The Low Countries are situated on the lower courses and mouths of three major European rivers (the heavily polluted Rhine, the Meuse and the Schelde), on the shores of the world's busiest sea (the North Sea), so that waterways and coastlines are subject to all forms of waterborne pollution. Furthermore, the Low Countries are in the vicinity of other densely populated industrial countries such as Great Britain, Germany and France. No matter from which direction the wind blows, it always carries with it pollutants from outside the region.

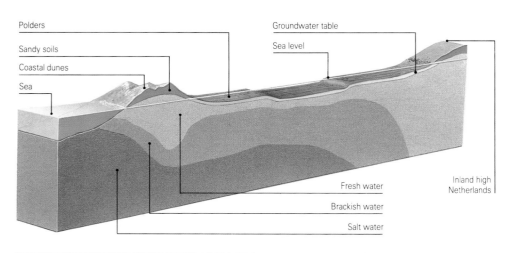

Polders
Sandy soils
Coastal dunes
Sea
Groundwater table
Sea level
Fresh water
Brackish water
Salt water
Inland high
Netherlands

THE POLLUTION OF RESOURCES

The fundamental resources of water, air and soil are under continual threat of contamination in the Low Countries. Water is at the same time both plentiful and scarce. *A Wet Country Short of Water* was the title of a publication contrasting the abundance of water in the Netherlands with its suitability for public use. But the demand for water of good quality from factories, farms and homes continues to rise. In the Netherlands, water consumption has increased from 505 million cu m (17,800 million cu ft) in 1960 to about 1,300 million cu m (45,900 million cu ft) in 1990.

The public supply of water comes from either ground or surface sources. The availability of groundwater is limited in the lower and more densely populated parts of Belgium and the Netherlands, because if too much fresh groundwater is pumped out, brackish and salt water takes its place. Instead, more and more surface supplies have to be used. Most of the surface water, however, comes into the region through rivers such as the Rhine and the Meuse, both of which are heavily polluted upstream. They carry high loads of organic waste, heavy metals, salts and other pollutants. The costs of producing clean drinking water from this mixture are very high, and have to be met by the public, who are faced with ever higher water charges.

Poisons in the air

The most serious sources of air pollution in the Low Countries are traffic and industry. Road traffic is extremely dense in the region, largely because of the high

Saltwater incursion (*above*) In the polder region of the southern Netherlands brackish water lies close to the surface of fertile land, a residue of the seawater that once flooded the area. Saltwater seeps upward as freshwater is extracted for irrigation.

number of private automobiles – almost 5.5 million in the Netherlands and some 4 million in Belgium and Luxembourg. Vehicle fumes contain a number of toxic chemicals including nitrogen oxides, carbon monoxide and lead. In the Netherlands, the growth in automobile ownership caused an 18 percent increase in emissions of nitrogen oxides during the period 1975–1984. Nitrogen oxides are associated with the problems of acid rain and global warming, and long exposure to these gases can cause lung damage.

The highest levels of vehicle emissions are in the major urban areas. In 1981, a

comparison of the blood content of people from 10 major cities around the world showed that Brussels' residents had the second highest level of lead in their bloodstreams (only in Mexico City were conditions worse), largely because of exposure to exhaust fumes.

Traffic – road, rail and air – also generates high levels of noise pollution, especially in the large metropolitan areas, with their dense road and rail networks and major airports. Between 20 and 30 percent of the Dutch population is seriously affected by noise nuisance. Residents of southwestern Amsterdam live only a few kilometers from the national airport of Schiphol, where an average of 646 planes land or take off every day. In addition, the railroad that connects the airport with Amsterdam carries over 150 trains per day in each direction, and a dense network of expressways and other heavily traveled roads crosses the area.

Contaminated soils

Industrial pollution in the Low Countries takes several forms, including water pollution by liquid effluents, and gaseous emissions such as sulfur dioxide, which contributes to the region's serious problems of acid rain. Soil pollution caused by the dumping of industrial and urban waste is more restricted in area, but its consequences can be extremely grave.

A notorious pollution site was discovered in 1980 in the municipality of Lekkerkerk near Rotterdam. Toxic waste was found in the soil beneath a new housing estate – the result of industrial dumping on the site in the past. The estate had to be evacuated temporarily to enable the total replacement of the soil under and around the houses. The incident initiated an intensive search for other dump sites in the Netherlands. Over 7,000 have since been discovered.

Pollution of the soil by agriculture is not so localized. Farmland occupies about two-thirds of the Low Countries, and is used very intensively. Artificial fertilizers and pesticides are applied to the land along with enormous quantities of manure. Belgium and the Netherlands add more nitrogen and phosphate to the soil per hectare than almost all other European countries. In many cases this leads to the build-up of chemicals harmful both to the continued productivity of the soil and to groundwater supplies.

A rich scum of effluent (*left*) washes onto the shore near Amsterdam. Dumping of sewage and chemicals in the North Sea went on throughout the 1980s despite new regulations. The province of Holland plans to designate a huge conservation area in the North Sea, the first to apply to such a large area of water.

Space at a premium (*above*), on the beach and in the parking lot. Bicycles are widely used for local travel, but the Dutch drive more cars per square kilometer than any European nation. Per capita consumption of fossil fuel is highest in tiny Luxembourg, however, which ranks sixth worldwide in emissions per capita.

DISPOSING OF POLLUTED MUD

In the harbors of Rotterdam-Europoort in the southwestern Netherlands a total of 23 million cu m (812 million cu ft) of mud is dredged annually from the rivers and docks to maintain sufficient depth for shipping. Almost 50 percent of the mud is too polluted to be spread on land or dumped at sea. The bulk of the pollution comes from the towns, factories, mines and farms of other countries – particularly Germany – within the Rhine catchment area.

In 1987 the Dutch authorities constructed a huge dump in the extreme west of the Rotterdam-Europoort area for the disposal of the polluted mud.

Known as the "Slufter", it consists of a 28 m (92 ft) deep hole covering an area of 260 ha (640 acres), encircled by a dike rising to 24 m (79 ft) above sea level. This gives the dump a total depth of over 50 m (164 ft), sufficient for the storage of 150 million cu m (5,300 million cu ft) of polluted mud.

Despite its great size, it is predicted that the Slufter will be full to the brim by the year 2002. The authorities hope that by that date the water quality of the Rhine will have been improved to such a degree that they will not be forced to construct a second dump to contain further polluted mud.

POLICIES FOR IMPROVEMENT

The governments of Luxembourg, Belgium and the Netherlands have all made it clear since the late 1980s that they intend to confront some of the long-standing environmental issues within the region. In Luxembourg, a series of environmental plans and regulations were launched in 1990, taking special account of the inevitable concentration of polluting sources in such a small country. In Belgium, a 460-page report entitled *The Environment in Belgium Today and Tomorrow* was published in the same year. It endorsed the so-called "precautionary principle" that wherever a threat to the environment exists, counter-measures have to be taken to ensure that environmental conditions remain acceptable for future generations.

A plan for the Netherlands

The Netherlands has a good reputation for its stance on environmental issues. Since 1970 a series of government acts have addressed most forms of pollution in the country, including surface and groundwater contamination, air quality, chemical waste, noise nuisance and fertilizer use. All are based on the principle "the polluter pays": the citizen or company that causes the pollution is liable for

the costs incurred in dealing with its effects. During the 1980s pressure increased for the creation of an overall environmental policy, an "umbrella" for the various sectoral laws. The result was the publication in 1989 of the National Environmental Policy Plan (NEPP).

The NEPP is based on two fundamental principles, the first of which is the need to deal with pollution at its source. In the past, environmental measures were frequently aimed at combating the effects of pollution rather than the sources. Such efforts included the removal and disposal of contaminated soil and silt, expensive

A sound barrier (*above*) along an urban expressway in Amsterdam. Noise pollution is taken increasingly seriously in congested urban areas, where life can be made unbearable by traffic. Noise reduction is part of the Dutch National Environmental Policy Plan of 1989.

purification processes for drinking water, building taller chimneys – some over 200 m (650 ft) high – to disperse the pollution from oil refineries, and the placing of noise screens along expressways. Some of these measures will continue to be necessary for the foreseeable future, but for better and more permanent results in the long term it is necessary to restrict the output of pollution at its source.

Stages in the creation of a polder (*below*) Reclaiming low-lying coastal land has been part of Dutch environmental practice since the 10th century, providing room for a burgeoning population as well as land for agriculture. On land surrounding polders the risk of flooding is greatly reduced and the water quality improved. Roads built on the larger dikes have improved communication links between isolated areas.

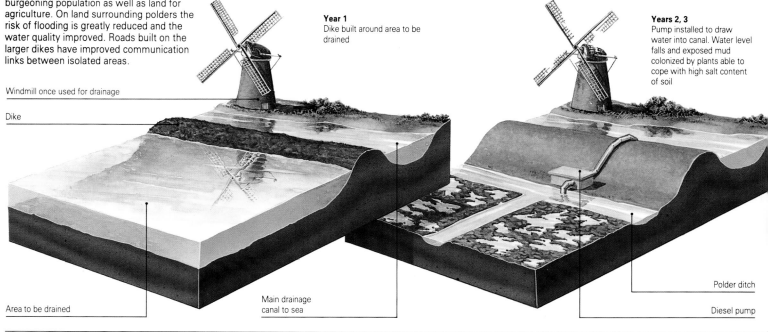

Windmill once used for drainage

Dike

Area to be drained

Main drainage canal to sea

Year 1
Dike built around area to be drained

Years 2, 3
Pump installed to draw water into canal. Water level falls and exposed mud colonized by plants able to cope with high salt content of soil

Polder ditch

Diesel pump

RECYCLING IN THE NETHERLANDS

The Netherlands has the best recycling record of any country in the world. In 1988 the country recycled 35 percent of its waste, including nearly 60 percent of the paper and cardboard, 36 percent of the glass and 28 percent of the textiles. Paper is often collected in order to raise money for charities and sports clubs, and several municipalities guarantee a minimum price for waste paper by making up the difference if the market price falls too low.

Beesel, in the southern province of Limburg, achieved a remarkable reduction of 60 percent in the amount of unsorted waste collected from its residents. Owing to a positive attitude among the population of 12,700 – the principle of "internalization" described in the National Environmental Policy Plan of 1989 – it took just a few months in 1991 for a set of municipal measures to make a noticeable difference. The measures included an increase in the number of bottle banks, a subsidy for waste paper, and collection schemes for chemical waste and textiles. The municipality also distributes free barrels for converting organic waste into compost. The total costs of waste management in Beesel are now slightly lower than they would have been if the original amount of waste had been transported to an official dumping site – a luxury that this crowded country, with its limited space, can ill afford.

Pollution could be dramatically reduced at source, the report suggests, if people switched from using private cars to traveling by public transport or bicycle. Similarly, farmers should decrease the size of their herds and curb their use of chemical fertilizers and pesticides. Factories and power stations should stop discharging waste into rivers and seas, and use "cleaner" raw materials. Following the discovery of large reserves of natural gas in the northern Netherlands, many factories and power stations in all three of the Low Countries were able to improve their environmental record by using gas as fuel instead of coal. When gas burns it produces far less pollution than either coal or oil.

The second principle of the NEPP concerns "internalization". This involves a change in mentality in government, business and the ordinary citizen, so that the need and responsibility for conservation of the environment is accepted by everyone. Environmental laws and measures, the report claims, will only be truly effective if people and companies are prepared to make sacrifices to ensure their success. With a little extra effort, citizens can contribute to systems of separated waste collection and thus increase the already high percentage of waste that is recycled in the Netherlands.

Cross-border negotiations

Because the source of much of the environmental pollution in the region lies in neighboring states, consultation with other countries plays an important role in the solution of environmental problems. Negotiations between the Netherlands, Belgium, Germany, France and Switzerland have taken place for many years, particularly over the issue of river pollution. Over 60 percent of the total water supply in the Netherlands comes from the river Rhine, Europe's most polluted river.

So far, agreements between the countries that share the river have helped to reduce the load of heavy metals in the water, but in many respects the situation has yet to be improved. Apart from the regular discharge of pollutants, there is also the threat of incidental pollution caused by industrial or river traffic accidents. In November 1986, large quantities of chemicals poured into the Rhine during a fire at a Swiss chemical plant at Basel. Following the accident, water supply outlets along the Rhine in the Netherlands had to be closed for 10 days.

Cooperation in environmental management is also important for the regional authorities within each country. In the Netherlands, for example, some neighboring municipalities carry out joint collection and processing of waste. But joint environmental policy in Belgium is hampered by the country's division into three more or less autonomous regions – Flanders, Wallonia and Brussels. The regions differ not only in the measures they adopt, but even in the way they define environmental issues.

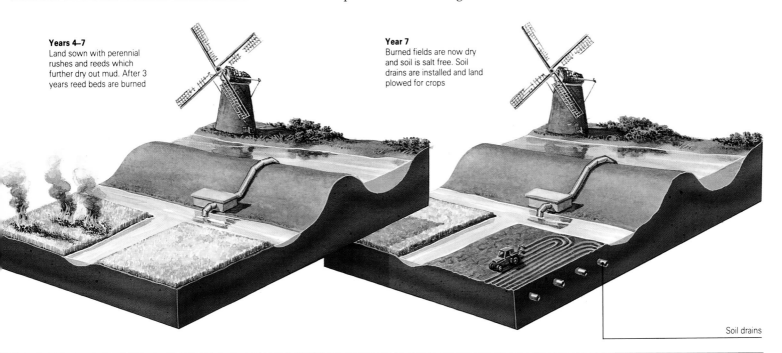

Years 4–7
Land sown with perennial rushes and reeds which further dry out mud. After 3 years reed beds are burned

Year 7
Burned fields are now dry and soil is salt free. Soil drains are installed and land plowed for crops

Soil drains

"The manure mountain"

Intensive livestock farming in the east and south of the Netherlands and northern Belgium is placing a severe strain on the environment of both countries. The principal problem is manure. The huge concentration of pigs, cattle and poultry generates far too much waste for safe disposal in the local environment.

The droppings from livestock accumulate in large storage tanks underneath the intensive feeding sheds. In the past, the normal way to dispose of the decomposing manure was to spread it over the farmer's land, where it served to fertilize the soil by returning nutrients – nitrogen, phosphorous and potassium – to the ground. But since 1970 the number of intensive livestock holdings has rapidly increased, as has the sheer volume of manure in the tanks. Consequently the amount that has to be dumped on the fields has exceeded the carrying capacity of the land. Dutch farmland can safely take about 50 million tonnes of manure per year, but the livestock farms produce nearly twice that amount.

Excessive manuring of arable fields and meadows supplies the soil with more minerals than the plants can absorb. The surplus contaminates both groundwater and surface water. Unfortunately, the sandy districts where pig-farming predominates contain the greatest number of sites from which groundwater is extracted for drinking water. Over-manuring also causes local and regional air pollution. The concentration of manure creates a powerful stench and releases high levels of ammonia, a gas that exacerbates the problem of acid rain. About 20 percent of the acid depositions in the Netherlands originate from agricultural sources.

Coping with the surplus

So far, research and management efforts have focused on techniques for reducing the impact of "the manure mountain". Changes in the composition of fodder can reduce the production of manure per animal or influence its chemical composition. The transportation of manure to arable farming areas results in a better regional distribution of the effluent, and manure can also be processed into pellets and then sold for export. Ammonia emissions can be reduced greatly by covering manure silos and by using machinery to inject liquid manure into the ground to a

More pigs than people (*right*) The Netherlands and Belgium have 7.5 percent of western Europe's population but 20 percent of its pigs. Disposal of the vast quantities of manure produced by pigs, calves and poultry is a major problem.

Spreading it thickly (*below*) A dark cloud of manure follows a tractor through a field. The high-yielding farmland of the Low Countries is loaded with overly generous applications of commercial fertilizer as well as local manure. Pesticides are also widely used.

depth of about 18 cm (7 in) instead of merely spraying it onto the surface.

As well as technical measures, a series of regulations have come into force in an attempt to control manure-related problems. In 1987, a Fertilizer Act in the Netherlands required farmers to keep a

record of manure production and to pay a levy if they exceeded a fixed quota. In Belgium, the Flemish authorities issued a decree forbidding the import of liquid manure from the Netherlands, and some municipalities have now placed restrictions on the number of months when manure can be applied to the land.

In both countries, however, opinion is growing that such limiting measures and regulations will never be sufficient. The real solution lies in being able to achieve a state of balance, with no more minerals applied to the land than the plants can absorb. More and more people believe that a drastic reduction in the livestock population will be unavoidable. In agricultural circles, there is understandable opposition to such an idea, but in the long term, agriculture as a whole can only benefit from the change.

Combating airport noise

Amsterdam's Schiphol International Airport is linked to the city by a mere 8 km (5 mi) of railroad. This close proximity eliminates the long commute to the airport that is typical of other major cities, and offers tourists with short stopovers an ideal opportunity to explore the city. However, for residents of Amsterdam, Schiphol's convenient location has a major drawback: high levels of noise pollution even in residential neighborhoods.

The noise produced by jet engines is a random assortment of air particles and motion waves. Most of the noise that planes make occurs during takeoff and landing, when the engine is straining the hardest – although low-altitude cruising by small craft can be surprisingly loud. Regulations adopted by the International Civil Aviation Authority have helped to reduce the noise around airports by enforcing steeper ascent rates – planes travel a shorter distance over an airport's immediate area as they climb quickly to a higher elevation out of earshot. On the ground, enclosed boarding ramps protect passengers from the scream of engines, while ground crews have worn padded earmuffs for protection since long before communication headsets were invented.

Restrictions on night flights also help to reduce overall amounts of noise, as does new technology that muffles engines. Acoustic mufflers with thousands of honeycombed cells are attached to the inside of the engine and soak up most of the jet roar. Almost all of the major international airports now have strictly controlled noise levels – microphones are planted around the airfield to record any transgressions of the noise guidelines.

At Amsterdam's Schiphol Airport airplanes refuel and wait for takeoff. Noise and land pollution levels have been lowered by the introduction of stricter environmental controls, which in turn have encouraged industrial innovations to reduce noise at source. The development of "silent engines" is one example.

COPING WITH RAPID CHANGES

LOST LANDSCAPES · ENVIRONMENTAL ILLS · PROTECTING THE ENVIRONMENT

The Iberian Peninsula is one of the most rapidly changing regions of Europe and one where pressure on the environment is great. Most affected are the industrial parts of northern Spain, as well as the larger cities and the extensive coastal areas that have been developed for tourism. But even remote parts of the rural interior are threatened with deteriorating air quality. Problems of water supply are likely to be a major cause for concern throughout much of Iberia. However, encouragement can be drawn from the growing awareness of the need to protect the environment, as shown by the increasing effectiveness of environmental movements. Popular concern over "green" issues has grown since Spain and Portugal developed their democratic institutions and joined the European Community (EC).

COUNTRIES IN THE REGION

Portugal, Spain

POPULATION AND WEALTH

	Portugal	Spain
Population (millions)	10.3	39.2
Population increase (annual population growth rate, % 1960–90)	0.5	0.8
Energy use (gigajoules/person)	39	62
Real purchasing power (US$/person)	4,190	8,250

ENVIRONMENTAL INDICATORS

CO_2 emissions (million tonnes carbon/annum)	17	73
Municipal waste (kg/person/annum)	221	275
Nuclear waste (cumulative tonnes heavy metal)	0	2,800
Artificial fertilizer use (kg/ha/annum)	103	99
Automobiles (per 1,000 population)	125	78
Access to safe drinking water (% population)	95	100

MAJOR ENVIRONMENTAL PROBLEMS AND SOURCES

Air pollution: urban high
Marine/coastal pollution: medium; *sources*: industrial, agricultural, sewage, oil
Land degradation: soil erosion; salinization; habitat destruction; *causes*: agriculture, industry, population pressure
Population problems: tourism
Major event: San Carlos de la Rapita (1978), transportation accident

Spain and Portugal

Key environmental issues

- • major town or city
- ⬤ heavily polluted town or city
- ◗ major pollution event
- 〰 heavily polluted river
- ▦ main area of coastal tourism

soil degradation

▓	severe
▒	high
░	moderate
☐	low

Map of environmental problems Overexploitation of the land has caused soil degradation and has contributed to the threat of desertification. Air and water pollution is common near major towns, and sea pollution is prevalent along the coasts, which are overrun by tourist developments.

LOST LANDSCAPES

People have had a dramatic impact on the landscape of Iberia. Only 3,000 years ago the peninsula was almost entirely covered with forests of oak, cedars and cypresses. Little of these forests remains today, except in a few small areas such as Portugal's Geres National Park in the northwest of the country. The forests were first cleared for farming from about 1000 BC onward. This process continued rapidly under Roman rule from 211 BC, when widespread irrigation systems were introduced, altering the volume and paths of rivers. Irrigation was further developed by the Arabs from the 8th century AD, especially in the regions of Andalucia and Valencia in southern and eastern Spain.

Environmental change accelerated during and after the period of reconquest and consolidation under the crowns of Castile and Aragon. Trees were cut down at a tremendous rate for use as fuel and for

Old solution to a continuing problem The Romans, famous for their engineering skills, built aqueducts such as this one in southern Portugal to direct water from rivers to supply their towns and farms. Water scarcity remains a problem in this arid region.

building, and topsoil that had been held in place for centuries by the trees' roots was quickly washed away. Once wooded mountains were reduced to bare stone and the high plateaus of the interior lost their vegetation.

In Andalucia, the landowners who replaced the Arabs neglected the irrigation systems, preferring to concentrate on pastoralism. Enormous herds of sheep and goats consumed sapling trees and other forms of vegetation across the southern half of the peninsula, reducing large areas of once fertile land to maquis – a habitat of tangled scrub, aromatic shrubs, stunted trees and thorny bushes – or garigue, a harsh environment only a little removed from desert. Much of this environment is little changed today.

An interest in crop farming and irrigation techniques gradually revived over the next few centuries, especially in Valencia. Water systems that had survived from Arab times were enlarged and extended, and in some places arid areas were transformed once again into fertile farmland. Although the gradual move away from sheep farming reduced some of the pressures on the land, the effort to increase agricultural productivity only created new problems.

Unreasonable demands

The regimes of General Francisco Franco (1892–1975) in Spain and Antonio de Oliveira Salazar (1889–1970) in Portugal both pursued policies of national self-sufficiency in food production. As a result the area of land under cultivation was greatly extended, even into areas unable to sustain the pressures of farming. In 1987 a survey by the Spanish Ministry of Agriculture estimated that as much as 46 percent of cultivated land was unfavorable to agriculture. In sheep-farming areas, the drive for self-sufficiency encouraged over-grazing, especially in the already degraded south.

More recently the region's growing prosperity, particularly in Spain, has brought new threats to the environment. The rapid expansion of large industrial cities such as Madrid, Barcelona and Valencia in Spain and Lisbon in Portugal has engulfed huge areas of countryside, while the growth of tourism has transformed the Spanish Mediterranean coastline into an almost continuous urban strip. Portugal's Algarve has suffered similar tourist-based urbanization. New landscapes of semidesert are emerging, the product not of deforestation and soil erosion, but of the concrete mixer.

ENVIRONMENTAL ILLS

The comparatively recent and rapid industrialization of Spain and Portugal has inevitably been accompanied by environmental deterioration. Air pollution has increased, and though the problem is presently concentrated in relatively few areas, leaving most of the interior little affected, there is concern that without proper control of industrial emissions air quality will fall throughout the peninsula.

A number of Spain's cities already have very serious air pollution problems. Unregulated urban expansion and rapidly increasing car ownership are leading causes of the trouble. This is most acute during long spells of windless, high-pressure weather when chemicals in the atmosphere are slow to disperse. Madrid, Bilbao, Huelva and Cartagena all suffer from high levels of sulfur dioxide and nitrogen oxides. Madrid also has high carbon monoxide levels from vehicle emissions. More acutely affected are four smaller Spanish towns: Aviles and Langreo on the north coast, and Badalona and San Adriano del Besos (both near Barcelona). These have all been officially declared air pollution zones; when pollution levels are high hospitals are filled with patients suffering breathing difficulties.

Bilbao in the industrial Basque area of northern Spain also has severe problems. The town's steel and shipbuilding works

Industry's backyard (*above*) A tiny patch of grass amid the smokestacks of Bilbao in northern Spain is the scene for a family picnic. Spanish cities have the highest air pollution levels in Europe.

Damage by interlopers (*right*) The native broadleaf trees of Spain's Extremadura region have been felled on a large scale to feed the timber industry. The imported eucalyptus trees that were planted as fast-growing replacements, however, require more water and make the ground around them sterile.

– only a few kilometers from the city center – are located in a narrow valley, which traps the pollutants and slows their dispersal. A similar alarming proximity of industry to population is found in Portugal. Steel, shipbuilding and chemical works crowd the southern bank of the Tagus river in the industrial suburb of Barreiro, opposite the heart of Lisbon. The high levels of sulfur dioxide from the industrial complex not only threaten the country's capital, but also the nearby ecologically sensitive areas of the Arrabida mountains and Tagus estuary.

Troubled waters

Water supplies are scanty throughout much of the region and water pollution is a major concern. Industry is a leading contributor to the problem. Wood pulp factories (part of the papermaking industry especially important in Portugal) cause serious damage by discharging untreated waste directly into the rivers, which in turn pollute irrigated agricultural land. Watercourses in the Nabão

and Almonda valleys in central Portugal, and parts of Catalonia in northeastern Spain, have been particularly badly affected by this kind of pollution.

Some of the worst pollution in the region, however, is caused by the oil and petrochemical industries, such as at Tarragona in northeastern Spain. These damage the environment directly by discharging polluted waste into rivers, and indirectly by consuming huge quantities of water – an increasingly scarce resource throughout the region. In some lowland coastal areas water extraction has caused the freshwater table to fall, allowing seawater to permeate inland and damage agricultural and wilderness areas.

The Ria de Aveiro in Portugal – a unique complex of estuarine lagoons, with a rich variety of plants and birds – has suffered particularly serious damage since the 1960s from industrial pollution emanating from nearby towns, such as Aveiro. The traditional salt extraction and seaweed fertilizer industries of the Ria, reliant on clean water flowing into the estuary, have also suffered.

Agriculture creates even more widespread water pollution than industry. Olives are one of the leading crops in Iberia, and an estimated 19,800 cu m (700,000 cu ft) of olive dregs are washed from presses into streams and rivers every year. In the Guadalquivir valley in Andalucia the waste from olive oil presses pollutes the waters over a wide catchment area. As the olive dregs in the water are broken down by decomposing bacteria, oxygen levels in the water fall, killing aquatic life, and methane gas is produced. Settling pools have been created to try and limit the pollution caused, but as yet the problem persists.

New agricultural technology – such as plastic greenhouses and plastic sheeting spread over the soil – has dramatically increased the demand for water, especially in Almeria in the southeast and in Valencia. As a consequence land in the interior, where the water is drawn from, may soon begin to suffer severe problems of salinity as the water table falls, and desertification may become a danger.

Poisonous forests

Soil erosion resulting from deforestation is one of Iberia's oldest ecological problems – but in recent decades the creation of new forests has, ironically, also had a damaging effect on soil. Since the 1950s there has been a rapid increase in the planting of eucalyptus trees to supply the wood pulp industry: between 1970 and 1985 the area in Spain planted with these trees more than doubled.

The eucalyptus is an unusually soil-destructive tree. The deep plowing that precedes planting makes the ground susceptible to erosion and also destroys the soil structure. In addition, the fallen leaves from the eucalyptus trees are poisonous, containing oils that prevent normal bacterial activity in the soil and undergrowth from growing. This discourages wildlife and enables the sun to dry out the ground. Iberia's new woodland is in many respects sterile.

SPAIN'S COOKING OIL SCANDAL

In the spring and summer of 1981 a mysterious illness swept through Spain. The symptoms were distinctive and terrible: shriveled muscles, deformities, frequently death. An estimated 700 people lost their lives and some 25,000 were taken ill. Government scientists at first thought toxic cooking oil was the cause. A well organized racket was discovered in which industrial oil was being sold – at great profit – as cooking oil. To do this the perpetrators had to remove an aniline dye, which had been added by the manufacturers to ensure that the oil would only be used in industry. It was believed that during the process of removing the dye, the oil had become toxic.

Several people were tried and imprisoned, but scientists never managed to discover any deadly toxins in the cooking oil. More recently a different explanation for the disaster has been suggested. Many Spanish farmers use large quantities of fertilizers and pesticides on their land, though they are often poorly informed as to the possible consequences. It is thought that a lethal cocktail of these chemicals may have caused the 1981 epidemic. A likely source is organophosphate pesticides used on tomato crops in Almeria.

PROTECTING THE ENVIRONMENT

Environmental awareness in Spain and Portugal grew steadily during the 1970s and 1980s, though it still remains some way behind some other European countries, such as Germany, Holland and the Nordic Countries. The demise of the Franco and Salazar regimes in the mid 1970s encouraged a new spirit of democracy in which environmental issues could be openly discussed. Although public involvement remains quite small in scale, a number of well-publicized issues, such as the "cooking oil" disaster, city air pollution and the plight of the Ebro wetlands in northeastern Spain have broadened the debate.

The crisis over the river Ebro's coastal wetlands began with the construction of major hydroelectric schemes upriver to provide electricity for the region. The dams across the Ebro have starved the river estuary of silt, allowing the sea gradually to penetrate inland, destroying vital rice fields, shellfish beds and one of Europe's main feeding grounds for mi-

grating birds. There is no obvious solution to the problem, but it has had a galvanizing effect on local groups in the delta. Farming, industry and conservation interests have all come together in close agreement over the need to save the Ebro wetlands.

The problems of the Coto de Doñana wetlands in Andalucia have, in contrast, revealed how high antienvironmental feelings can run when local economic needs come into conflict with those of the environment. Hundreds of thousands of birds – including Greylag geese, flamingoes, spoonbills, herons and egrets – depend on the seasonally flooded marshlands. But despite the area's status as a national park, it has been rapidly deteriorating. The water table has fallen dramatically – estimates suggest by up to 50 cm (20 in) a year – as water is extracted for use in nearby tourist resorts and for irrigation. As a result, the area of wetland has diminished by as much as 85 percent since 1972.

Local and central authorities, including the Institute for Nature Conservation, have tried a number of schemes to correct the damage, but have been hampered by

lack of money. In addition, they have come into conflict with the Ministry of Agriculture and the local farmers and developers, who use the high local unemployment figures to support their case for continued water extraction. Environmentalists have nevertheless gained some ground; since 1985 it has been obligatory to seek government approval for changes in wetland activities.

Legislative controls
Both the Spanish and Portuguese governments have enacted legislation to try and protect the region's environment. In the mid 1970s Portugal created a series of laws to conserve forests, soils and other natural resources. The 1980s saw further legislative changes in response to pressure groups, and in 1990 the National Ecological Reserve and Protected Areas Law came into force.

Over air pollution the Spanish government was much quicker to take action than the Portuguese. In 1972 and 1975 Spanish law set specific emissions standards and established a national system to monitor emission levels; they also established air pollution zones in the

Wetland protection (*above*) The authorities at La Albufera on Majorca, off the Mediterranean coast of Spain, have put a stop to the encroachment of tourism, farming, industry and water extraction to preserve a threatened coastal swamp.

SAVING A NATIONAL PARK

The Tablas de Daimiel National Park is an area of marshland beside the Guadiana river – near the confluence of the Ciguela and Zancara rivers – deep in the interior of southern Spain. It is fed by the La Mancha aquifer – an area of calcareous rock in which water collects, and in the past was covered by up to 1 m (3 ft) of water for most of the year. Since the early 1970s, however, the La Mancha aquifer has been heavily exploited for irrigation, with water extraction tripling between 1974 and 1987. The ecological consequences for the Tablas have been considerable. It is now waterless for most of the year and peatland that was previously submerged is now so dried out that it sometimes spontaneously bursts into flames in the heat.

The Spanish government has taken decisive action. Emergency wells have been drilled inside the park, while 60 million cu m (2,100 million cu ft) of water has been transferred from the Tagus aqueduct to the Zancara river, and a new reservoir has been built to improve control of surface water entering the marshland. Although these measures are unlikely to provide a permanent solution to the problems of the Tablas, they have bought the national park valuable time in which to try and change local irrigation methods and improve groundwater management. The incident also provides encouraging evidence of the willingness of the Spanish government to take prompt action when faced with an ecological crisis.

EC intervention

In recent years the European Community (EC) has been increasingly influential in the region, and has helped to reduce the disparity in environmental activity between the two governments. EC funds have been allocated to alleviate Portugal's water problems, and EC standards have been framed in new laws in both countries. It is hoped that the EC may inspire greater cooperation between the two Iberian governments over issues such as river management. Several of the region's largest rivers – the Tagus and the Douro, for example – are shared between Spain and Portugal. The EC may also play an important judicial role in resolving disputes between the two countries. For example, two Spanish nuclear power stations at Almarez and Sayago are viewed with some concern from neighboring eastern Portugal.

larger cities and the industrial area of the north. The Portuguese government's first law relating to air pollution was enacted in 1980. Air quality management schemes were introduced in cities with acute problems, such as Lisbon, Oporto in the northwest and Sines in the southwest.

The same disparity between the two governments was apparent over water pollution. In Portugal, until very recently, the only legislation on water use went back to the 19th century, and was hopelessly out of date. Spain, however, enacted a very important water law in 1980. This required authorization for any activity likely to cause contamination of water. It also established the key principle that both ground and surface water should be considered part of the public domain. The change has been very beneficial. In 1980 less than 18 percent of the Spanish urban population were served by water treatment plants, but by 1988 this had risen to 60 percent.

Limiting erosion in mountain valleys

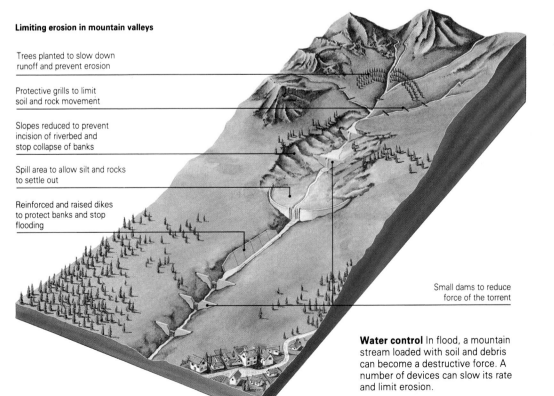

Trees planted to slow down runoff and prevent erosion

Protective grills to limit soil and rock movement

Slopes reduced to prevent incision of riverbed and stop collapse of banks

Spill area to allow silt and rocks to settle out

Reinforced and raised dikes to protect banks and stop flooding

Small dams to reduce force of the torrent

Water control In flood, a mountain stream loaded with soil and debris can become a destructive force. A number of devices can slow its rate and limit erosion.

Tourism: environmental costs and solutions

Some 54 million foreign tourists visited Spain in 1989, and 16 million visited Portugal. The growth of the low-cost, mass market tourist industry – the largest in the Mediterranean – has brought much needed wealth to the peninsula, but at great cost to the environment. Problems have been exacerbated by the speed of development over the last 30 years, and by its relatively confined area.

The earliest tourist centers grew up along picturesque stretches of coastline not too far from airports. In the late 1940s Torremolinos on the south coast was still a quiet fishing village attracting a few regular foreign visitors, but by the 1960s, peace and quiet had become a thing of the past. As the idea of low-cost tourism took off, huge hotels were built as cheaply as possible along the Mediterranean coast, much of which has now been completely transformed. Three-quarters of Spain's tourist accommodation is concentrated on the Mediterranean coast and Balearic Islands, and the Canary Islands, off northwest Africa; 31 percent of Portugal's is located in the Algarve, and Madeira also off northwest Africa.

Counting the cost

The effects on the environment have been profound. Coastal marshes were drained to provide suitable land for hotel construction, destroying important feeding grounds for birds. Sand dunes have been bulldozed. Water tables have fallen as a result of the huge demand for water from the tourist resorts and from agriculture. This has not only allowed sea water to encroach into freshwater lagoons, but also gives rise to severe and recurrent summer water shortages.

Litter is another problem, and noise, especially in the main tourist centers such as Torremolinos and Benidorm. Traffic congestion and air pollution from vehicle emissions have also increased dramatically, especially on the Algarve's main east–west road. In Spain there have been serious difficulties coping with sewage; one-third of the country's sewage is discharged into the Mediterranean without being treated.

Spain and Portugal's governments have responded to the increased environmental pressures in a number of ways, especially in Spain where public reaction against overdevelopment has been very strong. Attempts have been made to diversify and upgrade the tourist market

Hardy local plant life (*above*) A wild cactus grows above tennis courts, part of a resort on Tenerife in the Canary Islands. This species, unique to the islands, has managed to survive the plague of overdevelopment, but many less rugged unique species have not.

Mass market tourism (*right*) has made Benidorm one of the most popular vacation spots in Europe, but the intense concentration of highrise hotels, restaurants, bars and people adds to the problems of pollution along Spain's southeastern coast.

by attracting more affluent visitors, especially to inland areas, but these have had only limited success. Spain's 13 autonomous regions, each of which has its own planning powers, have made efforts to control future development in the main tourist areas. In addition, the Spanish government has passed a law that has brought the whole of the country's Mediterranean coastline under direct state control. The central government was also empowered to prevent any further developments within a fixed distance of the sea, and to ensure public access to the sea. It has made a number of efforts to improve existing tourist complexes.

Government action of this kind to protect both the tourist industry and the environment comes into direct conflict with the interests of developers. But perhaps the environment's best defense against further development is its present condition. Tourist numbers reached a peak in 1989, but by 1992 had begun to decline. As water shortages have grown more common, the sea has become increasingly polluted with sewage, and the land has been despoiled with cheaply built multi-storey hotels and apartment blocks, some of the tourists themselves have chosen to venture elsewhere, while others who were planning to visit the region for the first time were put off by the reports of environmental degradation.

A WELL-USED LAND

SHAPED BY HUMAN HAND · THE PRESSURES OF MODERNIZATION · LOOKING TO THE FUTURE

Environmental problems are nothing new to Italy and Greece. Deforestation and soil erosion have been evident for several centuries, while in recent decades both countries have witnessed industrial development, rapid urban growth and huge increases in car ownership. Such developments have led to pollution of the air, of rivers, and the sea. The environment has also come under increased pressure from tourism, which has transformed coastal areas, especially in Greece and on Italy's Adriatic shoreline. The region's scant water supplies have been stretched to the limit. However, there is growing public interest in environmental problems, and increased willingness by governments to intervene. The question is whether action can be taken soon enough to halt the region's increasing environmental degradation.

COUNTRIES IN THE REGION

Cyprus, Greece, Italy, Malta, San Marino, Vatican City

POPULATION AND WEALTH

	Greece	Italy	Malta
Population (millions)	10	57.1	0.4
Population increase (annual population growth rate, % 1960–90)	0.6	0.4	0.2
Energy use (gigajoules/person)	72	105	52
Real purchasing power (US$/person)	6,440	13,000	7,490

ENVIRONMENTAL INDICATORS

	Greece	Italy	Malta
CO_2 emissions (million tonnes carbon/annum)	20	120	0.2
Municipal waste (kg/person/annum)	259	263	n/a
Nuclear waste (cumulative tonnes heavy metal)	0	1,400	0
Artificial fertilizer use (kg/ha/annum)	171	190	46
Automobiles (per 1,000 population)	143	398	n/a
Access to safe drinking water (% population)	97	100	100

MAJOR ENVIRONMENTAL PROBLEMS AND SOURCES

Air pollution: locally high, in particular urban; acid rain prevalent; high greenhouse gas emissions
River pollution: medium; *sources*: agricultural, sewage
Marine/coastal pollution: medium/high; *sources*: industrial, agricultural, sewage, oil
Land degradation: *types*: soil erosion; *causes*: agriculture, industry, population pressure
Waste disposal problems: domestic; industrial; nuclear
Population problems: urban overcrowding; tourism
Major events: Seveso (1976), poisonous chemical leakage; Rhodes (1987), major forest fire; Adda Valley (1987), major floods; *Haven* (1991) oil tanker spill

SHAPED BY HUMAN HAND

The landscape of the region has long been vulnerable to changes brought about by human activities and exacerbated by weather conditions. The Mediterranean climate can be harsh, with hot dry summers and destructively torrential rainstorms. Some 3,000 years ago large areas of forest, which once covered much of Italy and Greece, began to be cleared for farmland. Trees were cut down for use as fuel, and for constructing houses and ships. On mountain slopes all over the region the soil – deprived of protective treecover – was washed away by the rain,

Barren hills and terraced slopes (*above*) on the Mani peninsula of southern mainland Greece. After hundreds of years of cultivation these terraces have fallen into neglect, leaving the thin topsoil on the mountain slopes vulnerable to erosion.

leaving just bare rock in many places.

Many coastal and lowland areas that were cleared of forest were transformed into scrubland, or maquis. The vegetation changed again as a result of overgrazing by sheep and goats, until some land became no more than an arid, thorny habitat known as garigue in Italy and phrygana in Greece. In some areas, such as Calabria in southern Italy, the environment has been reduced to near desert.

The human impact on the landscape

has not always been so destructive, however. The Arabs introduced terracing techniques to Sicily and southern Italy when they colonized the area in the 9th century, and terraces are still found on fertile hillslopes throughout the region. These are often planted with olive trees, which provide a useful barrier against soil erosion. The Arabs also introduced irrigation methods to bring farming to previously arid areas. These are now used extensively in both countries. Greece, in particular, has implemented widespread irrigation schemes over the last 50 years.

The draining of coastal marshlands has been another important agricultural innovation, bringing huge changes to the landscape through the destruction of wetland areas. It is estimated that in Roman times, some 2,000 years ago, Italy possessed about 3 million ha (7.4 million acres) of wetlands, of which no more than 200,000 ha (494,200 acres) remain.

Smothered by concrete

During the 19th and 20th centuries urban growth caused the greatest environmental change. When Greece achieved independence in 1830 Athens was hardly bigger than a large village. But by 1992 the metropolitan area of Athens had more than 3 million inhabitants – about 1 in 3 of the country's population – and covered some 430 sq km (170 sq mi). The story is the same in Italy, where the growth of Rome, Milan and other cities has caused the loss of an estimated 1.8 million ha (4.4 million acres) of farmland since 1980.

Tourism has also left its mark. Most of Malta's coast north of Valetta is now a continuous urban strip of hotels, villas, bungalows and restaurants. The same is true of many of the Greek islands, as well as beach areas on the mainland, especially along the northern Peloponnese and western Chalcidice. Development has also been exacerbated by poor planning laws.

In both Italy and Greece a new building cannot be opposed by local government once it has been constructed to a certain height, regardless of whether planning permission has been gained. This has encouraged people to build secretly – often at night – and at great speed, dotting the landscape with urban sprawl.

Mountainous areas of Italy have also been affected by tourism. Woodland too remote to be cleared for farmland has been cut down to make way for ski runs and elevators, as well as for hotel and road building. In 1987, deforestation contributed to the disaster in the Adda valley in the Italian Alps. Torrential rainstorms caused millions of tonnes of mud and rock to slide down unprotected hillsides, killing 50 people and engulfing two villages. As the wilderness becomes ever more desirable as a tourist destination, it is only too easy for it to be damaged by the amenities that make it accessible.

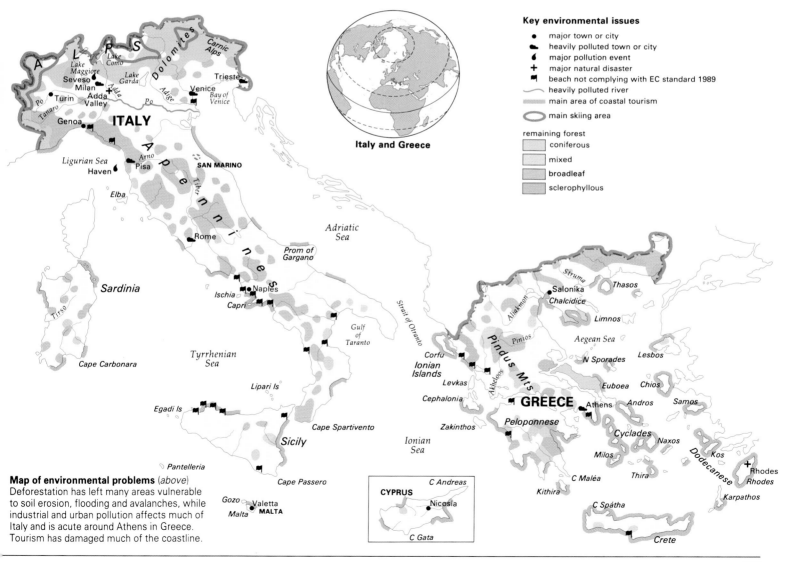

Key environmental issues

- ● major town or city
- ◖ heavily polluted town or city
- ◗ major pollution event
- ✛ major natural disaster
- ◪ beach not complying with EC standard 1989
- ∿ heavily polluted river
- ▓ main area of coastal tourism
- ◯ main skiing area

remaining forest
- coniferous
- mixed
- broadleaf
- sclerophyllous

Italy and Greece

Map of environmental problems (above)
Deforestation has left many areas vulnerable to soil erosion, flooding and avalanches, while industrial and urban pollution affects much of Italy and is acute around Athens in Greece. Tourism has damaged much of the coastline.

THE PRESSURES OF MODERNIZATION

One of the region's oldest and thorniest environmental problems – soil erosion – is a major current concern. Erosion now affects one-third of all cultivated land, with Greece suffering an average soil loss each year of 10 cu m for every 1 ha (350 cu ft for every 1 acre) of land. Much of the land is further damaged by intensive farming practices. The excessive use of sprinklers and other forms of irrigation can lead to water shortages and to a build-up of salts in the soil through evaporation. In Greece it is estimated that nearly one-half of all irrigated land – or one-quarter of cultivated land – has been seriously affected by salinity.

In many areas efforts have been made to stem soil erosion by planting trees, but elsewhere forests are frequently destroyed by fires. In Greece it is estimated that 4,000 ha (10,000 acres) are planted with new trees each year, but 30 times this area is lost through fires. The Greek island of Rhodes suffered a disastrous fire in 1987, when 65,000 ha (160,000 acres) of forests and olive groves were destroyed, and thousands of livestock killed.

Water pollution affects a large part of the region. The river Po passes through northern Italy's industrial heartland and some of the country's most intensively farmed areas. On its course it is polluted by industrial oil, toxic chemicals and heavy metals, as well as fertilizers and pesticides. These flow into the northern Adriatic, where the water circulates slowly, allowing pollutants to build up in the Gulf of Venice. Since the 1970s there has been an alarming increase in the amount of seaweed and algae that thrive on the nutrients in the water. When the algae die, oxygen levels in the water are reduced by the bacteria that break them down, killing other aquatic wildlife. In the summer months in Venice the smell of rotting algae can be overpowering. Attempts to clean up the lagoon have so far met with little success, as no sites have been found in which to dump the stinking algae.

Venetian decay

Flooding in Venice, though not a new problem, has become more frequent in recent decades, and the damage to the ancient city has been devastating. The

city – founded in the 5th century on 118 islands in the middle of the Venice lagoon – has been sinking. One of the causes can be attributed to excessive water extraction from the freshwater aquifers below the lagoon in the 1960s. The water was pumped from the aquifers to the mainland for use in the huge industrial complex at Marghera, causing Venice's foundations to subside. In November 1966 unusually high tides inundated the city, and polluted water lapped the wood and brick buildings at just over 1 m (3 ft) above the sidewalk. Since the early 1970s

Inundated with problems St Mark's Square, Venice's lowest point, floods when the tide reaches a mere 60 cm (23 in) above sea level. Tidal barriers have been planned around the city's lagoon, but haggling over the estimated $1.8 billion cost has delayed action.

no more water has been pumped from the aquifers, but severe damage has already been done and the chemicals in the water continually eat away at the stonework. Above the water level, air pollution has caused equally serious damage to the fabric of buildings.

Water shortage is increasingly common in the region, particularly in the vicinity

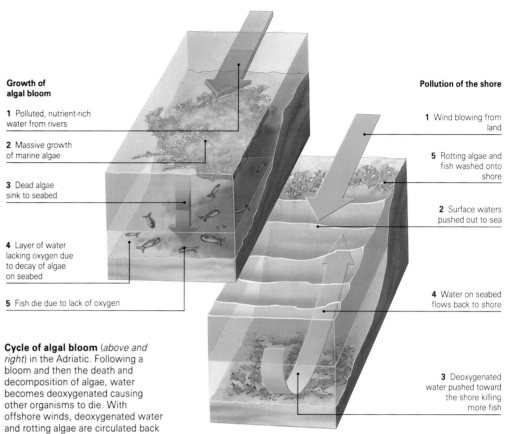

Growth of algal bloom

1 Polluted, nutrient-rich water from rivers

2 Massive growth of marine algae

3 Dead algae sink to seabed

4 Layer of water lacking oxygen due to decay of algae on seabed

5 Fish die due to lack of oxygen

Cycle of algal bloom (*above and right*) in the Adriatic. Following a bloom and then the death and decomposition of algae, water becomes deoxygenated causing other organisms to die. With offshore winds, deoxygenated water and rotting algae are circulated back to shore.

Pollution of the shore

1 Wind blowing from land

5 Rotting algae and fish washed onto shore

2 Surface waters pushed out to sea

4 Water on seabed flows back to shore

3 Deoxygenated water pushed toward the shore killing more fish

COUNTING THE COST

In July 1976, the dangers of poorly regulated industry became devastatingly apparent to many Italians. The Icmesa chemical plant at Seveso near Milan had been producing the weedkiller 2,4,5-T. This product is not particularly dangerous, but during its manufacture an extremely toxic waste chemical was also created: a type of dioxin known as TCDD. On 10th July a valve failed to operate and a cloud of this substance shot 50 m (164 ft) into the air, gradually settling on some 1,800 ha (4,450 acres) of surrounding land. A terrible disaster had occurred. More than 700 people had to be evacuated, most of them suffering from a disfiguring skin complaint called chloracne.

Pregnant women were advised to have abortions because of the damage that would have been caused to their unborn children. All domestic animals had to be slaughtered, and the top 20 cm (8 in) of soil in the affected area had to be removed and buried in special pits. The factory itself was dismantled.

The longterm consequences are hard to estimate. Certainly those exposed to the gas are considered to be highly at risk from cancer. At the same time the incident attracted much public attention, adding to national awareness of the dangers posed by such plants, and encouraging interest in environmental issues. Some good may have resulted from the disaster.

of new coastal resorts. If too much water is removed from underground reserves, the water table falls and seawater is able to seep in, contaminating freshwater supplies. In Malta expensive desalination plants have had to be built, as the island is completely dependent on its depleting underground reserves. In Palermo in Sicily piped water has for some years been obtainable for only a few hours each day during summer months. Such problems are set to grow worse. It is estimated that demand for water in the region will double during the next 20 years.

Sewage is a health threat in many areas, as the region has few adequate treatment plants. Tourist areas are particularly susceptible: it is not uncommon for sewage to be discharged into the sea very close to beaches, as at Cefalù in Sicily. Such arrangements contributed to the cholera outbreak in Naples in 1973, when the disease passed from discharged effluent to edible shellfish in the bay of Naples.

Smog over the Acropolis
Air pollution is another growing concern in both Italy and Greece. Industrial waste

and vehicle exhaust emissions have combined to cause problems in most of the region's large cities. The trouble is most severe in the summer months when Italy and Greece often experience long periods of hot, windless weather, allowing fumes to build up and react with sunlight, creating photochemical smog. Milan is frequently submerged beneath a white haze, while in Athens – one of the most polluted cities of Europe – the problem became so severe in September 1991 that hundreds of people were admitted to hospital with respiratory problems.

LOOKING TO THE FUTURE

Protecting the environment requires action both by governments and the local population, and in Italy and Greece there are encouraging signs from both quarters. However, although there is much concern about the quality of the environment, there is also confusion and conflict about how best to tackle the problems. On environmental matters, the region still lags far behind other parts of the developed world, such as Germany, the Nordic Countries and the Low Countries.

Green movements

Environmental movements in the region have grown considerably in recent years, especially in Italy. The Italian League for the Environment (Lega per l'Ambiente) was founded in 1980, and within 12 years had acquired a membership of more than 50,000 people. The league has sufficient resources to fund a "Green Boat", which sails around the coast each year monitoring pollution levels. The Worldwide Fund for Nature (WWF) is also a significant force in the region; it has guaranteed protection to parts of the coastline of Elba

and Sardinia by purchasing land itself. In Cyprus the Green Echo group is actively seeking ways to protect the environment from destructive development.

In Malta two organizations – the Maltese Ornithological Society and the Maltese Association for Hunting and Conservation – have put aside their differences in order to try and prevent problems of excessive shooting of wildlife on the island. (The Greek government on Cyprus has also recognized the need to take action, and has declared its intention to ban the hunting of birds – estimated to result in the killing of as many as 25 million migrating birds each year – though it is doubtful how effective this will prove.) Pressure groups with quite contrary aims remain influential in the region. In Italy, for example, proposals in a national referendum in June 1990 to curb hunting were defeated after the hunting lobby mounted a strong campaign to persuade voters to abstain. As a result hunters can continue their annual slaughter of between 140 and 170 million animals, which has pushed some species close to extinction.

Even though environmental pressure groups are growing, they are still finding

that large numbers of the population are indifferent to the region's environmental problems, especially in areas where resources are limited and unemployment is high. The tourist industry, often responsible – either directly or indirectly – for creating environmental problems, was itself only alerted after suffering a severe setback in 1989, when vast quantities of rotting algae were washed up on beaches along the Adriatic coast. The number of tourists visiting the area immediately fell by two-thirds. In 1990 a 35 km (22 mi) long barrier was erected at great cost to protect resorts from the yellowish-brown slime. Meanwhile scientists are conducting research into ways of controlling the scum using naturally occurring viruses.

Government action

Legislation is essential for the protection of the environment, and governments in the region have recognized the need to adopt environmentally friendly policies. Greece, for example, has recently passed laws that restrict tourist developments in sensitive coastal areas. In Italy towns suffering from vehicle pollution may ban all traffic from their central areas if supported in a local referendum: Bologna was the first to do so in 1984. In Athens, where traffic pollution is particularly acute, the government has adopted the novel approach of controlling access to the city center by banning automobiles with even and odd registration numbers on alternate days. This measure was intended to reduce traffic by half, though some Athenians have circumvented the ban by purchasing a second car, so adding to the city's parking problems.

All countries in the region are signatories to the Barcelona Convention of 1975, which sought to clean and protect the Mediterranean, and consequently governments have been obliged to introduce legislation to try and control pollution discharges. In Italy the Merli Law requires that authorization must be given for all waste discharged into the sea, with

Holding back danger (*left*) An anti-avalanche fence exposed in summer, close to the popular ski resort of Cortina d'Ampezzo, Italy. Careful management is needed to preserve the Alps' fragile ecology from human encroachment and keep it as a wildlife haven.

Protecting national treasures (*right*) from the corrosion and grime of traffic fumes is a never-ending process. This copy of Michelangelo's *David* shows signs of decay; behind it, the Piazza della Signoria is benefiting from new techniques in restoration.

specific limits set on quantities. Governments have also established a number of marine nature reserves – such as one off the coast of Crete to protect the endangered Mediterranean monk seal – with more in the planning stage.

The European Community (EC) has also begun to play an important role in environmental matters. It has provided funding for forest reserves in southern Italy that combine careful conservation practices with efforts to create employment through the exploitation of forest products. However, both Italy and Greece have been noted for their failure to implement EC environmental directives to the full. In Greece, for example, 60 percent of automobiles do not meet current EC emission standards.

SAVING THE PAST

There is increasing concern about the decaying stonework of many of the region's most valued historic buildings, especially in the larger and more polluted cities, such as Athens and Rome. The Scientific Laboratory of the Misericordia (LSdM) in Venice has become a leading center for research into the problem, and has shown it to be caused largely by emissions from car exhausts and oil-fired power stations, which pollute the air with nitrogen oxides and sulfur dioxide. These often form acid rain, but can also prove a menace in dry climates. In Athens they accumulate on stonework as dry "acid depositions", which turn into destructive acids when rainwater falls. Salts and "black crusts" build up, disfiguring and eroding buildings.

The LSdM in Venice has made significant progress on methods to halt such destruction. A new microblasting cleaning method has been developed, using aluminum powder blown in a narrow jet. The technique, though costly, causes less abrasion than traditional sand blasting with silica powder, and is more effective than water spraying, which often washes pollutants deeper into the stonework. Once cleaned the buildings must be protected against further attack. Research into this field of conservation is less advanced, but there is optimism that resins and polymers will prove effective. The task of restoration is huge, especially in cities such as Venice where almost all the buildings have been affected. It is recognized, however, as being vital both for safeguarding tourism and national prestige.

The Mediterranean: sink or swim?

The Mediterranean is well known for its natural beauty. Warm seas and dramatic coastlines make it one of the most visited areas, catering each year for as many as one-third of all the world's tourists: the annual number of visitors almost equals the 130 million people who live permanently around its shores.

Yet the Mediterranean is in trouble, suffering increasingly from pollution. Part of the problem is tourism; most of the coastal resorts lack the proper facilities needed to deal with the ever-increasing quantities of sewage and garbage. The main cause of the pollution, however, is the increasing quantity of chemical waste from new manufacturing and intensive-farming methods. Treatment plants have not kept pace with output, and it is estimated that four-fifths of discharges into the Mediterranean receive only minimal processing. In 1982 alone the Mediterranean was polluted by some 800,000 tonnes of nitrogen, 320,000 tonnes of phosphorus, 120,000 tonnes of oil, 60,000 tonnes of detergents, 3,800 tonnes of lead and 100 tonnes of mercury.

Oil pollution is now a serious hazard. One-third of international oil cargoes are carried through the Mediterranean, and every year as much as 850 million liters (187 million gallons) of crude oil are disgorged from tankers through small spills and tank cleaning. This is equal to 17 times the quantity of oil leaked in the notorious *Exxon Valdez* disaster in Alaska in 1989. The Mediterranean has also had disasters of its own. In April 1991 the oil tanker *Haven* collided with a ferry off Leghorn in the Gulf of Genoa, spilling 10,000 tonnes of oil and creating an oil slick that damaged tourism and wildlife in the area.

Pollution problems in the Mediterranean are worsened by its geography. The only connection with the wider oceans is through the narrow Strait of Gibraltar, and it is estimated that the Mediterranean's waters are only fully exchanged with those outside every 80 years. The warm climate adds to problems by causing a high rate of evaporation, which concentrates pollutants. This is especially so in shallow waters, where the results can be very destructive. In September 1975, 7,000 tonnes of dead or dying fish were washed up at Cesena on Italy's northern Adriatic coast. This was the result of an algal bloom that had deoxygenated the coastal waters.

The Mediterranean Sea: a polluted paradise Sixteen cities of more than half a million people line the shores of the Mediterranean, which is visited by some 100 million tourists every year. The waste produced by so many people is further increased by discharges from major industrial areas, oil refineries and petrochemical plants, drainage from farmland, and waste from the numerous ships and smaller craft that sail from one end of the sea to the other. Stocks of fish have been depleted by 50 years of overfishing, and safe drinking water is at a premium. In 1990 the European Community, the European Investment Bank and the World Bank committed themselves to providing $1.5 billion in immediate cleanup funding. Further regional cooperation is required if the Mediterranean is to escape total environmental ruin.

1 Gibraltar
2 Cartagena
3 Barcelona
4 Marseille
5 Genoa
6 Rome
7 Naples
8 Athens
9 Istanbul
10 Beirut
11 Tel Aviv
12 Alexandria
13 Benghazi
14 Tripoli
15 Tunis
16 Algiers
17 Tangier

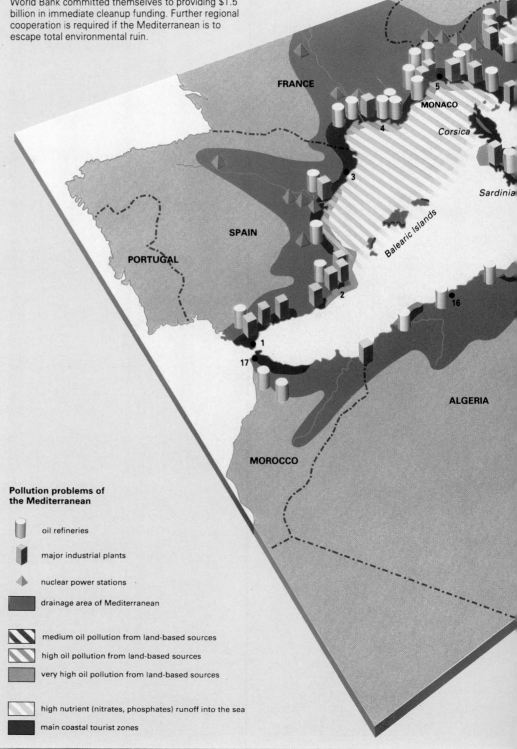

Pollution problems of the Mediterranean

- oil refineries
- major industrial plants
- nuclear power stations
- drainage area of Mediterranean
- medium oil pollution from land-based sources
- high oil pollution from land-based sources
- very high oil pollution from land-based sources
- high nutrient (nitrates, phosphates) runoff into the sea
- main coastal tourist zones

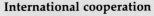

International cooperation

The Mediterranean is bordered by no fewer than 20 countries, so protection from pollution can only be achieved through international cooperation. In 1975 the Barcelona Convention agreed to an international research program to collect data on the Mediterranean's environment from about 80 national centers, so problems could be identified and solutions suggested. Then in 1984 the EC recommended building sewage treatment plants for all cities over 100,000 people, as well as requiring that companies take responsibility for processing their own waste. It ordered a complete halt on discharges of toxic chemicals into rivers and the sea, and demanded that oil tankers be fitted with extra tanks to prevent contamination of their water ballast with oil. However, with no sanctions to use against them, violations continue.

In 1990 the Nicosia Charter proposed that action be taken to eliminate all significant environmental problems in the Mediterranean by the year 2025. It set up new legal frameworks and adopted financial incentives to encourage good environmental management, offering hope that international cooperation will eventually solve the problems of the Mediterranean.

Oil sink (*left*) The wreck of the tanker *Haven* spilled 10,000 tonnes of oil off the Italian coast in April 1991. The Mediterranean is heavily traveled by tankers and other ships carrying potentially hazardous cargoes, increasing the risk of accidents.

A RACE AGAINST TIME

THE CHANGING LANDSCAPE · POLLUTION'S DARK SHADOW · A CHALLENGE FOR THE RICH

Central Europe was one of the first regions to recognize that conservation and economic growth must go hand in hand. A vocal environmental lobby has consistently pressed the governments of the region to find an acceptable balance between what is environmentally and economically desirable. The two main causes of environmental degradation have been the rapid expansion of industry and the intensification of agriculture. Air pollution from industry has caused health problems, particularly in former East Germany, where pollution controls ran counter to the socialist aim of industrial self-sufficiency and high productivity. Acid rain has damaged large areas of forest; agrochemicals have poisoned wildlife habitats; and in Austria and Switzerland the Alps are being harmed by tourism.

The largest open-cast iron-ore mine in Europe (*above*) cuts a swathe through an Austrian forest. Industrialization, new roads, tourism and other forms of human exploitation have all left their mark on the landscape. Deforested areas are at risk from erosion, flooding and landslides.

COUNTRIES IN THE REGION

Austria, Germany, Liechtenstein, Switzerland

POPULATION AND WEALTH

	Austria	Germany	Switzerland
Population (millions)	7.6	77.7	6.6
Population increase (annual population growth rate, % 1960–90)	0.2	0.2	0.7
Energy use (gigajoules/person)	118	231* 165*	111
Real purchasing power (US$/person)	12,350	14,620	17,220

ENVIRONMENTAL INDICATORS

CO₂ emissions (million tonnes carbon/annum)	17	222	16
Municipal waste (kg/person/annum)	231	317	386
Nuclear waste (cumulative tonnes heavy metal)	0	3,300	700
Artificial fertilizer use (kg/ha/annum)	198	421* 337*	431
Automobiles (per 1,000 population)	358	412	420
Access to safe drinking water (% population)	100	100	100

MAJOR ENVIRONMENTAL PROBLEMS AND SOURCES

Air pollution: generally high, urban very high; acid rain prevalent; high greenhouse gas emissions
River/lake pollution: medium/high; *sources:* industrial, agricultural, sewage, acid deposition
Land pollution: medium/high; *sources:* industrial, agricultural, urban/household, nuclear
Waste disposal problems: domestic; industrial; nuclear
Resource problems: land use competition
Population problems: urban overcrowding; tourism
Major event: Sandoz near Basel (1987), chemical spill

* Former West Germany (top) and East Germany (bottom)

THE CHANGING LANDSCAPE

In prehistoric times, dense forests almost entirely covered Central Europe. These were gradually cleared – initially along the fertile river valleys – for farmland. Peasant smallholdings dominated the lowlands, while alpine pastures were cleared to provide seasonal grazing for sheep, cattle and goats. In mountainous areas, roads and settlements were protected from floods, avalanches and soil erosion by the natural forest barriers, or *Bannwälder*, on the mountain slopes.

Until the mid 19th century the region remained largely agricultural, with poor road systems and neglected waterways. Regionalism was largely the cause, as until 1871 Germany was divided into numerous kingdoms, principalities and free cities. Following unification the country began to industrialize, initially around Berlin, the new capital. But despite industrial growth, farming methods remained traditional and smallscale until World War II.

After 1945 environmental change came rapidly. The postwar mechanization of agriculture, combined with greater use of pesticides and fertilizers, increased farming output dramatically – especially on

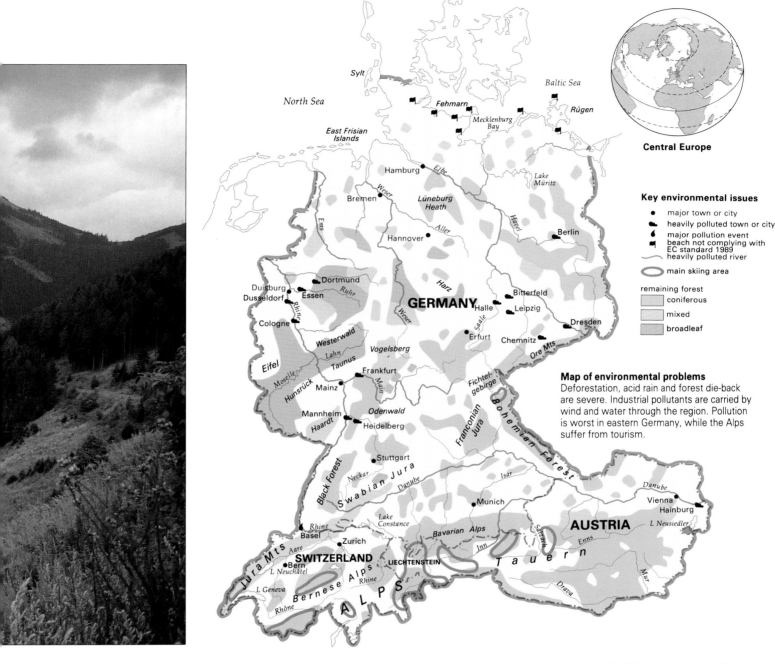

North Sea

Baltic Sea

Sylt

Fehmarn

Rügen

Mecklenburg Bay

East Frisian Islands

Central Europe

Key environmental issues

● major town or city

■ heavily polluted town or city

♪ major pollution event

⚐ beach not complying with EC standard 1989

〜 heavily polluted river

◯ main skiing area

remaining forest

☐ coniferous

☐ mixed

☐ broadleaf

Map of environmental problems
Deforestation, acid rain and forest die-back are severe. Industrial pollutants are carried by wind and water through the region. Pollution is worst in eastern Germany, while the Alps suffer from tourism.

Hamburg

Elbe

Bremen

Weser

Lake Müritz

Lüneburg Heath

Hannover

Aller

Ems

Havel

Berlin

Dortmund

Ruhr

Harz

GERMANY

Bitterfeld

Duisburg

Dusseldorf

Essen

Rhine

Halle

Leipzig

Cologne

Saale

Dresden

Westerwald

Weser

Erfurt

Chemnitz

Ore Mts

Eifel

Lahn

Taunus

Vogelsberg

Moselle

Hunsrück

Frankfurt

Main

Fichtel-gebirge

Mainz

Franconian Jura

Bohemian Forest

Mannheim

Haardt

Odenwald

Heidelberg

Stuttgart

Neckar

Swabian Jura

Danube

Isar

Danube

Vienna

Black Forest

Hainburg

L. Neusiedler

Lake Constance

Bavarian Alps

Munich

AUSTRIA

Rhine

Basel

Zurich

Inn

Salzach

Enns

Aare

Jura Mts

SWITZERLAND

LIECHTENSTEIN

Tauern

Mur

L. Neuchâtel

Bern

Bernese Alps

Rhine

Drava

L. Geneva

ALPS

Rhône

the fertile lowlands of northern and central Germany – but at the expense of widespread groundwater pollution and a loss of wildlife habitats.

The price of industrialization

From 1949 onward, German industry began to expand at an unprecedented rate, boosted by financial aid from the United States. Saarland and the Ruhr district of western Germany now have the greatest concentration of industry in the world, while in former East Germany sprawling industrial conurbations form a triangle around the cities of Leipzig, Dresden and Chemnitz.

In Austria and Switzerland widespread environmental change was brought about by hydroelectric schemes, with whole valleys being flooded and large areas deforested. The environment has also had to cope with the recreational demands of an increasingly affluent and highly mobile population. New leisure amenities near national beauty spots enable walkers, hunters, golfers, anglers and sailors to enjoy the beautiful mountain, forest and lake scenery. Even once-remote mountain areas of Switzerland and Austria are now highly developed, with ski resorts on many of the upper slopes.

Throughout the region awareness has grown of the great pressure that industrialization and an affluent lifestyle place on the environment. In the 1970s huge tracts of forest were discovered to be poisoned by acid rain. The Black Forest in the southwest, the Bavarian Forest in the

southeast and the Harz mountains forests of central Germany all suffered dead and dying trees. Alarm in West Germany at the extent of the damage resulted in the foundation of the Green Party in 1979. Its aims were to pressure central government to prevent irreparable damage being done to the environment, and in 1983 it became the first environmental party to win representation in any national government. Similar movements have developed elsewhere in the region. In 1985, the Austrian government withdrew its plans for a dam on the Danube at Hainburg after a public outcry and demonstrations in which 27 people were injured. Green issues are now firmly established on the political agendas of all the Central European countries.

POLLUTION'S DARK SHADOW

Most air pollution in Central Europe originates in Germany, where fossil fuels (coal, oil and gas) supply 70 percent of the energy used to power industry. When burnt, these produce noxious sulfur emissions that are damaging to health. Austria and Switzerland create far less air pollution, as most of their industry is powered by hydroelectricity, but both countries suffer the effects of wind-blown pollution from Germany.

The smog from factory and vehicle emissions (another major cause of air pollution) is so bad in some German towns – for example, in the heavily industrialized Ruhr – that when a smog alarm is triggered, large areas are closed to traffic and factories are forced to cease production. However, the pollution is not always so localized: people living in the ancient city of Heidelberg often suffer sore eyes and breathing difficulties when the wind is blowing from industrial Mannheim some 16 km (10 mi) away.

By far the greatest threats to human health are found in the heavily industrialized areas of former East Germany. The industrial area of Leipzig, Halle and Bitterfeld, with its dense concentration of energy and chemicals industries, is particularly heavily polluted. The extent of the environmental degradation in eastern Germany only came to the attention of the Western media in 1990, following German reunification. Industry there is fueled largely from the abundant – and therefore cheap – local reserves of brown coal (lignite), which burns less efficiently than bituminous coal and produces large quantities of highly polluting sulfur dioxide. Some 36 percent of the population in eastern Germany were found to be breathing air that contained more sulfur dioxide than regulations permit.

Ground level problems

The toxic waste generated by industry creates further pollution problems on the ground. There are some 30,000 waste sites in eastern Germany alone, and the number is growing rapidly. Since reunification, more and more toxic waste – 100,000 tonnes in 1991 – has been sent from western Germany to be dumped illegally and more cheaply in the east.

Rivers and groundwater are affected by industrial and agricultural chemicals (fer-

tilizers and pesticides), raw sewage and heavy metals. More than 1,000 noxious substances have been recorded in the environment in the Ruhr area, while in Westphalia nitrate fertilizers – spread onto the fields to boost crop yields – have seriously contaminated the groundwater: two-thirds of the 6,000 wells tested in the area exceeded the 50 mg per liter (0.008 oz per gallon) nitrate limit set by the European Community (EC).

The accumulation of nitrates in rivers causes freshwater algae to multiply rapidly, forming a green scum or "bloom" on the water's surface. When one nutrient becomes scarce the algae die, and they are broken down by bacteria that quickly consume the dissolved oxygen in the water on which the fish population depends (a process called eutrophication). Starved of oxygen, the fish die too. Some stretches of rivers including parts of the Rhine, have been pronounced "dead" as a result of this process.

Acid attack

Industrial pollution is not only damaging Central Europe's rivers. The forests have been badly affected by acid rain – the result of fumes from burning fossil fuels

Smog alert (*above*) Acute air pollution in the Ruhr region of western Germany triggers a local alarm, stopping traffic. However, sulfur dioxide emissions are 15 times higher in former East Germany. The closing of many obsolete noncompetitive factories in this part of the region will stop some pollution at the source.

Death of a temperate river (*right*) All rivers contain nutrients used by algae to grow; this is the normal eutrophic cycle (shown on the left). However, in rivers polluted by phosphates and nitrates (shown on the right), there is nutrient overload and the algae grow faster than normal. This abnormal growth reduces oxygen levels in the water, particularly at night. When these large masses of algae die and fall to the bottom, dissolved oxygen is also used up very rapidly by the bacteria breaking down the dead algae. The lack of oxygen in the water then kills fishes and other organisms. This abnormal cycle of growth and death in polluted waters is sometimes called cultural eutrophication.

Normal eutrophication cycle

1 Normal algal growth	**3** Decay uses some oxygen	**5** Oxygen level constant
2 Death of algae	**4** Oxygen from air	

TOURISM IN THE ALPS

Tourism in the Alps is a mixed blessing. Some 40 million people visit the area each year, bringing considerable revenue to all the countries in the region; but to environmentalists the situation has become a real nightmare. The combination of overdevelopment and industrial pollution has seriously endangered this fragile environment, making the clean air and unspoilt landscape a thing of the past.

In winter, skiing is the main attraction. To cater for the huge increase in numbers, large areas have been deforested and new developments built. New ski runs, lined with ski lifts, leave the mountain slopes denuded of trees and vulnerable to soil erosion and avalanches. Garbage disposal is an

acute problem, too: the rivers Navigenze and the upper Rhône are so littered with waste that the shape of the river beds has been altered and plant life has largely died off.

The railroads and highways that connect the resorts bring further problems. On the Brenner Pass between Austria and Italy, the volume of traffic increased from 1,500 vehicles a day in 1955 to 50,000 on peak days in 1988. The air pollution that this causes is severe, exacerbated by temperature inversions in the narrow valleys that prevent the fumes from blowing away. In addition, large quantities of salt are spread on the roads to thaw the ice, corroding bridges, polluting streams and damaging plant and wildlife habitats.

slopes vulnerable to soil erosion and avalanches, and valleys susceptible to flooding. Heavy rain in 1987 caused the river Reuss to flood in the sparsely forested Andermatt region of southern Switzerland. The water inundated fertile farmland, severely damaged the Gotthard highway and washed away railroad tracks, causing an estimated $320 million worth of damage.

At the other extreme is the threatened environment of the Wattenmeer, a shallow stretch of water between 5 and 20 km (3 and 12 mi) wide running between the German mainland and the East Frisian islands. The flat beaches, sand dunes and small islands support a complex ecology and are host to millions of migrant birds each year. However, tourists also flock to the area in their thousands to enjoy the spectacle. Their feet trample the dunes, causing serious damage to native plants and disturbing wildlife. Toxic chemicals and heavy metals – carried by the Ems and Elbe rivers from industrial and urban areas inland – pollute the waters. There are also threats to the area from the building of sea walls, airports and factories, and from oil drilling.

reacting with water vapor in the atmosphere. Hydrogen chloride, sulfur dioxide and nitrogen oxide gases can be blown hundreds of kilometers before they fall in a dilute solution as rain onto the forests and seep into water courses. Half of the forests in Germany, 34 percent in

Switzerland and 16 percent in Austria have trees that are dying from the effects of air pollution.

In the Alps, the combination of acid rain, poor forest management in the past, human alteration of land surfaces and climatic extremes has left many mountain

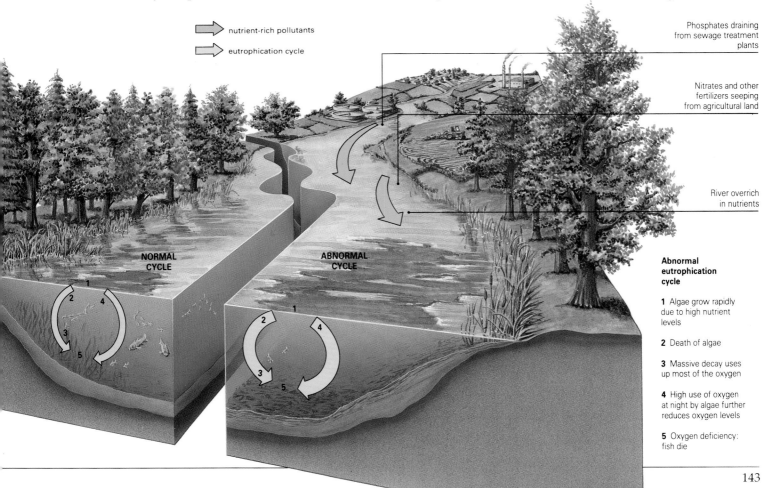

nutrient-rich pollutants

eutrophication cycle

Phosphates draining from sewage treatment plants

Nitrates and other fertilizers seeping from agricultural land

River overrich in nutrients

NORMAL CYCLE

ABNORMAL CYCLE

Abnormal eutrophication cycle

1 Algae grow rapidly due to high nutrient levels

2 Death of algae

3 Massive decay uses up most of the oxygen

4 High use of oxygen at night by algae further reduces oxygen levels

5 Oxygen deficiency: fish die

Walther-Rathenau-Straße

FRIEDEN ERNST NEHMEN
Keine Bundeswehr in unsere Goitsche!
JETZT ABRÜSTEN
DIE GRÜNEN

rettet den wald
DIE GRÜNEN

TAG FÜR TAG STIRBT EIN STÜCK NATUR!
DIE INDUSTRIE MACHT KASSE!

Zukunft – ökologisch sozial gewaltfrei
Grüne Liste / Neues Forum
Es wird Herbst Bürger bewegt Euch!

UMWELTSCHUTZ

A CHALLENGE FOR THE RICH

Central Europe – unlike its Eastern European neighbors – is in the enviable position of having both the expertise and the money (as well as the desire) to tackle most of its environmental problems. Germany probably leads the world in this respect, and is pressing the European Community to introduce higher environmental standards.

All the countries in the region have numerous protected areas: there are over 120 nature parks and reserves in Austria alone, covering some 19 percent of the country. Germany has plans to increase its protected area – at present covering about 8 percent of the country – by taking advantage of agricultural overproduction within the European Community. It aims to turn 10 percent of its agricultural land into wild corridors that will link together a chain of nature reserves, enabling wildlife to move freely between them.

Strict controls

Protecting the forests by reducing air pollution is a high priority throughout the region. One of Austria's first pieces of environmental legislation was the Forestry Act of 1975, which declared that new industrial plant would only be granted a license to operate on condition that protective devices such as scrubbers were installed. The most commonly used type of scrubber sprays water over industrial exhaust gases until the solid or liquid pollutants – such as sulfur dioxide – have been washed away. Between 1980 and

The role of political pressure (*above*) The West German Green party raised public awareness of environmental issues during the 1980s, but their electoral power has declined as ecology has been incorporated into mainstream politics.

1988 Austria's sulfur dioxide emissions were successfully reduced by two-thirds.

During the mid 1980s the Austrian government introduced the most stringent exhaust standards for cars in Europe. Legislation made the use of unleaded gasoline compulsory, and all new automobiles now have to be fitted with catalytic converters, which reduce emissions of carbon monoxide, nitrogen oxides and noxious hydrocarbons. The Swiss army has agreed to buy new vehicles fitted with catalytic converters.

In western Germany, traffic calming schemes – chicanes and mounds that slow down traffic – have not only reduced air pollution from vehicles by up to 70 percent; they have also halved the number of road accidents. Improvements to public transport have also reduced air pollution (though the majority of people are reluctant to give up the freedom and privacy afforded by their own vehicles): in Cologne, for example, trams now run on reserved sections of the highway so that they can avoid traffic jams; traffic lights change automatically to favor trams; and tram interchanges have been built at railroad stations.

Tackling former East Germany

East Germany only became subject to the environmental laws of the European Community in 1990, when it officially joined West Germany. An emergency

CUTTING BACK ON FOSSIL FUELS

Germany has reached an ideological crossroads. Industry – the basis of economic prosperity in the country – demands massive supplies of energy, but the fossil fuels that provide most of it are expensive, highly polluting for the whole region, and will eventually run out. It makes sense for the country to reduce consumption of fossil fuels and explore alternative energy sources.

Germany has plans to build the world's largest solar power plant at Bad Langensalza, near Erfurt in eastern Germany. The project will cost about one-sixth of the budget allocated for Germany's annual renewable energy program. Domestic rubbish such as paper, plastics and other combustible materials can also be used as fuel. Combustible waste from Herten in the Ruhr area is made into fuel pellets and sold as ECO-BRIQ, mostly to a nearby cement kiln. The plant handles 300,000 tonnes of combustible waste a year.

Saving energy is the other option. Even the most efficient power stations convert only 40 percent of the fuel's energy into electricity. Much of the rest is released into the atmosphere as heat. Combined heat and power schemes are designed to make use of this waste heat. At the Dow chemical factory at Stade in northern Germany, exhaust gases from a jet engine drive a turbine that generates electricity. Heat from the gases is collected to make steam, which is used in the chemical plant. Some 40 percent of the energy in the original fuel is converted to electricity and 47 percent is collected as heat.

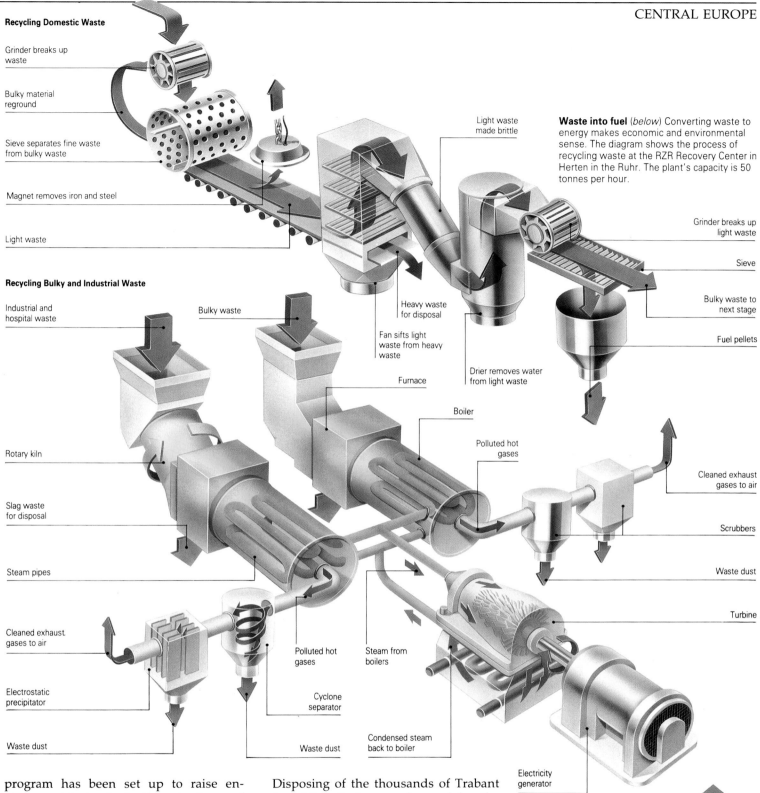

Recycling Domestic Waste

Grinder breaks up waste

Bulky material reground

Sieve separates fine waste from bulky waste

Magnet removes iron and steel

Light waste

Recycling Bulky and Industrial Waste

Industrial and hospital waste

Bulky waste

Rotary kiln

Slag waste for disposal

Steam pipes

Cleaned exhaust gases to air

Electrostatic precipitator

Waste dust

Light waste made brittle

Heavy waste for disposal

Fan sifts light waste from heavy waste

Furnace

Drier removes water from light waste

Boiler

Polluted hot gases

Polluted hot gases

Steam from boilers

Cyclone separator

Waste dust

Condensed steam back to boiler

Electricity generator

Electricity

Grinder breaks up light waste

Sieve

Bulky waste to next stage

Fuel pellets

Cleaned exhaust gases to air

Scrubbers

Waste dust

Turbine

Waste into fuel (*below*) Converting waste to energy makes economic and environmental sense. The diagram shows the process of recycling waste at the RZR Recovery Center in Herten in the Ruhr. The plant's capacity is 50 tonnes per hour.

program has been set up to raise environmental standards there to those of the western half of the country by the year 2000. As part of this program, the EC's provisions on waste management, water quality and air pollution will come into force in former East Germany in 1996. In addition, the five nuclear power stations in eastern Germany are to be closed down and replaced with one oil-fired station; some 27 water treatment plants are to be built on the Baltic coast; and 6,200 km (3,900 mi) of new sewers are to be laid. There are also plans to renovate a further 5,000 km (3,100 mi) of sewers.

Disposing of the thousands of Trabant automobiles that were dumped by the East Germans as soon as they could buy a Western automobile has proved a major problem. However, German scientists have now developed a strain of bacteria that can degrade plastic. The bacteria are able to reduce the phenol-formaldehyde resin body of a Trabant to 2 percent of its original weight.

Eventually, it is hoped, environmental protection will become as much a part of everyday life in former East Germany as it is in the rest of Central Europe. Government legislation on its own is not enough.

In Germany, Austria and Switzerland the growing number of "green" consumers has had a big influence on manufacturers. A modern washing machine from a German company such as Bosch or AEG, for example, consumes far less energy, water and soap powder than a 10-year-old model, and parts are especially designed to be reused or recycled. More such products are needed for "green" lifestyles.

The Rhine: a poisonous cocktail

The river Rhine is a truly international waterway, and one of the most polluted in the world. Barges carrying freight ply their way continually from Basel in Switzerland all the way to the North Sea, passing through Germany and the Netherlands. In Germany the river runs parallel to major rail and road routes, and dissects one of the most heavily industrialized areas of Central Europe. All along the river's course, huge quantities of industrial effluent are discharged daily into the water, producing a poisonous cocktail of waste that flows downstream to Rotterdam and out into the North Sea.

The effects of the pollution have been devastating, particularly to aquatic wildlife. For long stretches the water is now "dead", deoxygenated by the decomposition of algae. Salmon no longer migrate up the river as they once did, and with fewer fish and less insect life in the water,

many riverside birds have also disappeared. The wilderness wetlands along the North Sea coast, home to millions of shorebirds and seabirds, are polluted by the river's outflow, and in the North Sea itself seal populations have declined.

Following the rise of environmental concern in Europe, the River Rhine Commission was set up by those countries that share the river: Switzerland, Austria, Liechtenstein, Germany, France and the Netherlands. In 1986 the commission agreed to measures for combating the river's pollution, especially from industrial accidents: five serious accidents were reported in November alone of that year. The commission's aim is to make the river clean enough to support salmon again by the year 2000. As part of the program, the Netherlands national water authority,

An industrial waterway (*right*) Industrial plants line the banks of the Rhine near Duisburg. Much pollution comes from traffic on the water. Barges and ships are less supervised than cars, and carry thousands of tonnes of toxic chemicals and waste.

Going with the flow (*below*) Pollution on the Rhine has wide-reaching effects. The river carries a nutrient-rich and toxic cocktail as far as the North Sea, where it is mixed with pollutants from other European rivers including the Thames.

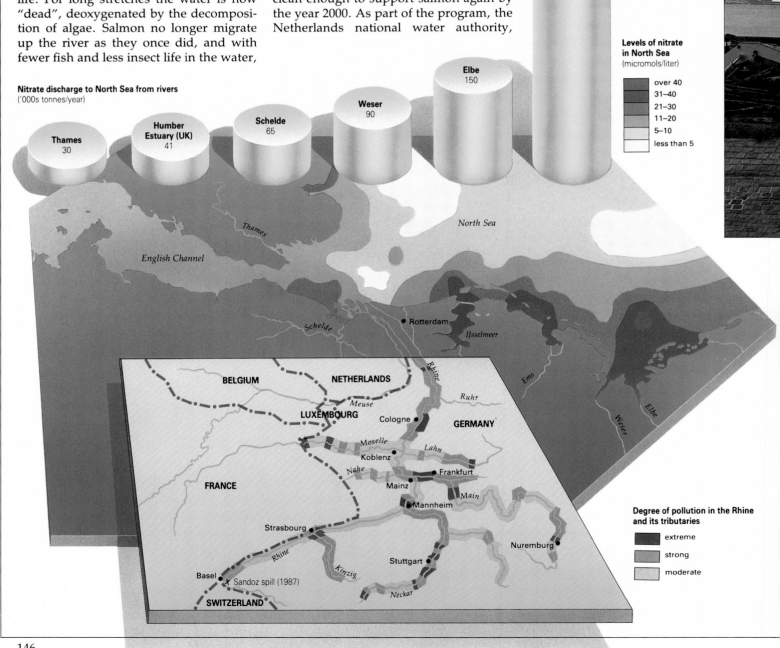

Nitrate discharge to North Sea from rivers
('000s tonnes/year)

Thames 30
Humber Estuary (UK) 41
Schelde 65
Weser 90
Elbe 150
Rhine/Meuse Estuary 420

Levels of nitrate in North Sea
(micromols/liter)

- over 40
- 31–40
- 21–30
- 11–20
- 5–10
- less than 5

Degree of pollution in the Rhine and its tributaries

- extreme
- strong
- moderate

English Channel
North Sea
Thames
Schelde
Rotterdam
IJsselmeer
Ems
Weser
Elbe

BELGIUM
NETHERLANDS
Rhine
Ruhr
Meuse
LUXEMBOURG
Cologne
GERMANY
Moselle
Lahn
Koblenz
Nahe
Frankfurt
FRANCE
Mainz
Main
Mannheim
Strasbourg
Nuremburg
Stuttgart
Rhine
Kinzig
Basel
Sandoz spill (1987)
Neckar
SWITZERLAND

Rijkswaterstaat, monitors pollution at 12 stations along the Rhine and its tributaries. Computers measure the water's oxygen, acidity, mineral and chemical levels, and remedial action is taken when thresholds are breached.

Environmental setback

In 1987 a single accident put back the plans of the River Rhine Commission possibly by a decade. At Basel, a fire broke out one night in a chemical warehouse belonging to the company Sandoz AG. The highly toxic chemicals included 12 tonnes of the fungicide ethoxyethyl, containing 1.5 tonnes of pure mercury. Fire fighters doused the flames with water, which – heavily polluted by the chemicals – washed straight into the Rhine. From Basel to Mainz, a distance of

Washing the rocks (*left*) A film of oil, bacteria and effluent clings to the riverbank, resulting from the contamination of the water. Hosing down the rocks may make them less slippery and dirty-looking, but the runoff pours straight back into the river.

some 320 km (200 mi), almost all living organisms, including 500,000 fish, were destroyed. It will take years for the river to recover and wildlife to return, and the threat of mercury having seeped into underground water supplies remains.

By chance, the chemists analyzing the water after the Sandoz spill discovered traces of the herbicide atrazine, which had not been released by the fire. Another company, Ciba–Geigy, later admitted that they had accidentally spilled 470 liters (103 gallons) of atrazine into the Rhine the day before the fire broke out. They had made no attempt to report the spill to the authorities. The Executive Director of UNEP (the United Nations Environment Programme) later summed up the situation when he declared that "these accidents reveal the full extent of the apathy, confusion and general unpreparedness of the world's most developed nations. They also illustrate the deplorable inadequacy of international legislation."

Stoking the fires

Technological advances in steel making in the Western world have helped the industry to become safe, less wasteful and more ecologically sound. However, in former East Germany and all across the former communist countries of Eastern Europe many steel-making factories still use 1950s technology in coke-burning blast furnaces that are dangerously lacking in safety measures. Workers are exposed to pollution and working conditions no longer acceptable in the West. In 1991, pollution controls in eastern Germany were nonexistent in many places.

The brown coal from which eastern Germany's coke is derived is highly polluting, producing tonnes of sulfur dioxide when burnt in power stations, and releasing various impurities in the form of gases when burnt to produce coke for the region's steel factories. Modern technology in the West has developed coke ovens to a high standard. The unit is self-contained and sealed so that the gases given off by the cooking coal are collected internally. Only a tiny amount of the dangerous gas escapes. But in many eastern German factories, smoke and gases – including cancer-causing benzene and ammonia – swirl out of the oven tops during the cooking process.

Lack of technology means that the workers have to stoke the coal and shut oven hatches manually. They continually face the risk of falling into the hot ovens. Although western German technology and pollution controls – such as a newly developed coke filter system that absorbs toxic pollutants – are being slowly introduced throughout the country, it is a slow and expensive process. Eastern Germany's dependency on brown coal is expected to drop by at least one-third by 2000, but at present the area remains the industrial blackspot of Central Europe.

Poisonous, swirling smoke and gases envelop a worker as he closes a hatch above a coke oven at the Gross Gasserei steel plant in Magdeburg, former East Germany.

THE RAVAGES OF PROGRESS

AN ALTERED LANDSCAPE · SOURCES OF POLLUTION · REVERSING THE DAMAGE

Eastern Europe's natural environment has been stricken by the onslaughts of state-directed industrialization, deforestation and intensive agriculture. Communist governments, relying on out-of-date technology, began their push to industrialize in the 1950s – ignoring pollution controls and the consequences for the health and viability of the land and its people. Throughout the region, power stations emit huge amounts of sulfur dioxide, and acid rain has killed vast stretches of conifer forests. Heavy metals, chemicals and untreated sewage are poured into waterways, especially the Danube. Cleaning all this up is bound to be costly and, although the need to do so is recognized, the new governments involved will need both Western aid and greater commitment if they are to make an impact.

AN ALTERED LANDSCAPE

Thousands of years of human settlement, deforestation and farming have dramatically altered the original landscapes of Eastern Europe. There are still lush remnants of the forests that covered much of the region in prehistoric times; Slovakia, for example, is one of the most densely forested areas of Europe, with great tracts of fir, spruce, beech and oak. However, these are now mostly found only on hills and mountains. The rest has been cleared for lowland agriculture.

Early instances of farmers clearing forested land for cultivation sometimes had unexpected repercussions: in the late Middle Ages, heavier-than-usual rainfall combined with the widespread removal of trees caused massive flooding in the lowlands of Poland, and forced many villages to be relocated. Even today, the logging of forests on steep-sloping land has led to serious soil erosion.

Widespread changes to the environment came in the mid 20th century with the modernization – and mechanization – of farming. With modern field systems, crop specialization and new schemes for drainage and irrigation came the heavy use of fertilizers and pesticides, which pollute land, rivers and air. In Hungary, as in other areas, nitrates and phosphates from fertilizers, human sewage and pig farms are a serious concern. The nitrates cause longterm contamination of the groundwater, and encourage the abundant growth of algae in rivers and lakes. When nutrients become scarce the algae die, and as they decompose oxygen levels in the water are reduced by bacteria. Without sufficient oxygen, fishes and other aquatic creatures also die.

Land reclamation, too, has destroyed some precious habitats. Under Romania's

---

COUNTRIES IN THE REGION

Albania, Bosnia and Hercegovina, Bulgaria, Croatia, Czechoslovakia, Hungary, Macedonia, Poland, Romania, Slovenia, Yugoslavia

POPULATION AND WEALTH

	Highest	Middle	Lowest
Population (millions)	38.4 (Poland)	15.7 (Czecho)	3.3 (Albania)
Population increase (annual population growth rate, % 1960–90)	2.4 (Albania)	0.8 (Romania)	0.2 (Hungary)
Energy use (gigajoules/person)	185 (Czecho)	136 (Romania)	38 (Albania)
*Real purchasing power (US$/person)	5,920 (Hungary)	4,860 (Yugoslavia)	4,190 (Poland)

ENVIRONMENTAL INDICATORS

CO_2 emissions (million tonnes carbon/annum)	56 (Poland)	25 (Romania)	1.2 (Albania)
Municipal waste (kg/person/annum)	756 (Bulgaria)	657 (Hungary)	212 (Poland)
Nuclear waste (cumulative tonnes heavy metal)	100 (Yugoslavia)	n/a	0 (Albania)
Artificial fertilizer use (kg/ha/annum)	303 (Czecho)	180 (Bulgaria)	130 (Romania)
Automobiles (per 1,000 population)	174 (Czecho)	122 (Bulgaria)	11 (Romania)
Access to safe drinking water (% population)	100 (Czecho)	97 (Hungary)	73 (Yugoslavia)

MAJOR ENVIRONMENTAL PROBLEMS AND SOURCES

Air pollution: generally high, urban very high; acid rain prevalent; high greenhouse gas emissions
River/lake pollution: high; *sources*: industrial, agricultural, sewage, acid deposition
Land pollution: high; *sources*: industrial, agricultural, urban/household, nuclear

* Only figures available. Note: figures for Yugoslavia relate to the former country now divided into individual states

dictatorial president Nikolae Ceausescu (1918–89), there was an attempt to turn 101,170 ha (250,000 acres) of the Danube Delta – one of Europe's leading wildlife sanctuaries – into farmland.

The industrial legacy

Pollution caused by toxic waste material from industrial activities has a much longer history. Iron-smelting, for example, was carried out in Silesia, in southwestern Poland, as early as 400 BC, and by the 17th century there were some

Map of environmental problems (*right*) Exploitation has ravaged Eastern Europe. Acid rain has defoliated 70 percent of trees in Czechoslovakia and up to 50 percent in Poland. Most water is contaminated, and air quality and industrial pollution are among the world's worst.

Struggling to survive (*below*) A woman searches for fuelwood among the rubble of bombed houses in Dubrovnik, Croatia, where prolonged civil war has brought a different meaning to environmental catastrophe. Basic resources are very scarce.

Eastern Europe

Key environmental issues

- major town or city
- heavily polluted town or city
- heavily polluted river

acidity of rain (pH units)

- 4.2 (most acidic)
- 4.4
- 4.6
- 4.8
- 5.0 (least acidic)

7,000 coal shafts in the area. However, the extent of the pollution generated then pales beside that caused by the industrialization of Eastern Europe in the mid and late 20th century. Acid rain (exacerbated by atmospheric pollution drifting in from western Europe) and pollution of the river Danube are two of the greatest problems. The 2,850 km (1,770 mi) Danube is already polluted by the time it enters Hungary from Austria; today the river is a sewer for the eight countries and

70 million people who live within its drainage area.

The true dimensions of the region's environmental problems were for many years obscured by state control of the media, the low priority given to environmental education, and propaganda claiming the superiority of the communist system (in which mistakes could not be acknowledged). With the collapse of communism in 1989, however, and the new climate of democratic debate, Eastern Europeans have become increasingly vocal about the dangers of pollution. The explosion in April 1986 of the nuclear reactor at Chernobyl in the Ukraine – lying just east of the region – was a major turning point in people's awareness of the environment's vulnerability.

SOURCES OF POLLUTION

Eastern Europe's environment and its peoples suffer most damage from air and water pollution, though land degradation creates its own problems. All types of heavy industry – especially chemical and metallurgical – contribute to air pollution. Ferrous (iron and steel) and nonferrous (for example, aluminum and copper) industries emit a wide range of toxic pollutants. Many of these substances such as cadmium and lead – both carcinogenic – are caused by industrial processes, particularly smelting.

The highest levels of air pollution occur along the northern border of Czechoslovakia and in adjacent areas of Poland. One of the worst affected areas is the Upper Silesian Industrial District, which sits on Poland's richest coal deposit. Here, industrial dust falling on towns such as Bytom, Gliwice and Zabrze reaches a staggering 1,000 tonnes on 1 sq km (0.4 sq mi) each year. The cement industry is responsible for giving off "white dust", particularly from furnaces with poorly maintained filtering systems. Air pollution also affects smallholdings in the vicinity of large factories: some have been abandoned because the vegetables raised in the plots are contaminated and unfit for consumption.

Industrial emissions may be worse locally if chimneys are too low or if pollution is not dispersed quickly by winds. However, tall chimneys are no solution, as they only spread pollution farther away. Some cities, such as Budapest and Prague, the Hungarian and Czech capitals, often experience dense smog caused by pollution from domestic and factory chimneys, and by automobile exhaust fumes. Although there is less traffic in Eastern Europe than in western Europe, cars such as Trabants and Wartburgs (made in former East Germany and imported into the region in great numbers) have simple two-stroke engines that individually emit much richer – and therefore fouler – exhaust gases than four-stroke engines.

Poisoned waters
Many rivers and lakes in Eastern Europe are severely polluted with industrial effluent, untreated sewage, and chemical fertilizer that has drained off farmland. A major threat comes from heavy metals –

chrome, iron, lead and mercury – released into the rivers by factories. (In 1989, there was 200 times the legal level of mercury in Poland's chief river, the Vistula.) A high proportion of Eastern Europe's rivers are, in fact, "dead" – they are unable to support any aquatic life. In Romania, 2,800 km (1,740 mi) of rivers – out of a total system of 20,000 km (12,420 mi) – are considered dead.

Toxic materials spilled or dumped into the water also cause serious pollution. In 1987, for example, authorities at the Romanian port of Sulina, at the mouth of the Danube, discovered that the paint and dye waste that it was storing was toxic and had leaked from the container drums into the Black Sea, threatening the local tourist beaches.

Polluted water or chemical waste seeping into the earth can also contaminate groundwater – the source of drinking water. One in ten Hungarians in 1992 lacked access to safe drinking water. Even Hungary's Lake Balaton – the country's greatest tourist attraction after Budapest – was heavily polluted by sewage and agricultural runoff until the government

initiated a massive cleanup program.

Lack of sewage treatment is a serious cause of water pollution everywhere. Some 50 percent of Hungarians have no sewerage system, and only a quarter of Budapest's sewage is treated before being discharged into the Danube. Romania's capital, Bucharest, had totally inadequate sewage treatment until new facilities were provided at nearby Glina.

Problems on the ground
Eastern Europeans have made a serious impact on the land. Intensive agriculture, with its heavy use of artificial fertilizers, has reduced the soil's natural fertility; while much land has been lost through mining, quarrying and tipping. Acid rain – the fallout of industrial pollutants as dry particles, rain, snow or fog – has caused widespread damage. By harming the leaves and leaching nutrients from the soil, it has killed vast areas of woodland – over 400,000 ha (nearly 1 million acres) in Czechoslovakia, and 242,808 ha (600,000 acres) in Poland. Acid rain also corrodes buildings and monuments, and in Upper Silesia – one of the most industrialized

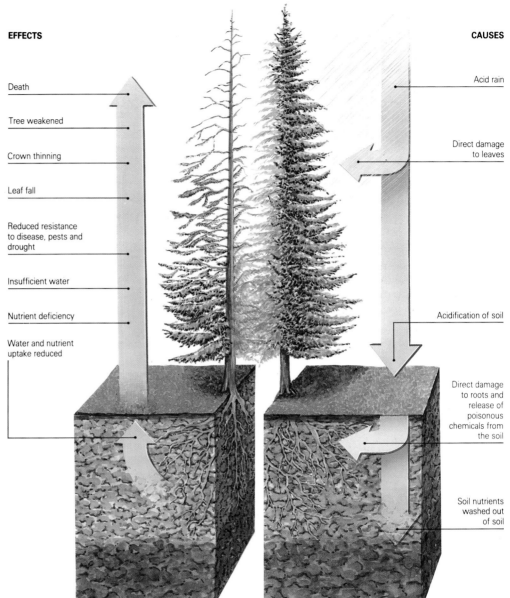

EFFECTS

Death

Tree weakened

Crown thinning

Leaf fall

Reduced resistance
to disease, pests and
drought

Insufficient water

Nutrient deficiency

Water and nutrient
uptake reduced

CAUSES

Acid rain

Direct damage
to leaves

Acidification of soil

Direct damage
to roots and
release of
poisonous
chemicals from
the soil

Soil nutrients
washed out
of soil

Smokestack pollution (*above*) Coal-fired industries account for about 70 percent of sulfur dioxide emissions in Poland. Even with relatively few motor vehicles, the country has some of the worst air pollution in the world.

Killed by acid rain (*right*) Several factors combine to cause the death of trees, including drought, disease and pest infestation. Air pollution contributes by reducing tolerance to other environmental stresses.

A DIRTY BROWN COAL

Although sulfur dioxide – generally considered the worst air pollutant – is produced in nature (for example, through volcanic eruptions), in industrialized parts of Eastern Europe most of it is generated by power stations and industrial boilers burning low quality brown coal (lignite). In Poland, in 1989, sulfur dioxide accounted for 2.79 million tonnes out of a total of 6.63 million tonnes of dust and gas entering the atmosphere. A further 0.90 million tonnes of sulfur dioxide was carried by prevailing winds from neighboring Czechoslovakia and East Germany. This gave Poland 9 tonnes of sulfur dioxide per sq km (23 tonnes per sq mi), compared with the western European average of 2.3 tonnes per sq km (6 tonnes per sq mi).

The sulfur dioxide emitted by lignite is a major contributor to acid rain. Much of the sulfur falls on the ground as dust. In northwestern Czechoslovakia – the region's most sulfur-polluted country – smoke from lignite-burning industries is often trapped between the mountains where it forms a thick smog that endangers health: old people and children are particularly vulnerable. The sulfur dioxide is an especial threat to sufferers of chronic respiratory complaints, such as bronchitis. On "sulfur-alert days", schoolchildren in the Czechoslovak town of Most are obliged to wear face-masks.

parts of Poland – even the railroad lines have been affected.

So much pollution has had devastating effects on people's health throughout Eastern Europe. Pollution-related health problems affect about a third of people in the Czech lands and some 15 percent in Slovakia. As many as one-third of the 38 million Poles are said to live in "ecological hazard" areas. Children suffer most of all, and the numbers of disabled children are especially high in areas with high heavy metal levels in air and water. In parts of Upper Silesia, for example, 35 percent of children show evidence of lead poisoning. Behavioral and emotional problems, anemia and kidney damage are just some of the possible illnesses related to having high levels of lead in the blood.

REVERSING THE DAMAGE

Eastern Europe's environmental concerns have been addressed for many years. Some policies, such as the conservation of the Tatra Mountains in Czechoslovakia, were put into effect in the late 19th century. Others, like the cleaning and restoration of historic cities, such as Prague and Krakow, have begun much more recently (after World War II). It has, however, taken a long time for the environmental damage caused by forced industrialization under the communist regimes to be adequately confronted.

The United Nations environmental conference in Stockholm in June 1972 – deemed a landmark in the rise of worldwide environmental concern – had some spin-off in Eastern Europe. Legislation stipulating acceptable pollution levels and encouraging environmental education was introduced; and new authorities were set up to coordinate and implement policy. One of these was Poland's State Council for Environmental Protection (established in 1981), out of which a State Environmental Protection Inspectorate emerged in 1984.

The 1980s saw growing regional concern for preserving national parks – even for creating new ones – and cleaning up the environment. But, in the politically unstable and economically desperate climates of Eastern Europe, antipollution measures were – and continue to be – inadequately enforced. Fines for pollution have been too low to encourage industries to reduce output or invest in cleaner technology; and no amount of legal protection can save nature reserves and national parks from the effects of air and water pollution.

The voice of concern
Following the political revolutions in Eastern Europe, however, deeper concern over environmental problems has developed. In most parts of the region the ecological movement is gathering momentum with local organizations campaigning for the abandonment of projects detrimental to the environment, and national organizations taking the fight into the wider political arena.

In Bulgaria, a country where industrial waste pollutes nearly 70 percent of the farmland and 65 percent of its rivers, an organization entitled "Ecoglasnost" is part

A child's view of ecology (*left*) A poster at an environmental rally in Prague expresses both a love of nature and a growing sense of optimism as Eastern Europe emerges from 40 years of totalitarianism and begins to clean up its act.

of a growing ecological lobby. In February 1990, opposition to a nuclear power station at Belene, which lies in an earthquake zone in northern Bulgaria, halted construction of the reactor. Unfortunately, existing antipollution regulations are usually ignored at industrial plants, such as the massively polluting – and antiquated – metallurgical complex at Kremikovtsi, near the capital Sofia.

The Polish Ecological Club (formed in 1980) made considerable headway in attracting members when the Solidarity trade union rose to prominence, while Poland's Green Party was formed in 1988. One marked success of the ecological lobby has been the closure of the aluminum smelter at Skawina, just to the southwest of Krakow.

Throughout the regime (1975–87) of Czechoslovakia's Gustav Husak, open debate about the environment – and in particular the environmental destruction caused by burning vast quantities of brown coal – was discouraged. Discussion could only be conducted through unofficial (*samizdat*) literature. One of the main sources for information was the protest group known as Charter 77, which was founded to protest against human rights abuses; however, during its 12 years of existence, "Chartists" produced a wide range of reports on many subjects, including nuclear power.

In the wake of Czechoslovakia's "velvet revolution" of 1989, which toppled the communists, the government has had to make major decisions about the environmental costs of the country's industry – which is said to use three times as much energy to produce one unit of net production as in Western Europe.

The way forward

Hungary is probably the one country in Eastern Europe where debate on the environment was effective even before the collapse of communist rule in 1989. Public pressure forced the introduction of enhanced safety provisions at the Paks nuclear power station, 100 km (60 mi) south of Budapest, which uses Chernobyl-type technology, while more recent pressure has led to the establishment of the Balaton Water Management Development Program, which is responsible for cleaning up the heavily polluted waters of Lake Balaton.

In addition, successful campaigning by Hungary's Green groups has stopped work on the Nagymaros barrage dam on the river Danube, originally part of a massive Czechoslovak-Hungarian hydroelectric project. Environmentalists – among them a scientific group called Duna Kor (the Danube Circle) – offered strong objections to the project. They warned that much of Hungary's drinking water, which comes from groundwater filtered from the Danube, would be adversely affected by excess organic material and low concentrations of oxygen caused by the dam system. They claimed, too, that the project would destroy the area as a fish breeding ground, and flood the historic and scenic stretch of the river known as the Danube Bend.

A MULTINATIONAL CLEAN-UP

Cleaning up Eastern Europe will need both international cooperation and also supervision by bodies such as the World Health Organization, which lays down guidelines for acceptable pollutant levels. As pollution does not respect borders, countries in the region cannot solve their problems in isolation. The emissions from a Romanian chlorine factory at Giurgiu on the Danube, for example, drift across the river – which serves as a national border – to affect the health of people living on the opposite banks, in the Bulgarian town of Ruse.

To deal with this "exportation" of pollution from one country to another, a Convention on Long-Range Transboundary Air Pollution, drawn up by the UN Economic Commission for Europe and ratified in 1987, required signatory states to reduce levels of sulfur dioxide by 30 percent by 1993. However, stricter measures to reduce nitrogen oxide emissions – an automobile pollutant – have been blocked by some countries, including Czechoslovakia. Eastern Europe also urgently needs Western antipollution equipment, such as scrubbers, which remove toxic gases from chimney emissions, and financial help in restoring historic cities such as Dubrovnik, damaged in the Yugoslav civil war in Croatia.

Romania's chemical apocalypse

The chemical industry of Romania – long-established and based on large reserves of oil and natural gas – is responsible for much of the pollution that now affects the country. Many of the offending chemical complexes sprang up over the last 25 years in response to the demands of Elena Ceausescu, the wife of Romania's former president, Nicolae Ceausescu. (Both were executed in 1989 during the Romanian revolution.) However, serious polluters also include older plants, such as the carbon factory of Copsa Mica, 240 km (150 mi) northwest of Bucharest.

The main problems concern air and water pollution. Much of this has been due to wasteful and obsolete technologies, the absence or inefficiency of pollution control equipment, and poor maintenance. In some instances, especially during the early 1980s when the Ceausescus ordered an austerity drive for the country's embattled economy, control equipment was simply switched off in order to save energy.

Horrifying evidence of air pollution can be seen in the town of Copsa Mica where fall-out of carbon black – a substance used in the manufacture of tires – from the Carbosin plant makes the town and the surrounding Mures valley one of Eastern Europe's worst pollution "black spots".

Not only are the town's buildings and vegetation blackened by the pollution, but the people and sheep are, too. One consequence is serious health problems, especially among babies and young children. No link has been shown between carbon black and cancer, but it is known to aggravate bronchitis and asthma. The high levels of lead, zinc and cadmium in the soil – derived from a local lead plant – are possibly more hazardous.

Some 260 km (160 mi) northeast of Copsa Mica, in the province of Moldavia, air pollution from an artificial fiber plant on the edge of the town of Suceava caused malformation in babies, and the factory was closed down after the 1989 revolution due to public pressure.

Poisoning the water

Water pollution is a common problem in many places. Petrochemical plants at Midia and Navoderi on the Black Sea pour effluent into the water, and this is then carried south by currents to threaten the tourist resort of Mamaia. In 1992 a chemical plant at Arad in western Romania was discharging ammonia, nitrates and phosphates at 10 times the permitted rate only 4 km (2.5 mi) from a major water source for the city. Fortunately, plans were made to avert poisoning of

Black sheep of the world (*above*) The industrial village of Copsa Mica is coated with a layer of carbon black from the Carbosin plant that produces the substance for making tires. Some 30,000 tonnes of soot fall on the town's 7,000 residents each year.

Toxic playground (*right*) Children play on the banks of the Tirnava Mare river opposite the Carbosin factory of Copsa Mica, unaware of the serious health risks they face from the dangerously high levels of heavy metals – lead, zinc and cadmium – in the soil and water.

the groundwater; polluted water will be pumped from the affected area and processed at a purification plant.

The estimated 6.9 million tonnes of pollutants – including chlorine, pesticides and detergents – that enter Romania's rivers and lakes each year have "killed" the lower reaches of rivers such as the Bistrita and Trotus (tributaries of the Siret in Moldavia), and almost 3,000 km (1,860 mi) of waterways in the Danube delta, disrupting habitats and aquatic life.

Special problems arise where treatment facilities have not been operating properly. At the Borzesti petrochemical complex in the Trotus valley, lack of spare parts affected the treatment plant, so that waste material from the production of rubber, fungicides, insecticides and pesticides was being dumped on land right next to maize fields, without any monitoring of the hazard.

Poland's devastating air pollution

The Nova Huta steel mill, just north of Krakow in southern Poland, was built in the 1950s using 1930s technology. It emits a staggering 10,000 tonnes of noxious gases every day, while each year it produces dust emissions containing 7 tonnes of cadmium, 170 tonnes of lead, 470 tonnes of zinc and 18,000 tonnes of iron. The consequences for the factory workers, the local population and the ancient city of Krakow have been devastating. Krakow lies in a valley that traps the pollutants, and in winter, a blanket of smog often covers the whole city. When it rains the sulfurous pollutants in the air dissolve and fall as acid rain. As a result, many of the ancient buildings – which escaped the ravages of World War II – are blackened and disintegrating, with stone friezes and statues flaking away.

In the Nova Huta steelworks itself, 80 percent of the workers retire on a disability pension, and 8 percent die while still working. The local population is affected by cadmium-laden dust from the smelting works, which falls on plots where vegetables are grown and enters the food chain, causing lung and kidney damage and weakening bones. According to members of the Polish Ecology Club, the scrubbers (filters) fitted to some of the factory chimneys to reduce dust emissions are turned off at night in order to save electricity.

Although Western companies have been eager to sell antipollution devices to Eastern Europe ever since the collapse of the communist regimes, the region's governments are more concerned with trying to keep their foundering economies afloat. Many economists argue that the only way that industries will be able to afford pollution control equipment will be to cut back on unprofitable activities, which would force layoffs of thousands of workers.

A disintegrating statue on Krakow's Cloth Hall – its porous limestone vulnerable to acid rain – bears testimony to the effects of uncontrolled industrial pollution in Eastern Europe.

MISUSE ON A VAST SCALE

MISMANAGED RESOURCES · FAR-FLUNG POLLUTION · A SLOW, COSTLY CLEAN-UP

Many of the environmental problems of the republics that comprised the former Soviet Union stem from the rapid industrialization that began after the Bolshevik Revolution of 1917. From then until the breakup of the Soviet Union in 1991 resources were exploited on a gigantic scale, controlled by a communist regime desperate to catch up economically with the West. Ecological destruction is now widespread: in Central Asia, the Aral Sea is shrinking because of excessive water extraction for irrigation from the rivers that feed it; Siberia's Lake Baikal is being fouled by industrial effluent; air pollution levels in Ukrainian and Russian industrial centers are 10 times above the official limit; while radioactive fallout from the 1986 Chernobyl nuclear accident still blights lives and land.

COUNTRIES IN THE REGION

Armenia, Azerbaijan, Belorussia, Estonia, Georgia, Kazakhstan, Kirghiz, Latvia, Lithuania, Moldavia, Mongolia, Russia, Tadzhikistan, Turkmenistan, Ukraine, Uzbekistan

POPULATION AND WEALTH

	*Former USSR	Mongolia
Population (millions)	288.6	2.2
Population increase (annual population growth rate, % 1960–90)	1.0	2.8
Energy use (gigajoules/person)	194	53
Real purchasing power (US$/person)	n/a	n/a

ENVIRONMENTAL INDICATORS

CO$_2$ emissions (million tonnes carbon/annum)	690	1.9
Municipal waste (kg/person/annum)	n/a	n/a
Nuclear waste (cumulative tonnes heavy metal)	n/a	0
Artificial fertilizer use (kg/ha/annum)	118	18
Automobiles (per 1,000 population)	55	n/a
Access to safe drinking water (% population)	100	65

MAJOR ENVIRONMENTAL PROBLEMS AND SOURCES

Air pollution: generally high, urban very high; acid rain prevalent; high greenhouse gas emissions
River/lake pollution: high; *sources:* industrial, agricultural, sewage, acid deposition, nuclear
Land pollution: high; *sources:* industrial, agricultural, urban/household, nuclear
Land degradation: *types:* desertification, soil erosion, salinization, deforestation; *causes:* agriculture, industry
Waste disposal problems: domestic; industrial; nuclear
Major events: Chernobyl (1986) and Sosnovyy Bor (1992), nuclear accidents; Kyshtym (1957) hazardous waste spill; Novosibirsk (1979) catastrophic industrial accident

* All figures relate to the former Soviet Union

MISMANAGED RESOURCES

The critical environmental situation of the former Soviet Union has its roots in the complex interplay of political, economic and social factors that has marked the region since the formation of the communist state in 1922. Long before the fall of Tsar Nicholas II (1868–1918), Russia had already opened up its Siberian territories – and the vast mineral wealth there – with the start of the Trans-Siberian railroad in 1891. On the overpopulated land to the west, agricultural techniques were backward, the soil was exhausted and there were frequent crop failures.

From 1924 onward, under the dictatorship of Joseph Stalin (1879–1953), a series of Five-Year Plans aimed to industrialize and modernize the Soviet Union. This included the building of gigantic new dams, power stations, industrial plants and huge factories. But the decisions of an overcentralized, bureaucratic government were made without thought for their environmental impact. The state's Central Planning Committee – influenced by the underlying Marxist ideology and the vast size of the country – believed that the Soviet Union's resources were inexhaustible. In most of the industrialized world, the volume of mineral and fuel extracted doubles every 15 to 18 years, but in the Soviet Union it doubled every 8 to 10 years.

Before the breakup of the Soviet Union – and with it the centralized bureaucracies – no one took responsibility for environmental problems. Ministries and institutions protected their own interests. Environmental control was either ignored or responsibility for it was shifted up and down the chain of command. Management was mainly interested in obtaining raw materials for their specific industry and meeting production targets, often regardless of cost and in the shortest time.

Unlimited exploitation

Irrigated areas were expanded on a vast scale, and there has been massive overuse of mineral fertilizers. Excess is also a characteristic of the region's many hydroelectric power (HEP) schemes. The harnessing of rivers for HEP provided a cheap means of increasing energy supplies. As a result of extensive dam building – the number of reservoirs in 1992 exceeded 3,500 – the former Soviet Union has lost 21 million ha (50.5 million acres) of arable land through flooding.

Burned out (*left*) Desolation reigns after a fire in a taiga forest in the northwest of the Ural Mountains in Russia. As many as 30,000 fires may rage through the forests of the former Soviet Union every year, contributing to pollution and destroying natural habitats.

Map of environmental problems (*below*) Air and water pollution are worst in the west of the region; 103 cities exceed by 10 times the air pollution limits set by the former Soviet Union. Even in the relatively cleaner east, industrial waste from pulp and paper mills is polluting Lake Baikal's freshwater. Permafrost constrains land use over a vast area of the far north.

The timber industry, too, reflects resource mismanagement. Although the former Soviet Union had over 810 million ha (over 2 billion acres) of forested land – 91.6 percent of it in the Russian Federation – it regarded these as inexhaustible. Loggers exceeded the permitted cutting quotas in many areas, and replanted too little. Since 1970 cedar forests in the far east of Russia have been reduced by 21.8 percent. For the Ministry of Timber Industry, the increased logging was a goal in itself, necessary for fulfilling the state plan, even though each year 85,000 cu m (3 million cu ft) of cut timber sank while being floated illegally down the rivers.

Key environmental issues

- major town or city
- heavily polluted town or city
- major pollution event
- heavily polluted river
- area affected by permafrost
- dead lake

annual air pollution (tonnes per square km)
- 20
- 10
- 5
- 2

Northern Eurasia

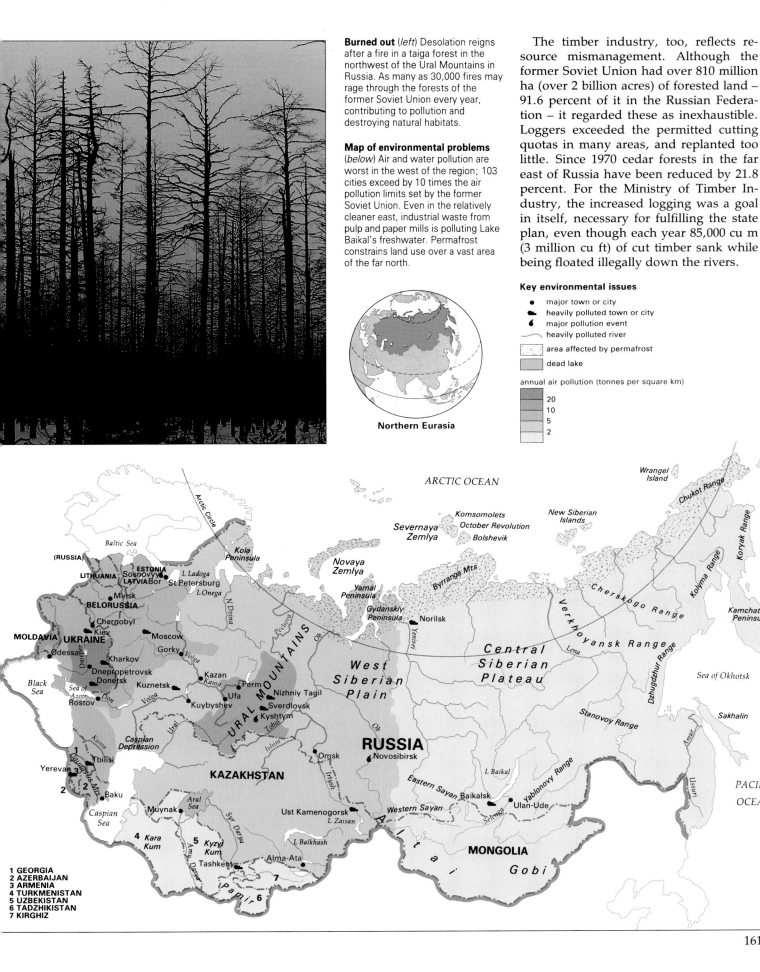

1 GEORGIA
2 AZERBAIJAN
3 ARMENIA
4 TURKMENISTAN
5 UZBEKISTAN
6 TADZHIKISTAN
7 KIRGHIZ

FAR-FLUNG POLLUTION

The most pressing environmental issues of Northern Eurasia relate to air and water pollution, radioactive contamination, excessive logging, and the slow death of rivers through salinization and stagnation. In Siberia and the far east, Russia's vast hinterlands stretching east from the Ural Mountains, destruction of permafrost (permanently frozen ground) is a growing problem.

The energy industry – just over 93 percent of which is run on fossil fuels (coal, oil and gas) – is the main polluter of air. Coal-fired power stations, which produce about 20 percent of the region's electricity, are the chief source of sulfur dioxide, nitrogen oxides and dust emissions. Levels of air pollution 10 times above the legal limit have been recorded in 103 cities – with a total population of nearly 40 million – across the region. The worst levels are reached in industrialized areas such as Donetsk–Dnieper in the Ukraine, some 250 km (200 mi) north of the Black Sea; in the Kuznetsk coal basin (Kuzbas) of southwestern Siberia; and in Norilsk in eastern Siberia, where sulfurous smoke pours out from smelters processing nickel, cobalt and copper.

Most of the pollutants – some 97.5 percent – are either solid particles or sulfur dioxide, carbon monoxide, nitrogen oxides and hydrocarbons. Each year, 15 million tonnes of sulfur fall as dust on the continental part of the region, 20 percent being carried by winds from Europe. However, 19 percent of world carbon dioxide emissions, which contribute to the "greenhouse effect" (atmospheric warming caused by so-called greenhouse gases trapping heat reflected from the Earth) are produced within the region itself, as is 20 percent of the sulfur dioxide that causes acid rain, and 10 percent of the CFCs (chlorofluorocarbons) that are helping to deplete the stratosphere's ozone layer, which filters harmful ultraviolet radiation from the sun.

Local pollution is often horrendous. In Nizhniy Tagil, a Russian town in the Ural Mountains, one industrial plant releases 700,000 tonnes of poisonous fumes annually – the equivalent of 2 tonnes for each of the town's inhabitants. In the Russian capital, Moscow, automobiles release over 840 thousand tonnes of pollutants each year, compared with just

Extent of radiation dispersal (*below*) seven days after the explosion at Chernobyl. The release of radioactive material lasted for 10 days in all, its distribution varying as the winds changed. Initially southeasterly winds blew it over Poland, Ukraine, Belorussia, Latvia, Lithuania, Finland and across Norway and Sweden. The winds then changed, leaving fallout over most of Europe. The extent of the fallout was determined by whether or not rain washed the contamination from the atmosphere, sometimes creating higher than expected levels as in the central British Isles. One of the most dangerous contaminants was caesium-137, which contaminated livestock that ate grass on which the chemical had settled. Restrictions on livestock movement and slaughter were introduced in the United Kingdom and the Nordic Countries.

Cause of the Chernobyl disaster (*right*) Prior to testing a generator's capacity to power emergency systems, engineers shut off the emergency core-cooling system so as not to affect the test results. In a further breach of safety standards, almost all the neutron-absorbing control rods were removed from the core to increase power output. However, coolant water was unable to reach the reactor as power output was too low. When the test started, power generation went out of control and an explosion of steam tore the reactor apart.

Radiation levels as multiples of normal level

- over 100
- 40–100
- 20–39.9
- 10–19.9
- 5–9.9
- 1–4.9
- up to 1
- no rise

over 150 thousand tonnes emitted in the Ukrainian city of Kharkov, 560 km (348 mi) to the south.

Troubled waters

The abuse and mismanagement of water throughout the region has led to many serious environmental situations. One of the most critical concerns is Lake Baikal, in southern Siberia. With a depth of 1,620 m (5,315 ft), it is the world's deepest freshwater lake. Its unique plants and aquatic animals – including freshwater seals – are seriously threatened by wastes pumped into it by pulp and paper mills on its shores. In addition, 198 million cu m (nearly 7,000 million cu ft) of sewage were pumped into the lake in 1989.

Such practices are not unique to Lake Baikal. Throughout the region untreated industrial effluent and raw sewage are pumped directly into rivers and lakes, killing fishes and poisoning water sup-

plies. The river Volga, flowing 3,690 km (2,293 mi) from northern Russia into the Caspian Sea, is so polluted with mercury, cadmium, petroleum residues and agricultural pesticides and fertilizers that the sturgeon that produce caviar have been brought close to extinction. The toxic chemicals either turn the prized eggs to mush or the females, losing their way in the polluted waters of the many hydro-electric reservoirs on the river's course, abort before they can reach their spawning grounds in the Caspian Sea.

Grass roots pollution

The land and its people are also under attack from pollution. Across Northern Eurasia, the overcutting of forests has led to severe soil erosion, the loss of animal habitats and soil fertility, and the silting of rivers from landslides. From Moldavia to Kazakhstan an indiscriminate use of agricultural chemicals, including

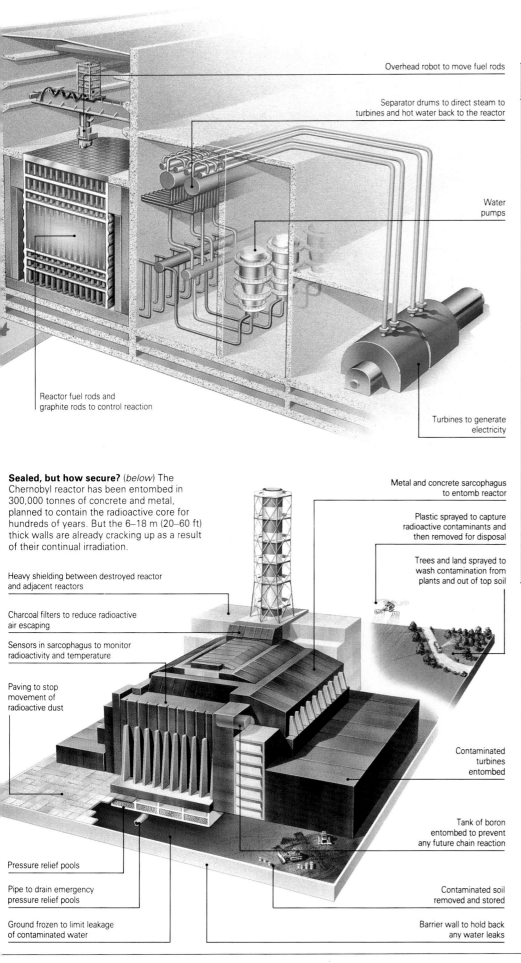

Overhead robot to move fuel rods

Separator drums to direct steam to
turbines and hot water back to the reactor

Water
pumps

Reactor fuel rods and
graphite rods to control reaction

Turbines to generate
electricity

Sealed, but how secure? (*below*) The
Chernobyl reactor has been entombed in
300,000 tonnes of concrete and metal,
planned to contain the radioactive core for
hundreds of years. But the 6–18 m (20–60 ft)
thick walls are already cracking up as a result
of their continual irradiation.

Heavy shielding between destroyed reactor
and adjacent reactors

Charcoal filters to reduce radioactive
air escaping

Sensors in sarcophagus to monitor
radioactivity and temperature

Paving to stop
movement of
radioactive dust

Pressure relief pools

Pipe to drain emergency
pressure relief pools

Ground frozen to limit leakage
of contaminated water

Metal and concrete sarcophagus
to entomb reactor

Plastic sprayed to capture
radioactive contaminants and
then removed for disposal

Trees and land sprayed to
wash contamination from
plants and out of top soil

Contaminated
turbines
entombed

Tank of boron
entombed to prevent
any future chain reaction

Contaminated soil
removed and stored

Barrier wall to hold back
any water leaks

CHERNOBYL – A NUCLEAR DISASTER

On Saturday April 26 1986, the world's
worst nuclear accident occurred in the
Ukrainian town of Chernobyl, 150 km
(80 mi) north of Kiev. At 1.23 am, a
powerful explosion in the No. 4 reactor
of the town's nuclear power plant rip-
ped aside the 1,000 tonne steel lid from
the reactor's core, and blasted through
the concrete containment shell. In the
raging fire that followed, a cloud of
highly radioactive substances – equivalent to 90 bombs of the type that
destroyed Hiroshima in Japan in 1945 –
escaped into the atmosphere.

For 10 days, radioactive fumes
poured out of the devastated plant,
carried by winds far beyond Ukraine.
The toxic gases included iodine-131,
which affects the thyroid glands; and
the longer-lasting – and therefore more
hazardous – caesium-137, which collects in muscle tissue. Contamination
from Chernobyl remains widespread:
in 1989, the area contaminated by
caesium-137 to levels dangerous for
human habitation amounted to some
1,500 sq km (579 sq mi) in Ukraine;
2,000 sq km (1,160 sq mi) in Russia; and
2,000 sq km (1,160 sq mi) in Belorussia.
Before the accident these lands comprised the agricultural heartland of the
region. By 1990, rates of cancer in high
fall-out areas had soared and animals
were being born without heads, while
in Belorussia alone 20 percent of cultivable land has been lost.

the banned pesticide DDT and the de-
foliant Butifos, has triggered a steep rise
in the incidence of cancer, mental illness
and birth defects.

Nuclear contamination from leaks or
open waste dumps is a major concern (in
the wake of Chernobyl). In March 1992
a leak of radioactive iodine from a
Chernobyl-style reactor at Sosnovyy Bor,
100 km (62 mi) west of St Petersburg,
created fears of a new disaster.

Exploitation of the ecologically sensitive tundra in the far north of Russia is
another worrying issue. In the search for
natural gas and oil, not only are crystal-
clear rivers and lakes being polluted,
forests cut down and the lifestyles of local
peoples dislocated, but also construction
work is destroying the moss and lichens that insulate the permafrost below
ground; the permafrost then melts, creating bogs that may have long-term effects
on Siberia's ecological health.

A SLOW, COSTLY CLEAN-UP

From the early 1970s, although economic goals remained the priority, authorities in the former Soviet Union began to wake up to the environmental costs of their industrial policies, and adopted many fundamental environmental laws. These included acts relating to land, water, the geological environment, forests and the atmosphere. Almost every year, the Communist Party and Council of Ministers issued directives drawing attention to nature conservation. Nevertheless, cleaning up an environment that has been abused for so long is both slow going and expensive. The dead hand of bureaucracy still weighs heavily, even after the demise of the Soviet Union.

Until the rise of the reform-minded President Gorbachev, environmental protest was regarded by the former Soviet authorities as dissident behavior – a form of political agitation. In the wake of *glasnost* (the period of "openness" and reform that preceded the collapse of communism), however, there has been considerable outspoken protest about the state of the environment. "Green" groups have proliferated. Among them are the All-Union Movement of Greens; the Russian Greenpeace movement; and a collection of protest groups such as the Baikal Fund, which battles to save Lake Baikal's waters from industrial pollution.

The fight for Lake Baikal, which has been going on for 30 years, has seen some successes. To prevent excessive logging, which can result in soil erosion and landslides – sending silt into the rivers and ultimately into the lake itself – all felling of trees along the lake's shoreline was banned in 1988. The greatest fight has been to stop untreated waste being poured into the lake from the 100 or so factories on Lake Baikal's shore and on the banks of the rivers that feed the lake. The worst pollution has come from two pulp and cellulose mills: one at Baikalsk at the lake's southern end and the other at Selenginsk on the Selenga river, near the lake's southeastern shore. In 1987, the former Soviet Union's Central Committee and Council of Ministers decided to phase out Baikalsk's hazardous operations by 1993, and ordered pollution control equipment to be installed at the Selenginsk plant. However, in a bid to close down the Baikalsk and Selenginsk

COPING WITH THE MISTAKES OF COMMUNISM

The environmental troubles facing the independent states of the former Soviet Union stemmed from two fundamental beliefs: that the natural resources of the region were "inexhaustible"; and that any environmental legislation was acceptable – as long as it did not threaten economic loss.

The idea of "inexhaustibility" lay behind the former Soviet Union's command economy, a system of production control dominated by the State Planning Committee, known as Gosplan. For six decades, its Marxist planners issued production targets for every factory and farm across the nation, irrespective of what products – or how much – were really needed, and of the cost to the environment. Its aim was to outstrip the West economically, and thereby to trumpet the victory of communism over capitalism.

After the breakup of the Soviet Union, and with it the collapse of the economy and the centralized planning authority, production levels of many of the region's resources fell dramatically. In 1991 coal production, for example, fell by as much as 11 percent. The combination of desperate shortages of food and fuel in the shops and growing public awareness of ecological issues has left people in no doubt of the devastating costs to the environment of 70 years of communism. It is hoped that with this knowledge, the newly independent republics will in future be able to make more rational demands on their energy, land, forest and water resources. Depending on priorities, they may – with the help of Western aid and more advanced technology – be able to start balancing the needs of the economy against those of the environment.

Early preparation (*above*)
Kindergarten students learn what to do in case of an accident at the gas factory in Astrakhan, 8 km (5 mi) from their village school. Faulty construction at the factory has increased the risk.

Follow-up to Chernobyl (*left*)
Radiologists from Kiev take soil samples in the village of Vylgvo near Chernobyl. Many of the 247 original residents have refused to leave in spite of the closure of local wells, a ban on fishing in the local river, and the contamination of 105 houses, now condemned.

pulp mills altogether, one activist went on what was perhaps the Soviet Union's first hunger strike in the cause of the environment. Encouraged by *glasnost*, in 1989 he went for five days without food as part of a protest campaign.

The Caspian controversy
An environmental problem of another kind has been tackled in the Caspian Sea. During the 1970s, lowering water levels prompted planners to dam the Kara-Bogaz-Gol – a large, almost land-locked bay adjoining the Caspian Sea in the former Soviet republic of Turkmenistan – which served as a shallow evaporation pan. It was hoped that by damming the bay, evaporation from the Caspian would be halted. The dam, completed in 1980, went against the advice of hydrologists who said that the level of the Caspian Sea rose and fell in cycles, and would rise

again. The results of the dam were disastrous: the Kara-Bogaz-Gol dried out and residual salt was blown inland to contaminate vast tracts of farmland, while the waters of the Caspian rose to flood fishing villages and other settlements on the sea's Turkmenistan shore. Finally, in March 1992 – after major environmental damage had been wreaked – work began on the extremely costly demolition of the dam.

Nuclear lessons
The 1986 Chernobyl nuclear accident deeply affected the former Soviet government's approach to nuclear power, which – before the union's disintegration – met about 3.3 percent of the country's electricity needs. A massive surge of environmental protest forced the authorities to stop the construction of a number of nuclear reactors, and plans for 10 nuclear-powered district heating stations in major

cities were cancelled. In addition, the nuclear industry was opened up, for the first time, to international scrutiny.

In the aftermath of Chernobyl, the state had to build new settlements for over 100,000 evacuees from the area, buy up condemned meat, and introduce new safety features and training at other nuclear reactors – all at vast expense. The reactor that exploded has been entombed in concrete walls 60 m (197 ft) high, 60 m long and 6–18 m (20–60 ft) thick. Unfortunately, the concrete has become brittle from continuous irradiation and the difference of temperature between the inner and outer walls, and the whole edifice is in danger of eventually breaking up. If that happens, yet more radioactivity will be released into the environment. Since 1989 the International Atomic Energy Agency has passed conventions to assure early notification in the event of an accident and prompt assistance on an international scale.

With the fragmentation of the Soviet Union, new arrangements have been made for tackling environmental problems. In January 1991, for example, the so-called "Kishinev agreement", was signed by all 15 of the Soviet Union's then republics, laying out a framework for environmental cooperation in a "post-Soviet" world. In addition a nongovernmental "Socio-Ecological Union" was founded to promote unity among all of the newly established environmental pressure groups in the region.

The Aral Sea crisis

The most obvious environmental disaster to strike Northern Eurasia – even worse, many say, than the Chernobyl nuclear accident – is the drying up of the Aral Sea. This vast salty lake lies some 400 km (250 mi) east of the Caspian Sea. In the early 1960s, a decision was taken by the former Soviet Union's central planning authorities to increase cotton production dramatically in Kazakhstan and the Central Asian republic of Uzbekistan bordering the Aral Sea. Water for irrigation was to be drawn from the two main rivers feeding the Aral, the Syr Darya and Amu Darya. This decision was to have catastrophic consequences for the area and its peoples. In the years that followed, so much water was drained from these rivers by irrigation canals for cotton, as well as other crops, that little remained to replenish the Aral and counteract the process of evaporation. The scale of the problem prompted the Soviet Union's ruling body, the Supreme Soviet, to declare the Aral Sea basin an ecological disaster area "beyond human control".

In 1960 the Aral's water level was just over 53 m (175 ft) and its area 66,820 sq km (25,800 sq mi). By 1987 the sea level had fallen 13 m (43 ft), and its area decreased by 40 percent – in other words, 27,000 sq km (10,424 sq mi) of the Aral's surface area had dried up, an area roughly the size of Belgium. The water is still receding, and at such a steady pace that, unless drastic action is taken, the Aral Sea could disappear by 2020. The consequences of the disappearing water have already been disastrous.

Fatal results

Each year, winds sweeping across the exposed, salt-encrusted seabed carry 75 million tonnes of salt, dust, sand and the residue of pesticides and other agricultural chemicals back onto the surrounding cottonfields to kill crops and the natural vegetation. The combination of this lethal, wind-blown brew and the huge quantities of pesticides and herbicides sprayed onto the cotton crops each year has taken a terrible toll on human health – especially of children. Infant mortality among the local Kara-kalpak people is four times the national average, and more than one in ten babies die before their first birthday. In the Central

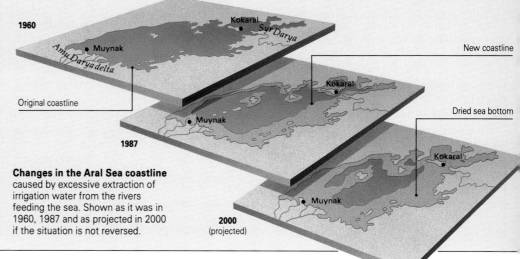

Changes in the Aral Sea coastline caused by excessive extraction of irrigation water from the rivers feeding the sea. Shown as it was in 1960, 1987 and as projected in 2000 if the situation is not reversed.

Ships of the desert (*above*) The rusty hulk of a boat litters the salty bed of what was once the world's fourth largest inland sea, but is now terrain more suitable for camels than ships. The fishing industry – formerly a major employer – has been decimated.

Asian cottonfields, some 83 percent of the children who survive longer suffer from serious illnesses.

With so much of the Aral Sea gone, the salinity of the remaining water has nearly tripled. Twenty species of fish have been killed off, and with them the sea's fishing industry, which once employed 60,000 people. The port of Muynak, which once stood on the Aral's southern shore, used to produce more than a tenth of the former Soviet Union's fish catch. This once-bustling city is now virtually a ghost town standing 48 km (30 mi) from the receding water.

The death of the Aral Sea has had other effects, too: the water once moderated the areas's climate, but extreme temperatures are now more common. Ironically, it was the sea's influence on the climate that made it possible to grow cotton so far north in the first place. The industry now suffers from shorter growing seasons and harsher conditions. It is not known what the ramifications will be for weather systems farther afield if the Aral Sea continues to dry up.

With the fragmentation of the Soviet Union, responsibility for decisions about the Aral Sea has passed from Moscow to the affected new republics (Kazakhstan, Uzbekistan and Turkmenistan). The simplest short-term solution would be to cut back on cotton production, but this would have disastrous effects on the fragile regional economies. An alternative would be to modernize the irrigation system – leaking and unlined canals could be replaced with pipes and computerized sluices and sprinklers – and thereby save enormous amounts of water. How to raise the money for such schemes remains a major question.

COPING WITH ARIDITY

LIVING OFF THE LAND · MODERN DILEMMAS · GRADUAL IMPROVEMENTS

Water is at the root of many of the Middle East's environmental problems. For thousands of years farmers have practiced irrigation to combat the region's aridity; as a result large areas of land, especially in Iraq and Syria, are now waterlogged and made infertile by a buildup of salt. In mountainous parts, where slopes have been cleared of trees, water runoff has caused severe soil erosion. Exploitation of oil reserves and rapid industrialization have led to uncontrolled urbanization and the attendant problems of air and water pollution. Motor traffic threatens fragile desert landscapes. Warfare in the region has left environmental scars: an orgy of destruction in the oilfields of Kuwait, a deliberately flooded border between Iraq and Iran, and a countryside studded with land mines in Afghanistan and Lebanon.

COUNTRIES IN THE REGION

Afghanistan, Bahrain, Iran, Iraq, Israel, Jordan, Kuwait, Lebanon, Oman, Qatar, Saudi Arabia, Syria, Turkey, United Arab Emirates, Yemen

POPULATION AND WEALTH

	Highest	Middle	Lowest
Population (millions)	54.6 (Iran)	4.0 (Jordan)	0.4 (Qatar)
Population increase (annual population growth rate, % 1960–90)	10.0 (UAE)	3.4 (Iraq)	1.3 (Lebanon)
Energy use (gigajoules/person)	642 (Qatar)	39 (Lebanon)	4 (Afghanistan)
Real purchasing power (US$/person)	19,440 (Qatar)	9,290 (Oman)	710 (Afghanistan)

ENVIRONMENTAL INDICATORS

	Highest	Middle	Lowest
CO$_2$ emissions (million tonnes carbon/annum)	42 (S Arabia)	4.7 (Oman)	0.6 (Yemen)
Artificial fertilizer use (kg/ha/annum)	750 (Bahrain)	92 (Oman)	6 (Yemen)
Automobiles (per 1,000 population)	207 (Kuwait)	48 (Turkey)	2 (Yemen)
Access to safe drinking water (% population)	100 (Bahrain)	83 (Turkey)	21 (Afghanistan)

MAJOR ENVIRONMENTAL PROBLEMS AND SOURCES

Air pollution: urban high
Coastal pollution: medium/high; *sources:* oil; war
Land degradation: *types:* desertification, salinization; oil pollution, *causes:* agriculture; war
Resource problems: fuelwood shortage; inadequate drinking water and sanitation
Population problems: population explosion; war
Major events: Gulf (1991), oil spills and oil well fires during and after the Gulf war

---

LIVING OFF THE LAND

The age-old struggle to squeeze a living from the arid environment of the Middle East by controlling limited water resources and irrigating the land has left its mark on rural landscapes throughout the region. In the fertile alluvial floodplains of the Tigris and Euphrates rivers in Mesopotamia (present-day Iraq), where farming began more than 8,000 years ago, complex irrigation systems were used to tap the river waters, leading gradually to the silting up of channels and the salinization of irrigated fields.

Rainfall throughout most of the Middle East is limited to less than 600 mm (24 in) a year and occurs mostly in winter. Rainstorms are frequently violent, transforming dry gullies into roaring torrents. The practice of rain harvesting – clearing an area of stones to enhance the runoff of rainwater, which is then channeled into canals leading to underground cisterns or used to irrigate agricultural plots – has been employed throughout the region for thousands of years.

Terraces to catch the sparse rainfall in desert areas such as the Negev, Lebanon and Yemen are another ancient technique still widely used. However, they quickly fall into disrepair if not maintained, and serious flooding can result. In Yemen, where more than 95 percent of the country's cultivable land is comprised of mountain terraces, many farmers have abandoned their holdings for more profitable activities. Without regular maintenance, rainwater washes over the terraces, stripping away fertile soil built up over thousands of years and gouging deep gullies.

Elsewhere, groundwater is channeled along subterranean canals known as *qanats*. These gently sloping tunnels tap the water from aquifers (underground reservoirs) beneath the alluvial fans at the edge of mountain ranges. Gravity carries the water downhill along the *qanats* to irrigate fields as much as 50 km (30 mi) from the source. Extensive *qanat* systems, some of them several hundred years old, still provide an efficient water supply in parts of Iran, Afghanistan and Oman. However, the increasingly heavy demand for water from urban areas means that groundwater from some of the aquifers has been extracted by pumping, leaving the *qanats* to run dry.

Until recently, large areas of the Middle East, including nearly all the Arabian Peninsula, were too arid for arable farming. However, mechanization and large-scale irrigation projects have greatly expanded the area of cultivable land. In Saudi Arabia, drip irrigation systems and the heavy application of fertilizers allows wheat to be grown in the desert, though at great cost. Similar techniques have boosted agricultural production in Israel.

Erosion and desertification

The expansion of cultivation into increasingly marginal areas has been the cause of widespread land degradation. The use of heavy machinery is inappropriate to dryland terrain. In parts of eastern Iraq and in the steppelands of eastern Syria – where mechanized rainfed barley cultivation has replaced livestock grazing – the removal of trees and shrubs means there are no long roots to hold the light soil together. Loosened by plowing, it is blown by the prevailing winds, forming hummocks and spreading dunes that can

Irrigation circles (*below*) near Riyadh, Saudi Arabia. Intensive smallscale irrigation allows crops to be grown in the desert, but the scarcity of water means that its scope is limited. According to estimates, Saudi Arabia's aquifers will be pumped dry by 2010.

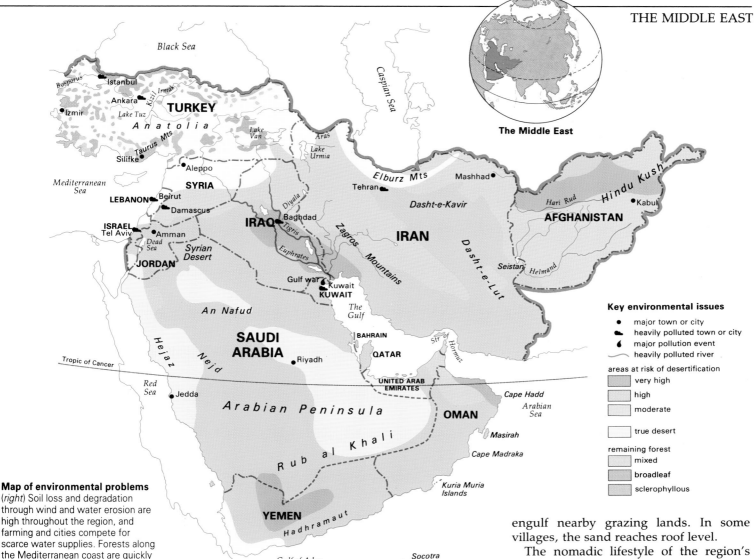

The Middle East

Key environmental issues

- major town or city
- heavily polluted town or city
- major pollution event
- heavily polluted river

areas at risk of desertification

- very high
- high
- moderate
- true desert

remaining forest

- mixed
- broadleaf
- sclerophyllous

Map of environmental problems
(*right*) Soil loss and degradation
through wind and water erosion are
high throughout the region, and
farming and cities compete for
scarce water supplies. Forests along
the Mediterranean coast are quickly
disappearing.

engulf nearby grazing lands. In some
villages, the sand reaches roof level.

The nomadic lifestyle of the region's
pastoralists has been severely curtailed by
the modern pace of change. Age-old
routes between seasonal grazing lands
have been disrupted by national boundaries, while cultivation and urbanization encroach onto former rangeland. As a
result herds are concentrated onto ever-
smaller areas of pastureland, leading to
overgrazing and land degradation. In
Jordan, livestock levels are now four
times higher than the land can sustain.
Attempts to improve conditions for surviving nomads by providing new wells
have often made things worse: livestock
congregate in huge numbers around the
waterholes, stripping the land virtually
bare and making it prone to erosion.

In northern Afghanistan and many
parts of Iran, overgrazing is exacerbated
by the cutting of wood for fuel, which
increases the power of wind and water to
erode fragile desert soils. Elsewhere in
the region, the construction industry's
demand for timber has helped to deplete
surviving areas of woodland. The mountains of Lebanon are now almost entirely
devoid of their famous cedars. Stripped of
their cover, hillsides become increasingly
vulnerable to erosion and gullying in
times of heavy rain or flood.

MODERN DILEMMAS

Until the discovery and exploitation of the Middle East's vast oil reserves in the mid 20th century, the impact from human activities on the land, particularly in the Arabian Peninsula, was only minimal. In many Middle Eastern states, oil wealth was the catalyst that helped to bring about the change from a traditional to a modern industrial society. The rapid pace of urbanization and industrialization throughout the region as a whole has placed enormous strain on fragile resources and given rise to severe environmental problems.

Most of the region's cities have grown extremely fast. Urban populations in Saudi Arabia, for example, more than doubled between 1960 and 1985. Air pollution from the cars that clog up city streets is an unwanted aspect of urban life that plagues many large cities in the region: Baghdad (Iraq), Beirut (Lebanon), Damascus (Syria), Istanbul (Turkey), Tehran (Iran) and Tel Aviv (Israel) are particularly badly affected.

City growth has brought with it acute new problems of sewage and industrial

TOO MUCH SALT

In dry regions of the world, soluble salts accumulate naturally at or near the soil surface because of high evaporation and insufficient annual rainfall to wash them out of the soil. An excessive buildup of salts can also be the result of poor irrigation techniques. It will occur if water is allowed to leak into the ground from supply canals, if too much water is applied, and if there is inadequate drainage, since this causes waterlogging. Too little water can also result in salinization, as excess salts are not washed away. Once irrigated land has become affected by salinity, crop yields fall significantly.

Salinization inhibits plant growth by altering the structure of the soil. As water evaporates from the soil, a concentration of sodium salts is left. These salts bind the clay particles together, reducing the spaces in the soil so that it cannot hold air and nutrients – essential for plant growth.

A high salt content in groundwater acts directly on the plant by preventing its roots from absorbing water, causing moisture stress – the plant starts to wilt and energy that would otherwise con-

tribute to growth is diverted to overcoming the increased stress of drawing moisture from the soil. In even quite low concentrations, some salts, notably boron, may be toxic to plants. A number of indigenous plants have adapted to living in saline conditions, but crop plants are much more susceptible, especially at the seedling stage.

Salinization and waterlogging (*below*) All water contains salts, which wash away when water is applied with sufficient drainage. When drainage is poor, salts accumulate (salinization) and the water level rises to cause waterlogging. Both problems affect nearly 30 percent of the world's irrigated land.

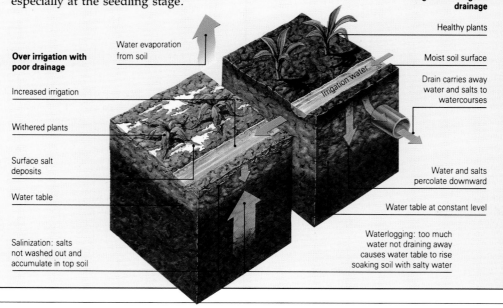

Over irrigation with poor drainage

Water evaporation from soil

Increased irrigation

Withered plants

Surface salt deposits

Water table

Salinization: salts not washed out and accumulate in top soil

Irrigation with good drainage

Healthy plants

Irrigation water

Moist soil surface

Drain carries away water and salts to watercourses

Water and salts percolate downward

Water table at constant level

Waterlogging: too much water not draining away causes water table to rise soaking soil with salty water

Deadly remains (*above*) Unexploded mortar bombs, relics of the long guerrilla war in Afghanistan, are found throughout the countryside. Various recent wars throughout the region have damaged the environment and left dangerous trash in their wake.

Deteriorating conditions (*left*) A flock of sheep in a sandstorm in the United Arab Emirates. Pastureland is diminishing due to overgrazing. Animals remove all the vegetation cover that holds the soil in place, causing erosion and desertification.

waste disposal. Most of the waste and effluent from the Istanbul area, for example, is discharged untreated into the narrow straits of the Sea of Marmara and the Bosporus, fouling them and making them unattractive for recreation. Local beaches and groundwater supplies are increasingly polluted.

Rivers suffer a similar fate. Syria's Barada river, whose waters are used for irrigation, is now heavily contaminated with Damascus' domestic and industrial effluent, which contains high levels of heavy metals. The scarcity of rainfall means that such wastes build up in high concentrations and are slow to disperse.

Increasingly in demand
Throughout the Middle East, water supplies for domestic and industrial use are under tremendous strain. To meet the ever-increasing demand, largescale projects involving canal building, well drilling and dam construction are undertaken. While easing the immediate need for water, these projects are likely to cause greater problems in the long term, leading ultimately to the exhaustion of aquifers or altering the flow pattern of rivers.

In the Iranian capital of Tehran, the largest city in the region with a population of over 6 million, a series of projects have been undertaken since the 1920s in the attempt to increase supplies to meet demand. Canals and dams have been built on the Karaj river to the west of the city, supplemented with the drilling of wells to tap groundwater supplies. By the 1970s supplies were again running low, so water was diverted from the Lar river, which flows into the Caspian Sea.

Rising consumption of water in urban areas has reduced the amount available for irrigation, bringing urban populations into conflict with the needs of rural communities. The construction of dams on the Karaj river, for example, limited the amount of water for farmland irrigation on the alluvial plains downstream. The potential for conflict is greatest when a river flows through more than one country: Dams built on the Euphrates in Turkey and Syria to provide water for irrigation have considerably reduced the quantity and quality of flow in Iraq. The waters of the Jordan river have similarly been a source of aggravation between Lebanon, Syria, Jordan and Israel.

Heavy irrigation has caused water tables to fall and increased soil salinity in many parts of Iran and Saudi Arabia, and on the flanks of the Hajar mountains in the United Arab Emirates. Nearly half of Syria's 670,000 ha (1.6 million acres) and up to 60 percent of Iraq's 1.8 million ha (4.4 million acres) of irrigated cropland are suffering from salt accumulation caused by the overapplication of irrigation water without adequate drainage.

Problems for wildlife
The enormous changes that have taken place in the rapidly modernizing countries of the Middle East put the region's wildlife habitats under increasing pressure. In the desert, cars, pick-up trucks and surfaced roads have all but replaced camels and traditional caravan routes, and threaten the delicate balance of a unique ecosystem. Quarrying to supply the booming construction industry is also a cause of disturbance to wildlife.

The marine life of the Gulf is constantly threatened by spillages from tankers – it is the most heavily used tanker highway in the world, carrying in 1987 more than a quarter of the world's oil supplies from market economy countries. War has added enormously to the size of the disaster. Even before the deliberate oil discharges of the Gulf War (1990–91), the conflict between Iraq and Iran throughout the 1980s had resulted in total spillages of nearly 2 million barrels of oil into the waters of the Gulf.

Elsewhere recent wars in the region have left deep scars on the environment. In Afghanistan, thousands of unexploded land mines buried during the Soviet occupation (1979–89) pose a constant danger to animals and people alike.

GRADUAL IMPROVEMENTS

As environmental awareness grows, most governments in the region have begun to take steps to tackle the pollution problems that go hand in hand with rapid urban growth and industrialization. A number of large cities, including Tehran and Istanbul, have been furnished with new and efficient sewage collection systems, though the waste is still often left untreated before it is discharged into the sea or rivers. Most smaller cities, however, may have to wait some time for such facilities to be built.

In Ankara, Turkey's capital, the problem of air pollution from vehicle exhausts is compounded by smoke from the city's thousands of inefficient domestic stoves that burn low-grade brown coal (lignite) and oil. This gives rise to a thick brown smog, especially in winter when visibility is often reduced to less than 50 m (165 ft). Better quality coal and more efficient combustion methods are currently being sought by experts at the city's Middle East Technical University.

Tackling water problems

All governments face the problem of providing a safe water supply for their growing urban populations without exhausting reserves for future generations. Where too much is extracted from aquifers, groundwater levels begin to fall dangerously low. In coastal areas, once the water table falls below sea level, salt water is able to penetrate the aquifers, contaminating supplies.

In its natural state, the coastal aquifer that supplies Tel Aviv, Israel's largest conurbation, is 3 to 5 m (10 to 16 ft) above sea level. By the 1950s so much water had been extracted for irrigation and for the new state of Israel's burgeoning industries that water levels over an area of 60 sq km

Reclaiming the desert (*right*) A former nomadic Bedouin waters tree seedlings in Syria. Tree-planting schemes have been launched to improve the land and halt the pace of degradation. The roots hold the soil together, and the trees provide fodder and fuelwood.

Capturing the sun's energy (*below*) with solar panels on the rooftops of Jerusalem. Lacking petroleum, but with plenty of free sun, Israel has been a pioneer in the development and use of solar energy, which provides hot water for households throughout the country.

Middle Eastern conservation issues do not always reach an international audience, but in 1988 plans for a major tourist development on the Goksu delta in Turkey were successfully opposed by a combination of local and international conservation groups. The delta, on the south coast of Turkey opposite the island of Cyprus, is the only Turkish breeding place of the rare Purple gallinule, among many other bird species, and is a key site for migrant and wintering birds – it has been rated by the International Council for Bird Preservation as "one of the most important bird areas in Europe and the Middle East". Its beaches are also among the main nesting sites for the two Mediterranean species of sea turtle, the Loggerhead sea turtle and the Green turtle.

The development plans for the delta, revealed by the Society for the Protection of Nature in Turkey (DHKD), included the building of a holiday resort as well as a new airport and a shrimp farm on its southwestern tip. The DHKD wasted no time in publicizing this threat to the delta's unique wildlife and in organizing international pressure against it. Their efforts were eventually successful in persuading the Turkish government to call a stop to the project. The decision was made to relocate the airport away from this important site, and on 1 December 1989 the Goksu delta was given the status of Specially Protected Area, only the fourth to be created in Turkey.

(23 sq mi) were below sea level, and the city's aquifer was being degraded by salt water intrusion. To meet this crisis, freshwater was delivered by a pipeline (known as the National Water Carrier) from the Sea of Galilee in the north of the country to fill a line of wells that had been constructed along the coast, thereby providing a freshwater barrier against the intrusion of seawater.

The National Water Carrier – which came fully into operation by 1969 – is central to Israel's use of water. Not only does the pipeline, some 3 m (10 ft) in diameter, supply the aquifers that feed the urban industrial complex of Tel Aviv, but it also delivers water to the agricultural settlements of the coastal plains and the Negev desert, transporting more than 1 million cu m (1.3 million cu yds) of water each day.

A severe drought in the early 1990s placed Israel's water supplies once again under strain. Levels in the Sea of Galilee reached an all-time low, reducing freshwater supplies to the south. Saltwater penetration of the aquifers again became a problem, causing the country's water experts to look carefully at ways of conserving water supplies and calling for stricter controls on the use of water for irrigation, especially of parks and gardens. Now most agricultural irrigation – whether by a sprinkler, trickle or dripfeed system – is carried out at night when evaporation

levels are reduced. Water saving schemes include the recycling of all urban waste water for irrigation.

Some of the Gulf countries, such as Kuwait, Oman, Saudi Arabia and particularly the United Arab Emirates, have met the rising demand for water by building desalinization plants to make seawater fit for drinking. However, these plants are expensive to build and to run, though they have the advantage of being fueled by surplus natural gas.

Improving the land
Land degradation in dryland areas has many causes, requiring many different solutions. In Iran a major effort has been made to tackle one of the most visible aspects of desertification – the movement of sand dunes onto productive agricultural land. Where the problem is most acute, immediate control of moving dunes is achieved by spraying them with a bitumen mulch. A method that takes longer to achieve results but is less harmful to the environment is to plant drifting dune fields with woody shrubs and grasses that hold down the sand.

In particularly windy areas, palisades are constructed to protect the young plants. No grazing or cultivation is allowed, and the spaces between the planted areas soon fill with native vegetation. In some areas where the project has been running longest, new plant growth has

been substantial enough to allow controlled grazing and fuelwood collection to take place once more.

Elsewhere, successful schemes have been launched to reverse the longterm trend of unsustainable forest use. Tree-planting programs in progress in Israel since before 1948 have helped to extend local timber resources while controlling erosion on hill slopes.

Similarly, on the coastal plains and mountains in the west Syrian province of Latakia – where water running off cleared slopes has caused particularly severe erosion – a major reforestation program was implemented with assistance from Germany at the end of the 1980s. Some 10,000 ha (24,700 acres) of neglected terraces, formerly used to grow winter vegetables and fodder crops, have been planted with trees.

Environmental terrorism – The Gulf War

The Gulf War (1990–91) that followed the Iraqi invasion of Kuwait precipitated an environmental disaster of enormous proportions – one of the worst, and certainly one of the most publicized, the world has ever known. Unforgettable pictures of oiled cormorants trapped in a sea of sludge, and of the desert ablaze with burning oil were seen on television screens around the world.

Along the Gulf, Kuwaiti storage tanks were deliberately opened by the Iraqi forces to release their oil directly into the sea. In all, about 7–8 million barrels were discharged, creating an oil slick that was nearly twice the size of the world's previously largest spill, caused by a blowout in the Ixtoc-1 well in the Bay of Campeche off the Mexican coast in 1979.

The bulk of the discharged oil from the Gulf slick came ashore as a heavy, sludgy mousse that contaminated a 460 km (290 mi) stretch of the northern coast of Saudi Arabia. A strip of oil up to 100 m (328 ft) in width was deposited along the sandy beaches that comprise 60 percent of the affected coastline, and extensive areas of the Gulf's unique salt marsh, mangrove and coral island habitats were also badly oiled. Among the sites affected were the island breeding shores of the rare Green and Hawksbill turtles.

As many as 30,000 seabirds, particularly wintering grebes and cormorants, died in the immediate aftermath of the spills, but the damage caused to intertidal feeding grounds affected far greater numbers than these – 1–2 million shorebirds are estimated to pass through the Gulf on migration. The depletion of fishing stocks

had a major impact on the Gulf's fishing industry, second only to the oil industry in economic importance.

Even before the war, the ecosystems of the Gulf – effectively a closed sea – were subject to major disturbance from high temperature variations and increased salinity, and from the persistent release of oil and other toxic pollutants. Although the full extent of the devastation caused by the wartime oil discharges will take many years to assess, experts fear that irreversible damage may have been done to the Gulf's biological productivity, leading in the long term to greatly reduced species diversity.

Desert fires

The retreating Iraqi forces set fire to some 850 of Kuwait's 1,116 oil wells. Dense black smoke choked the capital city of Kuwait and billowed into the atmosphere to spread still farther: soot from the raging well-fires was even reported to be falling in snow on the Himalayas in northern India. However, the immediate problems of soot particle inhalation, contamination of water and food supplies,

A vision of Hell on Earth (*right*) After the retreat of the combatants, oil wells burned 3 million barrels a day, spewing 5,000 tonnes of soot into the air at concentrations a thousand times above normal. On the ground, burned-out military vehicles together with live and spent ammunition littered the landscape. The fragile desert ecology will take years to recover.

A trail of deliberate pollution (*below*) The inset map shows the spread of some 7–8 million barrels of deliberately released oil. The larger map tracks the range of the smoke plume from oil well fires. This contained oil droplets and soot, which fell to the ground more than 100 km (60 mi) away.

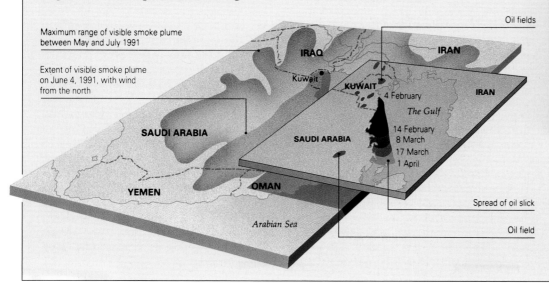

Maximum range of visible smoke plume between May and July 1991

Extent of visible smoke plume on June 4, 1991, with wind from the north

Oil fields

IRAQ
Kuwait
KUWAIT
IRAN
4 February
The Gulf
14 February
8 March
17 March
1 April
IRAN
SAUDI ARABIA
SAUDI ARABIA
OMAN
YEMEN
Spread of oil slick
Oil field
Arabian Sea

the risks of acid rain, photochemical smog and climate disruption from reduced solar radiation were not as severe as many environmentalists forecast and had mostly passed by early 1992.

A much greater hazard was created by the large lakes of oil emanating from nonburning, gushing wells. The oil did

not percolate into the sand but drained into wadis (river gullies) and steams, threatening to contaminate groundwater supplies. These oil lakes endanger wildlife. Birds – particularly swallows, bee-eaters and martins – that hunt for insects in the air attempt to drink from the black lakes, mistaking the oil for water.

The delicate desert ecology, relentlessly pounded by bombs, tanks and other military vehicles as well as soaked by oil spills, may take decades to recover from the aftermath of the war. Ruts in the ground made by heavy equipment will remain for many years, and disturbed gravel and compacted sand surfaces will enhance opportunities for wind erosion. The waste – including human sewage, garbage and toxic materials – left by the coalition forces has been buried in pits, but discarded military weaponry continues to litter the desert, and there remains a longterm risk of groundwater contamination.

THREATS TO A FRAGILE LAND

ALTERED LANDSCAPES · PRESSURES ON RESOURCES · EFFICIENT RESOURCE MANAGEMENT

The drought and famine that struck the Sahel countries, especially Ethiopia, in 1984–85 alerted the world to the serious environmental problems affecting Northern Africa. In such a dry region, water supply is a critical issue, exacerbated during the 20th century by rapid urban growth, particularly in the north, and the expansion of intensive agriculture. Water-related problems include the pollution of groundwater by excessive use of fertilizers and pesticides, diseases that are spread through communal water supplies, and the salinization of productive soil by overirrigation. On the edges of the Sahara land degradation and desert encroachment have been caused by widespread overgrazing and the clearance of trees to provide fuel, timber and agricultural land, exposing the fragile soils to wind and water erosion.

COUNTRIES IN THE REGION

Algeria, Chad, Djibouti, Egypt, Ethiopia, Libya, Mali, Mauritania, Morocco, Niger, Somalia, Sudan, Tunisia

POPULATION AND WEALTH

	Highest	Middle	Lowest
Population (millions)	53.2 (Egypt)	8.2 (Tunisia)	0.41 (Djibouti)
Population increase (annual population growth rate, % 1960–90)	5.6 (Djibouti)	2.6 (Morocco)	2.1 (Chad)
Energy use (gigajoules/person)	83 (Libya)	10 (Morocco)	1 (Ethiopia)
Real purchasing power (US$/person)	3,170 (Tunisia)	970 (Sudan)	500 (Mali)

ENVIRONMENTAL INDICATORS

	Highest	Middle	Lowest
CO$_2$ emissions (million tonnes carbon/annum)	25 (Algeria)	3 (Tunisia)	below 0.1 (Djibouti)
Artificial fertilizer use (kg/ha/annum)	351 (Egypt)	14 (Mali)	below 1.0 (Niger)
Automobiles (per 1,000 population)	90 (Libya)	8 (Mauritania)	below 1.0 (Ethiopia)
Access to safe drinking water (% population)	97 (Libya)	61 (Morocco)	19 (Ethiopia)

MAJOR ENVIRONMENTAL PROBLEMS AND SOURCES

Air pollution: urban high
Land degradation: *types*: desertification, soil erosion, salinization; *causes*: agriculture, industry, population pressure
Resource problems: fuelwood shortage; inadequate drinking water and sanitation
Population problems: population explosion; urban overcrowding; famine; war

ALTERED LANDSCAPES

Until some 6 million years ago, Northern Africa was covered with lush, tropical forests. As the climate slowly changed, becoming drier, the forests were replaced by grasslands. By about 10,000 BC, nomadic pastoralists had begun to graze their sheep and goats in the central Sahara; the plains were then still fertile enough to support a rich diversity of animals, including ostriches, elephants and antelope. Between 8000 and 2000 BC, severe climatic changes (leading to greater aridity), combined with increased livestock grazing and tree-felling, accelerated the spread of the desert, which now occupies most of the center of the North African region.

Whether, and to what extent, the Sahara continues to spread, is a complex matter. A United Nations conference in Kenya in 1977 asserted that the Sahara had moved south by about 100 km (62 mi) between 1958 and 1975 – an average of 6 km (3 mi) a year. However, satellite imagery taken between 1982 and 1987 shows that the sparse vegetation of the Sahel – the belt of savanna on the Sahara's southern border – first retreated south, and then advanced north (toward the Sahara) when the rains came.

Water for survival

Rainfall in the desert is spasmodic, and seasonal rainfall in the fertile Maghreb (Algeria, Morocco and Tunisia), the Sahel, the Ethiopian Highlands, Libya and Egypt is often unreliable. Droughts are a fact of life, while at the other extreme, sudden heavy rainfall causes severe flooding and landslides. Not surprisingly, water is the one resource that is most in demand throughout the region. Surface water storage is uncommon outside the Atlas Mountains and Ethiopian Highlands because of the limited amounts of runoff. But from very ancient times a number of techniques have been used to obtain such water.

In Libya, for example, the remains of desert farming communities from between the 1st and 5th centuries AD have been discovered more than 300 km (186 mi) from the coast, perched on the edges of wadis (rocky watercourses, normally dry in summer); here, walls and dikes trapped water and silt from flash floods, directing them into surrounding fields

where crops were grown. In the oases of southern Tunisia, date palms, peaches, pomegranates and vegetables have been cultivated since Roman times using water drawn from an aquifer (a layer of water-holding rock) near the surface. After Tunisia achieved its independence from France in 1956, government-funded boreholes allowed a deeper aquifer at about 200 m (655 ft) to be exploited, leading to widespread agricultural expansion. The demand for water in southern Tunisia is now so great that deeper wells (down to 2,800 m or 9,190 ft) had to be drilled in the 1980s. Careful management is needed to avoid overpumping.

Desert storm (*below*) A huge cloud of sand engulfs a Mauritanian village. Such storms are one result of desertification. Wind-blown clouds of loose sand may be deposited as far away as the Caribbean.

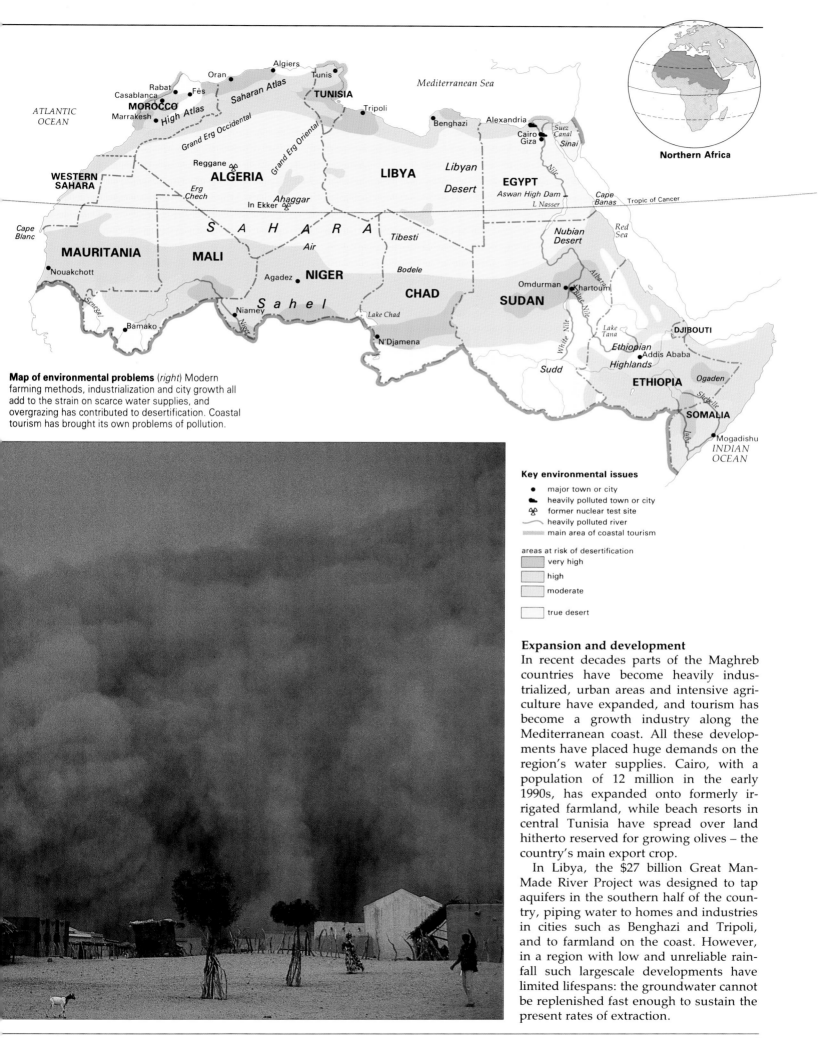

Map of environmental problems

Map of environmental problems (*right*) Modern farming methods, industrialization and city growth all add to the strain on scarce water supplies, and overgrazing has contributed to desertification. Coastal tourism has brought its own problems of pollution.

Northern Africa

Key environmental issues

- ● major town or city
- ◤ heavily polluted town or city
- ☢ former nuclear test site
- ◠ heavily polluted river
- ▒ main area of coastal tourism

areas at risk of desertification

- very high
- high
- moderate
- true desert

Expansion and development

In recent decades parts of the Maghreb countries have become heavily industrialized, urban areas and intensive agriculture have expanded, and tourism has become a growth industry along the Mediterranean coast. All these developments have placed huge demands on the region's water supplies. Cairo, with a population of 12 million in the early 1990s, has expanded onto formerly irrigated farmland, while beach resorts in central Tunisia have spread over land hitherto reserved for growing olives – the country's main export crop.

In Libya, the $27 billion Great Man-Made River Project was designed to tap aquifers in the southern half of the country, piping water to homes and industries in cities such as Benghazi and Tripoli, and to farmland on the coast. However, in a region with low and unreliable rainfall such largescale developments have limited lifespans: the groundwater cannot be replenished fast enough to sustain the present rates of extraction.

PRESSURES ON RESOURCES

The problems of water availability and water quality affect the whole of Northern Africa, and communities depending solely on rainfall for farming face crop failures and high levels of animal and human mortality during droughts. Because of the hot climate, much precious irrigation water is lost through evaporation. Large irrigation projects are also vulnerable to leaking canals and drainage ditches that have fallen into disrepair or been blocked by wind-blown sand.

Groundwater used for irrigation may contain high levels of soluble salts, particularly lime, gypsum and halite. As the water evaporates these salts build up in the soil, making it infertile. If there is inadequate drainage, the irrigated areas may become waterlogged, increasing salinization. In the Nile valley, the El Gezira region of Sudan (just south of Khartoum), and the inland delta of the Niger (in Mali), crop yields have been reduced because of waterlogging and salinization from unlined canals and flood irrigation. Salt-rich groundwater also causes limescale to build up and corrode industrial plant and domestic water pipes.

The quality of the region's drinking water is another concern. Most water used by people living outside towns and cities is drawn straight from wells and is untreated. In Egypt, for example, 88 percent of the urban population has access to safe drinking water compared to 64 percent in the rural areas. In Mali the proportions are much lower – 46 percent and 8 percent, respectively. Communal water supplies aid the spread of water-borne diseases such as typhoid, hepatitis and cholera, which are endemic.

Degraded environments

The region's soils are mainly very thin, nutrient-poor and susceptible to erosion. Some of the worst erosion in the world occurs in the Ethiopian Highlands, where it is estimated that 300 tonnes per ha (120 tonnes per acre) of topsoil are lost every year – largely the result of severe and extensive deforestation.

Fine, wind-blown soil particles lead to dust storms that can travel thousands of kilometers and disrupt radio communications. In Mauritania, about 100 million tonnes of soil are lost each year, much of it billowing across the Atlantic and ending up in the Caribbean and South America. Coarser wind-blown sand frequently banks up against roads and railroads, and buries villages, oases and farmland. Water-eroded soil washes into rivers and silts up reservoirs: each year some 10.3 million cu m (364 million cu ft) of sediment are deposited behind the Hasm-el-Girba dam in Sudan.

Much of the region's vegetation, crucial for binding the soil, has been removed through overgrazing and cutting for fuelwood. In southern Tunisia and southern Morocco and throughout the savanna belt

THE ENVIRONMENTAL EFFECTS OF WAR

Hundreds of thousands of people throughout the Sahel region of Northern Africa have died of starvation since the great famine of 1984–85. However, continuing poor harvests and food shortages are not simply the result of drought. The causes are more complex and are linked to the civil wars that have divided Sudan, Somalia and (until 1991) Ethiopia since the 1970s.

The effects of these conflicts have proved catastrophic for the environment. Money and energies that could have been spent on tree-planting, terracing or improving agricultural practices were devoted to military activities. Bombs and heavy military vehicles have torn up the land. The insecurities of war reduced people's ability to grow crops, blockades cut off the country's supply lines, and the recruitment of soldiers into national and rebel armies removed young men from the agricultural workforce.

In northwestern Ethiopia, these problems were compounded by pressure on the land from high population densities, inequitable land reforms in 1975 and the influx of more than 615,000 refugees from Sudan and Somalia. The government's attempt, abandoned in 1986, to move 25 million people from the north to the south of the country in a so-called "villagization" program proved a disaster. Apart from the human misery involved, farmers were encouraged to abandon growing staple food crops in favor of high-earning cash crops such as coffee in order to pay for Soviet arms' purchases. The government abandoned the scheme in 1986. This, too, contributed to famine.

Fueling deforestation (*above*) A timber depot outside Cairo, Egypt. Trees, which are the primary source of fuel throughout the region, are a vanishing resource, particularly in Sudan, Somalia and Ethiopia. Villagers often walk 16 km (10 mi) to gather firewood.

Eating everything in sight (*left*) Hardy goats and sheep – well adapted to the harsh conditions – forage in Morocco. Government-sponsored settlement programs for nomadic peoples have resulted in larger flocks and increased overgrazing, which contribute to desertification in the region.

of the Sahel, government schemes to settle nomads have resulted in greatly increased stock numbers. Most overgrazing occurs around wells, often drilled with government aid along main roads to markets. Large herds of cattle, sheep, goats and camels strip the vegetation bare around the wells and further exacerbate soil erosion with their trampling hooves.

Firewood and charcoal are the main domestic fuels in the savanna belt, Ethiopia and Somalia, and woodlands within hundreds of kilometers of major cities such as Khartoum (Sudan), Niamey (Niger) and Mogadishu (Somalia) have been degraded by wood cutting. In 1989, for example, Mogadishu consumed 156,000 tonnes of charcoal. To provide the wood needed to make that amount of charcoal, all of the woodland within 100

km (60 mi) of Mogadishu was destroyed. Now *Acacia* and riverine woodlands some 100 to 350 km (60 to 220 mi) away in the Bay region and Juba river valley, west of Mogadishu, are being intensively harvested. Stocks are estimated to last only until the end of the century.

Urban pollution

Population growth, industrial and urban expansion and increased tourism have brought a chain of environmental problems. Cairo is the largest conurbation in Africa. Its sewage system dates from 1914, and can only deal with half of the 2 million cu m (70 million cu ft) of sewage produced each day; some 3 million inhabitants living in the suburbs are not served by the system at all. Sewage often floods onto the streets, and much of it goes untreated directly into the Nile, causing groundwater contamination and posing a serious health hazard.

Air pollution is a problem in many cities, where industrial emissions and effluent are largely uncontrolled. In Cairo vehicle exhaust fumes and industrial pollutants have caused acid corrosion on the 4,600-year-old Sphinx, one of the most famous of the country's ancient monuments, at nearby Giza.

EFFICIENT RESOURCE MANAGEMENT

As water is the most precious commodity in the region, sound management by governments and farmers is imperative. The three key factors in achieving this are increased water supply, efficient use and conservation. Dam building has successfully increased supplies in some areas (despite huge losses from evaporation), while boreholes sunk into ever-deeper aquifers have increased the exploitation of groundwater: a well bored in an area receiving just 10 mm (0.4 in) of rain a year can yield more than 15,000 cu m (529,720 cu ft) a day for a limited period.

Irrigation consumes more water than any other activity in the region. In mountainous areas, traditional methods of collecting, storing and distributing rainwater runoff are still the most efficient ways of providing water for farmers. In Algeria, for example, hillslope runoff – collected and stored in 500 cu m (17,660 cu ft) capacity cisterns – is sufficient to grow fodder and foodcrops, and provide drinking water, for 30 sheep and 5 people. Similar systems are used in Tunisia and northern Somalia. The Tunisian government has also funded the construction of traditional floodwater diversion dams in gullies to supply water for agricultural development in the arid Matmata district in the south of the country.

Drip irrigation uses less water than other methods such as sprinkler and gravity systems, while maintaining crop productivity. By lining irrigation canals water loss from seepage is reduced, while piped water distribution helps prevent evaporation. Such measures not only help to conserve water, but also reduce soil salinization and hence maintain crop productivity levels.

Battling against erosion

Measures to halt wind and water erosion – thus helping to preserve the soil's fertility – are important throughout the region, especially on the desert fringes, in the semiarid parts of the Maghreb, the savanna zone and the Ethiopian Highlands. The most successful projects involve the local people and require little capital and little expertise. In Niger's Majia valley, near the town of Bouza in the southwest, villagers were encouraged by the United States voluntary relief

organization CARE to build windbreaks on their flat, millet-growing land in order to reduce wind erosion, which was causing serious soil loss, especially of the fertile topsoil. Between 1974 and 1985 the villagers built 330 km (205 mi) of tree windbreaks using a species of Indian tree called neem. As a result, crop production rose by 18 to 23 percent, compared with unprotected areas. The trees also provided a fuelwood harvest that could be sold in the markets.

In the Illeta district of Niger, about 350 km (217 mi) northeast of the capital Niamey, poor soils have been reclaimed for cultivation by using a soil management technique known as *tassa* – small pits that are dug in the soil and planted with sorghum. Simple improvements to the *tassa*, such as making them larger and adding dung for fertilizer, were introduced in 1989 after farmers observed similar systems in Burkina in Central Africa. The resulting increase in sorghum yields has been spectacular. Both the *tassa* and windbreak examples show that while reducing soil loss is an important objective of soil conservation, it is rarely successful unless it also addresses the issue of soil fertility.

Managing fuelwood resources

The loss of woodlands is a crisis that must be tackled throughout the region. For countries that are heavily dependent on fuelwood for their energy needs – for example Ethiopia (95 percent), Somalia

Holding down the desert (*above*) This plantation of young palms is part of a program to stabilize the extensive sand dunes in Tunisia. More than 75 percent of the land is threatened by erosion, with topsoil loss affecting 18,200 ha (45,000 acres) annually.

Providing the basics (*right*) Urgently needed water arrives by tanker truck at a refugee camp in Ethiopia. Millions of people threatened by famine – caused by war and drought – have been assisted through international efforts, but the twin crises continue.

MULTIPURPOSE AGROFORESTRY

Fast-growing, introduced trees, such as *Eucalyptus* and *Casuarina* species, are often grown in plantations to alleviate fuelwood and timber shortages. They are also planted to stabilize shifting sand dunes and prevent water erosion. However, studies of woodland plantations throughout Africa suggest that a shift away from such stands of fast-growing, single-species trees toward more integrated, multifunctional forestry would be better for the environment. It would also provide more opportunities for employment and a healthier living.

Single-species plantations have some serious drawbacks. Eucalypts, for example, rapidly deplete the soil's nutrients and moisture, while their leaves contain poisonous oils, making poor fodder for animals. Plantations of fast-

growing tree species located near cities often fail to answer the need for more wood, as the trees may have poor burning properties: some burn too fast, others give off too much smoke.

What is needed is multifunctional forestry, or agroforestry, an approach that recognizes the importance of local trees and the wide variety of uses that woodlands and trees can be put to. In this new system, farmers are encouraged to plant native trees that are beneficial to the environment, providing not only food in their leaves, pods, fruit and nuts, but also valuable fuelwood and fodder for animals. Livestock can graze beneath the trees either on undergrowth or low branches, and shade-loving crops can be cultivated in fertile soil protected from severe wind or water erosion.

(84 percent) and Sudan (80 percent) – more efficient fuel use and a switch to other types of fuel (such as solar energy) would go a long way to reducing woodland destruction. In Niger, a new type of metal stove has been designed that encloses the cooking pot; this reduces energy consumption by one-quarter, compared with the traditional stove. Some Algerian villages, in contrast, have installed solar cells that convert the energy in sunlight directly into electricity.

The clearance of trees for farmland is difficult to halt. In Kordofan and Darfur provinces in Sudan, 88,000 ha (220,000 acres) of *Acacia* woodland are cleared annually to grow sorghum. However, an estimated 48 percent of this cleared land is so degraded after 3 or 4 years – mainly due to poor cultivation practices – that it has to be abandoned. The solution to this problem, and to the related problem of the expansion of grazing lands at the expense of woodland, lies in the development of agroforestry, which combines forestry, grazing and cultivation.

In urban areas the solutions to the problems of urban sprawl and environmental pollution involve better urban planning and service provision. Cairo, for example, began a major overhaul of its inadequate sewage system in 1980 as part of a plan to improve the health of its citizens by the year 2000, and provide water from treated effluent to irrigate 400 sq km (150 sq mi) of reclaimed desert.

Legacy of the Aswan High Dam

The Aswan High Dam in southern Egypt serves as a striking example of how a massive dam and reservoir project can have unforeseen consequences for the environment, the local people and their livelihoods. Traditionally, farming along the Nile valley relied on the annual flooding of the river, caused by winter rainfall on its headwaters in the Ethiopian Highlands: each year the Egyptian *fellahin* – Arab for "tillers of the soil" – diverted the floodwater, with its load of fertile silt, onto their plots. But this ancient system was limited by the uncontrolled and seasonal nature of the flooding.

In the 1950s, in a drive to industrialize and modernize the country's agriculture, the Egyptian government decided it was necessary to control the Nile flood. To achieve this, a great High Dam (financed by the former Soviet Union) was begun in 1960, 11 km (7 mi) upstream from the city of Aswan. It was completed in 1970. Measuring 112 m (370 ft) in height and 3.6 km (over 2 mi) long, the dam holds back the 480 km (300 mi) long Lake Nasser, and provides Egypt with 7.5 billion cu m (9.8 billion cu yds) of water. Hydroelectric power from the dam also provides about one-third of the country's electricity.

Holding back the waters

The immediate impact of the dam was the filling of Lake Nasser, but the rising waters threatened the famous rock-cut temples built by pharaoh Rameses II at Abu Simbel over three thousand years ago. In 1963, in a massive salvage operation, the temples were sawn into pieces and raised up the hillside, above the

Creating as many problems as it solves (*below*) Although the Aswan High Dam provides water and hydroelectricity, Lake Nasser (with the resited temples of Abu Simbel), is filling with the silt that once provided rich soil for crops downstream. Now the Nile delta is receding because the silt no longer reaches it.

White elephant on the White Nile (*left*) The Jonglei Canal in southern Sudan was planned to divert the White Nile from entering the Sudd swamps so that more water would flow into the Nile. However, civil war has put an end to the project, which would have destroyed the swamp's ecosystem.

waters, where they were reassembled. Less well known was the partial flooding of the town of Wadi Halfa, farther south, across the border in Sudan.

Most of the effects of the dam, however, have been felt downstream. The fertile Nile silt, once vital to farmers, now fills Lake Nasser, reducing the reservoir's projected lifespan by half. Only 8 percent of the original silt load now travels downstream, a development that has had unfortunate side-effects. Without silt, the Nile flows faster: as a result, the river has cut a deeper channel immediately downstream of the dam, and the inlets of irrigation channels have had to be lowered in order to receive water.

Without silt building up at the delta, the sea is advancing inland, reducing the productivity of agricultural land. The lack of nutrient-rich silt has also devastated Egypt's fishing industry around the Nile's mouth, leading to the loss of about 30,000 jobs and millions of dollars income, and depriving the Egyptian people of a major source of protein.

Salinization of the soil has been extensive. The amount of salinized land in Egypt in the early 1980s was between 28 and 50 percent of all irrigated land. Now, 10 percent of agricultural production is lost each year through the decline in soil fertility that salinization causes. To offset the loss of the annual load of 60 to 180 million tonnes of fertile silt and to combat salinization, 13,000 tonnes of calcium nitrate fertilizer have to be put on the land annually, at enormous expense.

Diseases transmitted by aquatic organisms are common in areas where farmers wade in irrigation ditches, and have spread with the increase in irrigated land. In the early 1990s, about 30 percent of the population were infected by schistosomiasis and 50 percent by bilharzia. Furthermore, the water input to Lake Nasser is not assured. The Sudanese have built dams on the Atbara river and Blue Nile, while the Ethiopians may be planning a dam on Lake Tana, the source of the Blue Nile (which provides 85 percent of Egypt's water). This would deny Egypt control over its lifeblood – the waters of the river Nile.

From rags to riches

Cairo's unofficial refuse collectors, the Zabbaleen, live in seven squatter camps around the outskirts of the city. The shacks that make up their housing are nestled among the mounds of garbage collected daily from the city's streets. Only about 60 percent of Cairo's garbage is collected at all, about one-third of it by the city's official collection service and two-thirds of it by the Zabbaleen. The rest is left to blow about the streets.

The Zabbaleen bring the refuse back to their camps, where it is sorted into several categories. They recycle 80 percent – or 30 percent of the city's total garbage output. Different Zabbaleen families specialize in specific materials: some in paper, others in rags, plastic, metals, glass and bones. The separate piles are then taken by trucks back to the city for reprocessing – paper waste to the paper mills, bones to the glue factories. Even the family pigs are caught up in the recycling process. The pigs eat the edible waste, their manure is sold as fertilizer, and eventually the pigs are taken to the slaughterhouse.

It is an unhealthy business living amid flies, stench and disease, but for some of the Zabbaleen it has proved very profitable. In two months a Zabbaleen can accumulate one tonne of plastic and sell it for between seven and twelve times as much as a government employee earns in one month. As a result, a few Zabbaleen have become relatively wealthy by Cairo standards. The shacks have been replaced with brick buildings and the horse and cart has given way to the pick-up truck. Televisions, refrigerators and even video recorders are not uncommon sights in the Zabbaleen camps.

Waste not, want not Urban recycling is big business in Cairo, where there is insufficient landfill for the amount of garbage produced. The city's unofficial refuse collectors, the Zabbaleen, collect, sort and sell whatever they can find.

ON THE BRINK OF DISASTER

PRESSURE ON A FRAGILE LAND · ACCELERATING CRISIS · RISING TO THE CHALLENGE

For almost as long as the human species has existed, humans have been modifying the rainforest and savanna environments of Central Africa through burning, grazing, felling and cultivation. Today, however, environmental change is taking place faster than ever before – in part, the result of accelerating population growth, which places relentless pressure on fragile land resources. The effects, both for people, the wildlife and the environment, are little short of disastrous. In lowland tropical areas, rainforest is being cleared for fuelwood, grazing land or plantations, while in the savanna grasslands the introduction of intensive farming methods has led to severe soil erosion and in some places desertification. Measures are under way to tackle all these environmental ills, but success is slow in coming.

PRESSURE ON A FRAGILE LAND

Dense tropical rainforest, now confined to the Congo Basin and small parts of coastal west Africa, once covered a much larger area. For thousands of years the indigenous peoples of the region practiced shifting cultivation in these forests, which preserved soil fertility very efficiently. A family would clear a small area of the forest by slashing and burning the vegetation, leaving the ashes to act as fertilizer. After three years of growing forest crops such as yams, the soil would become exhausted and the family would move to another plot, the old patch being left for the forest to regrow.

With the arrival of European colonizers, however, large areas of the forests were permanently cleared to make way for plantations for commercial crops such as coffee and cocoa. Other areas were and are being logged for prized tropical hardwoods such as mahogany. For most people in the region, wood is the only source of energy for cooking and heating, and large tracts of prime forest are being cut down to produce fuelwood.

All these factors are helping to reduce the rainforests. Ivory Coast, for example, boasted 145,000 sq km (55,970 sq mi) of dense rainforest in 1900, but by 1980 this had been reduced by two-thirds. West Africa as a whole is losing its remaining forest at a rate of 4 percent a year. If this continues, by 2010 there will be none left.

Population growth, pressure on land and the introduction of commercial farming methods means that traditional shift-

COUNTRIES IN THE REGION

Benin, Burkina, Burundi, Cameroon, Cape Verde, Central African Republic, Congo, Equatorial Guinea, Gabon, Gambia, Ghana, Guinea, Guinea-Bissau, Ivory Coast, Kenya, Liberia, Nigeria, Rwanda, São Tomé and Príncipe, Senegal, Seychelles, Sierra Leone, Tanzania, Togo, Uganda, Zaire

POPULATION AND WEALTH

	Highest	Middle	Lowest
Population (millions)	108.5 (Nigeria)	5.8 (Guinea)	0.4 (Equ Guinea)
Population increase (annual population growth rate, % 1960–90)	3.6 (Kenya)	2.8 (Zaire)	1.1 (Equ Guinea)
Energy use (gigajoules/person)	34 (Gabon)	2 (Guinea)	below 0.1 (C Verde)
Real purchasing power (US$/person)	3,960 (Gabon)	910 (Guinea)	410 (Uganda)

ENVIRONMENTAL INDICATORS

	Highest	Middle	Lowest
CO₂ emissions (million tonnes carbon/annum)	53 (Nigeria)	24 (Burkina)	9 (C Verde)
Deforestation ('000s ha/annum 1980s)	510 (Ivory C)	55 (CAR)	1 (Burundi)
Artificial fertilizer use (kg/ha/annum)	41 (Kenya)	4 (Ghana)	below 1.0 (Guinea)
Automobiles (per 1,000 population)	15 (Tanzania)	5 (Kenya)	below 1.0 (Rwanda)
Access to safe drinking water (% population)	75 (Gambia)	34 (Zaire)	12 (CAR)

MAJOR ENVIRONMENTAL PROBLEMS AND SOURCES

Land degradation: *types:* desertification, soil erosion, salinization, deforestation, habitat destruction; *causes:* agriculture, population pressure
Resource problems: fuelwood shortage; inadequate drinking water and sanitation; land use competition
Population problems: population explosion; urban overcrowding; inadequate health facilities; disease; famine; war
Major event: Lake Nyos (1986), gas cloud released

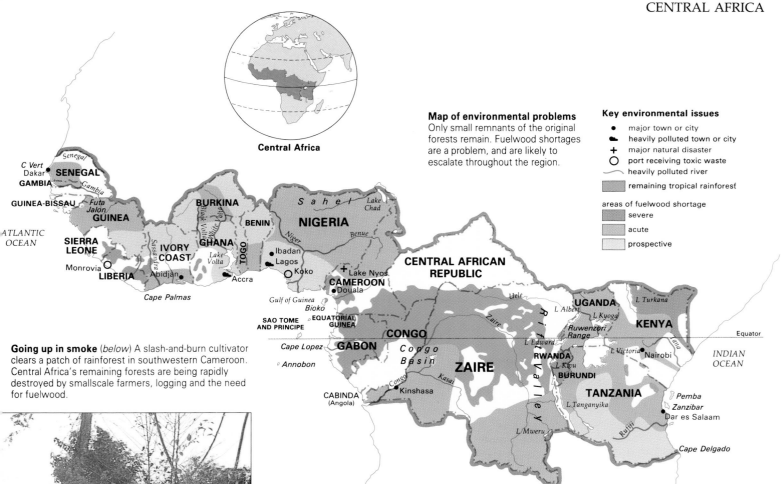

Central Africa

Map of environmental problems
Only small remnants of the original forests remain. Fuelwood shortages are a problem, and are likely to escalate throughout the region.

Key environmental issues
- • major town or city
- ⌐ heavily polluted town or city
- + major natural disaster
- ○ port receiving toxic waste
- 〜 heavily polluted river
- ▨ remaining tropical rainforest

areas of fuelwood shortage
- ▨ severe
- ▨ acute
- ▨ prospective

Going up in smoke (*below*) A slash-and-burn cultivator clears a patch of rainforest in southwestern Cameroon. Central Africa's remaining forests are being rapidly destroyed by smallscale farmers, logging and the need for fuelwood.

ing cultivation is being abandoned. Farmland is rarely allowed to remain fallow and uncultivated for any length of time. As a result, the fertility of forest soils has been lost, leading to severe problems of erosion in many areas.

The dwindling savannas
To the north, east and southeast of Central Africa's rainforest belt the vegetation gives way to expansive savanna woodlands and grasslands, which were once filled with great roaming herds of wildlife – wildebeeste, zebra, antelope, elephants and buffalo. However, these grasslands also provide good agricultural and grazing land for farmers and their livestock, and the conflict between people and the environment is therefore at its most intense in this zone.

In the past the nomadic pastoralists of the savannas, such as the Fulani of northern Nigeria, moved their cattle in the dry season to pastures in the south, where it was wetter. The cattle grazed on the stubble left in the fields after harvest, fertilizing the fields with their droppings. With the return of the rains, the southern woodlands would become infested with the disease-carrying tsetse fly, so the nomads would return with their herds to their northern pastures, which had by this time started to grow again.

The imposition of political boundaries by European administrators disrupted these traditional routes. Moreover, improved veterinary knowledge has led to an explosion in the number of animals. Cattle, in particular, are kept as much as a symbol of status and repository of wealth than as a source of meat: it is not unusual for a livestock owner to have 300 cattle, 200 goats and 60 camels. In some areas existing pressure on grazing lands and watering places has been exacerbated by large schemes to develop cash crop farming, such as in the inland delta of the Niger, where herds have been pushed into ecologically fragile environments.

To create new areas for grazing, trees are cleared and grass is deliberately set on fire to promote new growth. Subsequent overgrazing and bush fires then make tree growth more difficult. Deprived of tree roots to bind it together, the soil becomes eroded, opening the way to desert encroachment. The drilling of a few boreholes in a limited area of rangeland during periods of drought exacerbates the problem – large herds, attracted to the water source, trample and graze the surrounding pasture bare in rings up to 15 km (9 mi) wide. At present rates of loss, it is feared that as much as 30 percent of the region will have become desert by the year 2040.

ACCELERATING CRISIS

The greatest threat to the environment of Central Africa comes from the sheer weight of numbers of people. From an estimated 310 million people in 1990, the population is expected to double in the next 25 years. The fragile character of the region's soils, and the low carrying capacity of the land, make it particularly vulnerable to pressure from people.

Very few areas of Central Africa are blessed with fertile alluvial soils such as those around Lake Victoria in Uganda, Kenya and Tanzania. Generally, the region's soils have a high iron content and are unable to tolerate continuous cultivation without becoming compacted and losing their fertility. Once stripped of vegetation, the topsoil is washed away leaving the land vulnerable to gullying.

Bare earth reflects solar radiation back into the atmosphere and so hinders the formation of rain-giving clouds. The prolonged droughts that affected large parts of the region in the 1980s and 1990s were in part caused by the environmental degradation that came about from overuse and mismanagement of the land. Drought, in turn, accelerates the process of desertification.

Deterioration in the structure of the soil is often compounded by heavy machinery introduced under largescale development projects. In the 1970s, for example, a Canadian-funded aid project encouraged mechanized wheat farming on the Hanang plains of northern Tanzania, an area previously occupied by Barabaig pastoralists. For the first few years, the prairie-style cultivation with huge combine harvesters produced impressive yields. But there was a heavy ecological price to pay – heavy soil loss, erosion and extensive gullying.

Hazards to health

In the years after independence in the early 1960s, a number of large dams were built to tap the hydroelectric energy potential of several of Central Africa's major rivers, such as the Congo, Volta, Niger and Zambezi, and to control flooding and provide water for irrigation. Little thought was given at the time to the ecological consequences of these projects, such as the silting up of reservoirs and the increased salinization of soil in irrigated areas. One unexpected environmental con-

sequence was that the reservoirs behind the dams became favorable habitats for mosquitoes and disease-carrying water snails, responsible for malaria and schistosomiasis, a parasitic disease that infects the blood.

The warm tropical environment of Central Africa encourages a host of microorganisms, disease agents and insect carriers, which are responsible for killing or debilitating millions of cattle and people, especially children. Both trypanosomiasis, which causes sleeping sickness in humans, and nagana, a wasting disease in cattle, are transmitted by a protozoa carried by the tsetse fly. Onchocerciasis, or river blindness, is transmitted by the small fly *Simulium damnosum*, also known as the black fly. It breeds in fast-flowing river water. In places where it is rife, village communities are invariably forced to abandon the alluvial floodplains and move to higher land where water is scarce, droughts are frequent, and soil is less fertile and prone to erosion.

The most recent scourge in Central Africa is the fatal AIDS epidemic. The origin of the virus is still the subject of much dispute. Some people believe that it was transmitted to humans from the African green monkey, which inhabits the rainforest and savannas of Central Africa. Others disagree. Whatever the facts of the case, the disease is extremely common in the region, particularly in the cities. By 1995 it is likely to have affected more than 20 million people.

Mounting pressure on the environment

As more and more people throughout the region move to the rapidly growing cities, increasing amounts of untreated human and industrial waste are dumped directly into the ocean or nearby rivers, polluting water supplies, endangering health and destroying wildlife habitats. Offshore oil drilling operations and effluent from chemical industries located along the west African coast pose an increasing environmental menace to marine life. In addition, there is mounting concern about the illegal dumping of toxic waste – imported from Europe – on African land and in African waters.

Human activity is impinging more and more heavily on the unique wildlife of the region, bringing some species to the verge of extinction. The felling of forests for cultivation, fuelwood and charcoal, for example, is destroying the habitat of the

rare Mountain gorilla. Large mammals of the savannas, such as the Black rhinoceros and the elephant, are in great danger from poachers armed with sophisticated weaponry, despite international agreements to curb the export of ivory. Most surviving herds are confined to a few protected reserves in eastern Africa, where their numbers outstrip the carrying capacity of the limited area of grassland on which they are free to roam.

High risks to health (*right*) People in Africa are under constant threat from tropical diseases such as malaria, river blindness, sleeping sickness (transmitted by the mosquito, black fly and tsetse fly respectively) and bilharzia. Central regions are worst affected.

DANGEROUS LAKES

Central Africa's physical environment can be unpredictable and deadly. In areas of volcanic activity, such as the Great Rift Valley of eastern Africa and the highlands of Cameroon in the west of the region, high concentrations of dangerous gases may form in crater lakes. These can erupt without warning, killing animals and people over a wide area.

The worst of these disasters occurred at Lake Nyos in Cameroon on 21 August 1986 when more than 1,000 cu m (1,300 cu yd) of carbon dioxide, which had percolated through the volcanic rock into the deep waters of the lake, was released in a deadly cloud. In less than an hour, the ground-hugging gas drifted 16 km (10 mi), smothering some 1,700 villagers.

Two years before, a similar eruption had occurred on Lake Monoun, 95 km (60 mi) south of Lake Nyos, killing 37 people. Scientists fear that other lakes may be deadly time bombs, waiting to explode. Lake Kivu, on the border between Rwanda and Zaire, for example, is known to hold more than 63,000 million cu m (82,000 million cu yd) of methane gas and five times as much carbon dioxide. Meanwhile, urgent research is being carried out to determine what triggers the gas releases. Earth tremors, rockfalls, volcanic eruptions and storm winds are among possibilities being investigated.

Fragile earth (*left*) An erosion gully of enormous proportions in Nanka, Nigeria. It is estimated that half a million tonnes of soil are eroded from this area every year by heavy rains. Rivulets cut through the exposed topsoil – deprived of its protective cover by deforestation, overgrazing and the continuous cultivation of marginal lands – eventually joining together to form gushing torrents. Soil degradation has drastically reduced the region's ability to feed itself.

Life cycle of bilharzia (*right*) Bilharzia (schistosomiasis) is transmitted by water contaminated by sewage. Infected people or cattle pass eggs through urine or feces that hatch into flukes (parasitic flatworms). The larval flukes infest water snails before emerging into the water as cercariae (the final larval stage). The cercariae then penetrate the skin of people in the water. The disease is very debilitating and affects the bladder, reproduction organs, liver and intestines.

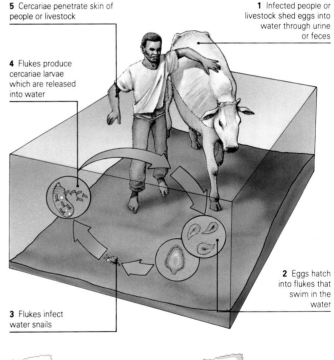

5 Cercariae penetrate skin of people or livestock

4 Flukes produce cercariae larvae which are released into water

1 Infected people or livestock shed eggs into water through urine or feces

3 Flukes infect water snails

2 Eggs hatch into flukes that swim in the water

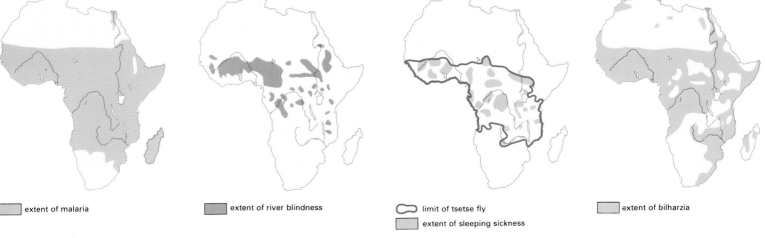

☐ extent of malaria

☐ extent of river blindness

⬭ limit of tsetse fly

☐ extent of sleeping sickness

☐ extent of bilharzia

RISING TO THE CHALLENGE

The realization of how crucially environmental degradation contributes to Central Africa's severe food shortages and endemic economic malaise is now spreading, and governments throughout the region are taking steps to tackle the root causes of the problem. Family planning is being encouraged to curb rapid population growth. Reforestation and land reclamation programs are being implemented. Sound agricultural techniques and proper rangeland management have become key issues.

However, there are formidable social and political obstacles to be overcome if such measures are to achieve widespread success. Technical skills are lacking in areas such as engineering, water control, ecology, and in the management of trees, field crops, soil and rangeland. The financial resources needed to fund training and research are severely limited.

Land reclamation and reforestation
Nevertheless, the countries of the semi-arid Sahel, stretching from Senegal to northern Kenya, have mobilized their limited resources to combat the urgent problem of desertification. The deep gullies and sand dunes formed by soil erosion are being stabilized by the planting of trees to hold their soil. Their leaves are a source of food for livestock, relieving pressure on overgrazed pasture. Elsewhere, irrigated areas laid waste by salinity are being replanted with salt-tolerant trees that reduce waterlogging.

Assistance from international organizations is vital to these projects, yet some of the most successful reclamation and reforestation schemes are those that involve the villagers themselves. For example, a nationwide tree-planting campaign was inaugurated by Kenya's Forestry Department in 1971. It teaches local farmers, most of whom have their own woodlots, technical skills in soil conservation and forestry management, and has won strong support at local level.

The Green Belt Movement, started by the National Council of Women of Kenya in 1977 to help villages and urban communities plant trees on any open space, has been hailed throughout Africa and the rest of the world as an exemplary success story. By the early 1990s it had established 21,000 green belts and set up

65 village-based plant nurseries. At the same time, efforts have been made in Kenya and other countries in the region to develop improved fuel-efficient clay stoves in order to reduce the demand for firewood and charcoal.

Environment-friendly farming
In forested areas, efforts have been made to develop corridor farming – first introduced by Belgian colonists in Rwanda, Burundi and Zaire in the 1940s – as a successor to the bush fallowing methods traditionally used by farmers in the region. The technique used is to clear several strips of land about 100 m (330 ft) wide. These are then cultivated in rotation, with one strip always being left fallow for a number of years. The strips are divided by wide borders of bush to encourage the quick growth of weeds and shrubs during the fallow period.

More recently, the idea of corridor farming has been adapted to alley cropping. In this, fast-growing, nitrogen-fixing trees that return nutrients to the soil are planted either side of alleys in which food crops are grown. The leaves of the trees provide fodder for livestock, and their branches are a source of fuel-

wood and building poles. Experiments by the Institute of Tropical Agriculture in Nigeria have demonstrated that alley cropping maintains soil fertility and also boosts crop yields, consequently allowing continuous cultivation of the same plot.

A number of solutions have been suggested to reduce pressure on overgrazed pastureland. The provision of many boreholes over a large area would distribute herds more thinly across the rangeland and prevent trampling. Of still greater ecological significance would be a plan that allowed pastoral nomads to move freely across political boundaries, thereby enabling them to return to their traditional practice of following seasonal rains with their herds.

Attempts to reduce the numbers of livestock persistently falter on traditional attitudes toward the keeping of large herds as a status symbol and as an insurance against drought. However, international organizations such as Oxfam are helping to set up programs to encourage pastoralists with large herds to cut back on numbers by enabling them to sell some of their cattle at profit. The meat is then used to relieve starvation in the famine-stricken areas of the Sahel.

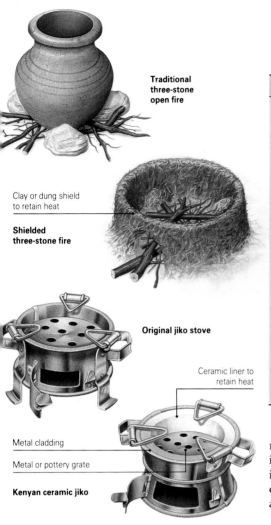

Traditional three-stone open fire

Clay or dung shield to retain heat

Shielded three-stone fire

Original jiko stove

Ceramic liner to retain heat

Metal cladding

Metal or pottery grate

Kenyan ceramic jiko

Fuel-efficient stoves (*above*) Cooking over an open fire (*top left*) allows most of the heat to escape, and sparks and smoke are dangerous. Building a simple shield of clay or dung (*top right*) retains some of the heat and saves fuel. The ceramic-lined jiko (*above right*) gives fuel savings of up to 50 percent over the original Kenyan charcoal-burning jiko (*above left*). More efficient stoves save time and ease pressure on scarce fuelwood resources; making them also provides work for local people.

Regreening initiatives (*left*) Local women plant out seedlings at a tree nursery in Kenya. Groups such as the Green Belt Movement have been very successful in promoting community tree-planting and reforestation schemes in hundreds of villages throughout the country.

Corridor farming (*right*) Fast-growing leguminous (nitrogen-fixing) trees are planted in alternate rows with a food crop such as maize. The trees continue to grow in the dry season and are cut back before a new annual crop is planted. The process is repeated, with the trees supplying a constant source of fuel and fodder while restoring fertility to the soil.

TSETSE FLY CONTROL

Control of the tsetse fly is held by many people to be of vital economic importance to Central Africa. The fly renders virtually useless immense tracts of land that could be used for grazing, and the trypanosomes – microscopic parasites – that it carries put at risk of disease more than 50 million people and 40 million cattle.

Several regional organizations are studying ways of reducing its incidence. A cost-effective means of controlling it is by the use of odor-baited traps laced with a potent pesticide. However, this and spraying massive areas from the air with the pesticide endosulfan and on the ground with DDT are vigorously opposed by environmentalists who warn that the pesticides will make their way up through the food chain, ultimately threatening many birds of prey.

Bush clearing, game shooting and game fencing have all been tried, but with only limited success. One of the most promising projects is the development of dwarf breeds of cattle, such as the ndama cattle in Gambia, that are resistant or tolerant to the trypanosomic parasite carried by the tsetse fly. Genetic engineering to produce strains of tsetse fly resistant to trypanosomes may also be a possibility in the future.

Tsetse fly control has been condemned by wildlife conservationists, who argue that wiping out the tsetse fly will destroy the few remaining pockets of wildlife refuge in Central Africa. The woodlands where the fly breeds will be cleared to provide new pastureland for livestock, leading – so they argue – to severe land degradation and the eventual desertification of virtually the entire continent.

With the help of newly developed remote sensing technology and monitoring equipment it is becoming increasingly possible to monitor changes to the environment over wide areas to give advance warning of impending degradation and land loss. Under the auspices of the United Nations Environment Program, a regional center in Nairobi, Kenya, has been set up to train professionals from all over Central Africa in the use of this equipment.

1 Annual crop (such as maize) and leguminous trees planted together

2 Annual crop ripens and is harvested; trees are left to grow

3 Trees continue to grow in dry season

Annual crop

Year 1

Leguminous tree

Year 2 (and subsequent years)

4 Trees cut for fuel and fodder. New annual crop planted and harvested

5 Trees grow on to produce new growth for harvesting

Toxic waste: a deadly import

One deadly byproduct of the modern industrial era has been the generation of millions of tonnes of hazardous wastes produced during the manufacture of chemical goods ranging from plastics, pesticides, soaps, medicines and fertilizers to ammunition. Among the most deadly of the waste materials are polychlorinated biphenyls (PCBs), heavy metals such as mercury and lead, organic chemicals, dioxins, cyanide, radioactive-tainted waste and discarded household goods that contain toxic material, for example, antifreeze (ethylene glycol) and drain cleaners (hydrochloric acid). Well over 90 percent of the toxic wastes are produced in the developed world.

Since the 1970s, the politicization of environmental issues in most developed countries has increased public awareness of the dangers of hazardous waste disposal. Most of the governments in Europe, North America, Japan and Australasia have passed stringent laws to ensure the safe disposal of toxic wastes.

As companies found it increasingly difficult – and expensive – to dump their deadly byproducts within the borders of the developed countries, they began to seek less restrictive and cheaper sites abroad. These they found in developing countries such as those of Central Africa, which – desperately in need of foreign currency – found it difficult to resist multimillion-dollar offers to provide locations for toxic-waste dumps.

Trading in lives (*above and right*) Toxic waste from Europe was dumped at the port of Koko, Nigeria in 1988, but the drums – some with false labels attached to hide their deadly contents – leaked. The Italian company responsible for trying to dispose of the toxic waste was forced to collect it. It chartered the ship *Karin B* (*right*), which for two months sailed the seas being refused entry to a number of European countries, including Italy, Germany, Spain and the United Kingdom. Eventually the ship was forced to return to Italy. Some African countries are now trying to enforce stricter regulations on the import of hazardous wastes.

Countries such as Nigeria, Senegal, Guinea, Benin, Cameroon and Congo were all approached. Some governments succumbed, while others such as Nigeria have, at least at government level, strongly resisted the offers. Despite this, toxic waste dumping in Central Africa – often done secretly by bribing government officials – intensified during the second half of the 1980s.

An agenda for disaster

Most countries in Central Africa lack the technology to dispose safely of toxic waste. The environmental consequences of mishandling a toxic load could be tragic: dangerous compounds such as PCBs dumped into the ocean or allowed to seep into the water table would contaminate the entire food chain. Thus the rich fishing grounds off the African coast are directly threatened by waste dumping, and the health of people living close to the sites is put at major risk.

To make matters worse, unscrupulous

dealers frequently mislabel toxic-waste drums, declaring them to contain fertilizer or harmless cleaning or degreasing fluid. Abandoned drums that contain deadly chemicals have sometimes been emptied by unsuspecting villagers and used for storing drinking water or brewing local millet or maize beer. Even if the drums are correctly labeled, the corrosive effects of Central Africa's high humidity

easily cause them to rust and rupture.

In one such incident a farmer in the river port of Koko, Nigeria, was paid $100 a month – a large fortune by African standards – to store 4,064 tonnes of toxic waste from Italy. When some of the drums ruptured, deadly PCBs entered the nearby river, and many villagers died after eating and drinking contaminated food and water.

Most developing and developed countries are bound by the Basel Convention on Control of Transboundary Movements of Hazardous Wastes, which was negotiated in 1989 between the exporters and importers of toxic wastes. Unfortunately, this convention permits the export of hazardous wastes to countries whose storage facilities are less advanced than those of the exporting country as long as

the importing state has detailed information on the waste shipment and gives prior written consent. Growing awareness of the dangers from the deadly trade led 12 African states to sign the Bamako Convention in January 1991 banning the import of toxic wastes from any country. It remains to be seen whether economic pressures to circumvent the Convention's agreement can be resisted.

193

Ivory trade goes up in smoke

Carved ivory, made from the tusks of African elephants, is highly prized by collectors for its beauty and durability. A male African elephant's tusk – a variety of dentin, or tooth enamel – averages about 2 m (6 ft) in length, and weighs about 23 kg (50 lbs). Europe was once the capital of the ivory trade, supplying the world with expensive trinkets, but the Far East took over as the carving and trading center early this century, with Japan, until 1989, accounting for 40 percent of all imports.

In many areas the African elephant has been hunted near to extinction by poachers who are increasingly well armed with automatic rifles and other weapons left over from the region's guerrilla wars. In order to obtain the ivory from an elephant the animal has to be killed, as one-third of each tusk is embedded in a bone socket in its skull. Profits from the trade go to the dealers, not the poachers.

By 1989 the situation was so grave that steps were taken by the Convention on International Trade in Endangered Species (CITES) to ban the ivory trade altogether. Some 100 countries, including Kenya, agreed to a worldwide ban on the sale of ivory in order to protect the endangered African elephant. By 1992, no countries were legally able to import ivory. Although illegal poaching is still a problem, by the early 1990s the World Wide Fund for Nature (WWF) announced that the world ivory trade had collapsed due to fall in demand. The African elephant population had increased to such an extent that some countries had to organize culls in order to keep populations viable.

Confiscated ivory stocks are publicly burned in the Nairobi National Park, Kenya, to prevent them entering the ivory trade and to demonstrate that poaching does not pay.

A DAMAGED EDEN

EXPLOITING THE LAND · FRAGILE ENVIRONMENTS UNDER STRESS · WORKING FOR THE FUTURE

Southern Africa's varied landscapes, which include tropical forests, savanna grasslands and desert, are beset by a range of environmental problems that stem from ever-increasing human pressure for more land. Overgrazing (particularly in Botswana and Namibia) and deforestation in Madagascar have led to severe soil erosion and gullying. Localized problems, such as those arising from mining, the introduction of alien species, or industrial air pollution (as in South Africa and Zimbabwe), can often be addressed. But the more complex environmental issues are harder to deal with, particularly in the face of the huge imbalances in wealth caused by hitherto harsh and inflexible racial and economic policies (especially in South Africa), and the ravages of civil wars such as in Angola and Mozambique.

COUNTRIES IN THE REGION

Angola, Botswana, Comoros, Lesotho, Madagascar, Malawi, Mauritius, Mozambique, Namibia, South Africa, Swaziland, Zambia, Zimbabwe

POPULATION AND WEALTH

	Highest	Middle	Lowest
Population (millions)	35.3 (S Africa)	8.5 (Zambia)	0.8 (Swaziland)
Population increase (annual population growth rate, % 1960–90)	3.4 (Botswana)	2.6 (Namibia)	1.7 (Mauritius)
Energy use (gigajoules/person)	83 (S Africa)	2 (Comoros)	1 (Madagascar)
Real purchasing power (US$/person)	5,480 (S Africa)	1,370 (Zimbabwe)	570 (Comoros)

ENVIRONMENTAL INDICATORS

CO₂ emissions (million tonnes carbon/annum)	47 (S Africa)	3.4 (Angola)	0.3 (Comoros)
Deforestation ('000s ha/ annum 1980s)	156 (Madagascar)	80 (Zimbabwe)	0.5 (Mauritius)
Artificial fertilizer use (kg/ha/annum)	307 (Mauritius)	18 (Zambia)	below 1 (Botswana)
Automobiles (per 1,000 population)	3,079 (S Africa)	49 (Madagascar)	6 (Lesotho)
Access to safe drinking water (% population)	98 (Mauritius)	50 (Swaziland)	24 (Mozambique)

MAJOR ENVIRONMENTAL PROBLEMS AND SOURCES

Air pollution: locally high
Land degradation: *types*: desertification, soil erosion, deforestation, habitat destruction; *causes*: agriculture, population pressure
Resource problems: fuelwood shortage; inadequate drinking water and sanitation
Population problems: population explosion; urban overcrowding; inadequate health facilities; famine; war

EXPLOITING THE LAND

The original inhabitants of southern Africa – nomadic hunter–gatherers such as the San (Bushmen) and larger numbers of pastoralist and farming Bantu-speaking peoples – lived in relative harmony with their environment. By contrast, the European settlers who arrived in increasing numbers from the end of the 17th century viewed the land and its resources as a limitless natural storehouse of commodities for profit and plunder.

Within a few decades of the initial Dutch settlement at Cape Town in 1652, the hardwood forests that covered the slopes of Table Mountain and the surrounding area had been cleared – the timber used for fuel, construction, and the building of wagons, ships and fur-

niture. Later, forests of endemic yellow-wood, stinkwood and cedar along the southern Cape coast were increasingly exploited, especially after the advent of the railroads in the 1880s created a vast new demand for hardwood timber as sleepers. On the grasslands, vast herds of grazing animals were shot for sport or food. By the late 19th century, large mammals had all but disappeared from much of the country.

The spread of British settlers and farmers of Dutch origin (Boers) during the 19th century brought further changes to the land. The introduction of European farming techniques to the very different environments of southern Africa damaged soil fertility and stability. Overgrazing removed vegetation cover, leaving the soil vulnerable to wind and water erosion and the formation of deep gullies.

Old and new together (*left*) Smoke from a charcoal-making kiln billows around an overhead power line in Malawi. Heavy reliance on wood and charcoal as a domestic fuel is putting southern Africa's surviving areas of woodland at increasing risk.

Map of environmental problems (*below*) The region faces severe problems of desertification and deforestation. The unique fynbos vegetation of the Cape and other wildlife habitats are vulnerable to the pressures of encroaching urbanization.

Key environmental issues

- • major town or city
- ⬝ heavily polluted town or city
- ⌒ heavily polluted river
- ◯ area of severe fuelwood shortage
- ◯ area of fynbos vegetation
- ▨ remaining tropical rainforest
- ▨ area of deforestation

areas at risk of desertification
- ▨ very high
- ▨ high
- ▨ moderate
- ☐ true desert

Mounting pressures

As more fertile land was taken over for white commercial farms, impoverished peasant farmers were forced to eke out a subsistence existence on marginal land in remote rural reserves. In South Africa's Native Reserves – the forerunners of the black "homelands" – the land was seriously overgrazed by pastoralists who were prevented from moving their cattle to fresh grazing lands. The forced relocation of people, restrictions on rural–urban migration and a rapid population growth rate have increased pressure on the land, accelerating the process of land degradation. South Africa's black "homelands" today are ravaged by drought, overgrazing, erosion and state neglect.

In Zimbabwe, the Land Apportionment Act of 1930 set aside half the country (then Southern Rhodesia) for whites. Much of this territory included the best and most fertile land; areas of poor land became "Native Reserves", where land pressure and infertile soil led to over-cultivation. Since Zimbabwean independence in 1980, there has been modest redistribution of land, although more is planned for the future. Here, too, population increase and sustained drought has put mounting pressure on fragile and diminishing land resources.

Exploitation of the region's mineral resources has had a major environmental impact. Copper, iron, gold and diamond mines have scarred the landscape and polluted waterways. The coalfields and power stations east of Johannesburg produce sulfurous air pollution on a par with parts of Eastern Europe. Urban sprawl and tourist developments also create local pressures on the environment. For example, sprawling holiday resorts discharge untreated or semitreated sewage into coastal estuaries, lagoons and inshore waters, damaging plant and animal life.

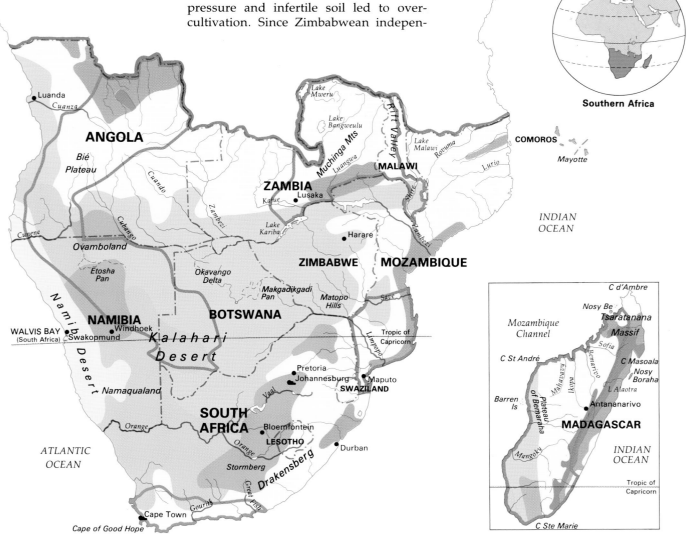

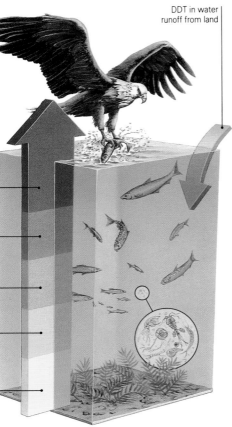

DDT in water
runoff from land

South Africa's "silent spring"
(*right*) Pesticides such as DDT and
dieldrin are still widely used in
southern Africa. They run off
farmland into rivers and enter the
food chain, accumulating in the fatty
tissues of animals. High levels of
toxicity make African fish eagle eggs
so thin they crack before hatching.

Fish-eating birds 50 ppm

Carnivorous fish 2 ppm

Small fish 0.5 ppm

Zooplankton (microscopic animals)
0.04 ppm

Water 0.000003 ppm

Levels of DDT   ppm = parts per million

FRAGILE ENVIRONMENTS UNDER STRESS

Overuse and misuse of productive land
have created wastelands right across
southern Africa. The most vulnerable
areas are island ecosystems (Madagascar
and the Comoros), marginal land on
mountain slopes such as the Drakensberg
range in South Africa, and the arid and
semiarid savanna lands such as those of
Botswana, but even the most tolerant
environments have been degraded by
overexploitation or pollution. Soil loss is a
major problem. In the mountain kingdom
of Lesotho, erosion affects nearly 90 percent of the land, with an annual loss of
18.5 million tonnes of soil from only 2.7
million ha (6.7 million acres) of land.

In Botswana and Namibia, cattle and
sheep ranches have expanded into marginal areas on the desert fringes, de-
nuding vast areas that already had only
sparse grass cover. In Namibia, thorn-
bushes have spread over some 40 percent
of the country's commercial grazing land.
These tough plants, unpalatable to cattle,
displace the remaining grasses in overgrazed areas, reducing even further the

A carpet of weeds (*below*) floats on
the huge lake created by the Kariba
dam on the Zambezi river between
Zimbabwe and Zambia. Many large
dams constructed in the 1960s have
become blocked by weeds or silt,
reducing their capacity and lifespan.
The lakes serve as breeding grounds
for insects carrying diseases such as
malaria and bilharzia.

THE IMPACT OF OPEN-CAST MINING

Open-cast mining for copper in Zambia, uranium in Namibia and iron, coal and diamonds in South Africa has had an enormous impact on the environment. Large slag heaps and mine facilities, such as crushing plant and reduction works, disfigure the landscape close to towns, and the constant movement of massive tipper trucks creates considerable noise and vibration. Poorly protected mineworkers are at risk from potentially harmful dust and discharges of fumes and liquids. Carried downwind, these toxic fumes may affect people living nearby, as well as agricultural activities.

The contamination of groundwater by toxic effluents is a serious side-effect of open-cast mining. Environmentalists claim, for example, that radioactive liquids from the immense uranium works at Rössing in Namibia's Namib Desert are entering the subterranean water-system of the nearby Khan and Swakop rivers. Although the mine-owners deny it, pollution on this scale would be disastrous, as the water supports wildlife over a large area, and supplies the coastal towns of Swakopmund and Walvis Bay, and the small market gardens of local oases.

Dirt and diamonds (*left*) A diamond mine dump at Kimberley Cape, South Africa. Open-cast mining is a major contributor to environmental damage throughout the region. Worse than an eyesore, dumps and abandoned sites leach toxic chemicals into the soil and water supply, with disastrous consequences.

and Namibia. The eradication of the tsetse fly in parts of the region has opened new areas of land to cattle herding, increasing pressure on scant water supplies. Several areas have been contaminated by toxic chemicals, such as dieldrin, used during tsetse fly extermination campaigns. The indiscriminate use of pesticides throughout southern Africa, together with deliberate poisoning and shooting, has drastically reduced birds of prey.

Armed conflict is another cause of severe land degradation in some of the region's environments. In Angola and Mozambique, agriculture and the economic infrastructure have collapsed, and poverty, malnutrition and environmental destruction are rife. In Mozambique, guerrilla forces slaughtered the country's elephants in order to sell the ivory for arms, while in Angola – prior to the ceasefire in mid 1991 – guerrilla troops felled and smuggled out vast amounts of precious hardwood timber in exchange for weapons and supplies.

The fuel crisis
Population pressure, fed by rising rural–urban migration, has created sprawling shanty towns on the urban fringes. In South Africa's Cape Province, the Cape Town metropolitan area (with a population of nearly 3 million people) has encroached onto historic vineyards, fertile farmland and previously open countryside. The majority of urban dwellers rely on wood and charcoal. Large areas of land around the cities and shanty towns are stripped bare of trees to provide fuelwood, and air in the cities is heavily polluted with woodsmoke. The vast South African township of Soweto suffers particularly badly from air pollution, and at dawn and dusk in winter is often submerged beneath choking clouds of coal smoke. In Lusaka (Zambia) and Harare (Zimbabwe), electricity has been introduced into some of the poorest areas to deter people from cutting down trees on the cities' edges. But few can afford electrical appliances and utility rates.

The poorest areas also suffer from a lack of sanitation and refuse disposal. Water supplies are often inadequate or nonexistent, thus posing serious health risks, particularly to young children and the elderly. All these problems are compounded by the rapidly rising population throughout the region.

land's ability to support livestock. Some 80 percent of southern Africa's 304 million ha (750 million acres) of productive drylands had suffered the effects of desertification by the early 1980s.

Threatened ecosystems
The pressure of human activities in Madagascar has turned the island into an environmental catastrophe. The forests, for example, are being cleared for farming, fuelwood and timber at a rate of 156,000 ha (385,000 acres) a year, while only 12,000 ha (30,000 acres) a year are being reforested. The island's habitats and unique plants and animals are highly endangered. On Mauritius to the east, 226 plant species out of an estimated total of 800–900 species are at risk.

The unique, Mediterranean-type fynbos floral kingdom of low-growing, ever-green shrubs on the southwestern Cape is at increasing risk from urbanization, bushfires and the invasive spread of *Acacia* tree species introduced from Australia to stabilize coastal dunes. As no native animals or insects feed on these *Acacias*, they are left unchecked to suffocate the native vegetation.

Many of southern Africa's coastal wetlands, including estuaries and mangrove swamps, which act as barriers to coastal erosion, have been cleared for tourist developments or have been polluted by sewage and industrial effluent. Coral reefs are fast disappearing, hacked away by souvenir hunters.

Interior wetlands are also under threat. Water is being extracted on a large scale for mining purposes from the swamps of the Okavango Delta, a vital supply for people and animals in Botswana, Angola

WORKING FOR THE FUTURE

Southern Africa's environments – its land, people, wildlife and resources – cannot be sustained in a healthy and viable way unless the basic issues of poverty and wealth are addressed. The redistribution of land between commercial and subsistence farmers, if properly managed, could reduce the pressures of overgrazing and erosion on the most degraded areas, while minimizing deterioration in newly resettled areas. At present, the majority of the rural population is forced to abuse the land in order to survive in the short term, and ingrained fear and distrust of officials in the region have hindered redistribution schemes.

The state, as the largest landowner in every country, can play a major part by improving land management practices on its plantations and in its forests. Retaining or planting suitable trees can help to reduce soil erosion, while providing windbreaks, shade, fruit and traditional medicines. If properly managed, they can also supply livestock fodder and fuelwood, both of which benefit small farmers. In Zimbabwe, the Rural Afforestation Project has established some 70 state nurseries that provide rural people with seedlings at subsidized prices. It has also encouraged widespread agroforestry, which includes the cultivation of trees within farming systems.

Cooperation at local level

Wildlife conservation programs will only succeed if the people living in and around the reserves benefit from them. If properly conserved, wildlife can be exploited as a renewable resource that will provide food, jobs, medicine and foreign exchange long after nonrenewable mineral resources such as copper have been exhausted. Until recently, however, many conservation areas, including wildlife parks, effectively excluded the rural population and provided recreation for the wealthy. Such a policy antagonized local pastoralists who were denied access to their traditional grazing lands. Many were forced to turn to poaching in order to make a living.

Both sides of the fence Tall grasses within a protected area of the Okavango Delta in Botswana contrast with closely cropped vegetation outside. The management aims of the Maremi reserve in the delta give local people a part in decision-making.

In experimental schemes in South Africa and Zimbabwe, local people have been allowed into the game and nature reserves to collect firewood, grass for thatching, fruit, tubers and traditional medicines. After they were allowed to use the meat taken from culled animals, there was a dramatic decline in poaching. Skilled local craftsmen have also been employed to produce handicrafts for sale to tourists, thereby generating income for themselves. In Zambia a scheme to integrate rural development and resource conservation was set up in 1987 in the Luangwa valley in the southeast of the country. In order to provide the local people and wildlife with a self-sustaining future a range of activities were developed, such as hippo cropping, game meat and forest cooperatives, safari hunting and tourism.

An innovative agreement, signed in July 1991, established the Richtersveld National Park, along the Orange river border with Namibia, as South Africa's newest reserve. Here, for the first time, the indigenous and impoverished people of Namaqualand – the Nama – will be allowed to stay on their traditional lands, and will receive financial assistance so that they may sustain their pastoral way of life. Tourism-related jobs will also be important. Such schemes point the way ahead. However, areas protected by the government cover only a small fraction of the region. Effective environmental conservation and management programs must therefore extend outside the national parks and game reserves to embrace particularly vulnerable areas.

Some large national and transnational corporations – eager to improve their public image – have proclaimed suitable areas of their land as nature reserves, restocking them with wildlife, and opening parts to the public. The South African Nature Foundation, which is affiliated to the World Wide Fund for Nature, has encouraged large landowners to register important areas as "natural heritage sites". Those who do so must undertake to follow expert conservation advice.

Working with nature

Farmers themselves have a key role to play in wildlife conservation. Livestock farmers have traditionally persecuted jackals, lynx and birds of prey as vermin that kill their lambs and calves. However, a state-sponsored scheme in parts of

THE PROBLEM OF SAFE WATER

Much of southern Africa's rural population lacks access to safe drinking water and proper sanitation. Consequently, the people face constant risks of gastrointestinal infection and life-threatening diseases, such as cholera. In addition, drought – such as the one that struck southern Africa in 1992 – is a recurring problem.

Zimbabwe has made concerted efforts both to conserve and to find new sources of drinkable water. There are over 8,000 storage dams in the country (with more being built all the time), providing domestic, agricultural and industrial water. Thousands of masonry weirs, just a few meters high, have been built by local communities. However, such dams are liable to siltation, which clogs reservoirs and, most importantly, deprives the land below the dams of fertile soil deposits.

Much of Zimbabwe is underlain by igneous rock, which means that there are few deep aquifers to be exploited. Environmentally controversial drilling has been carried out to tap shallower underground reserves – more than 35,000 boreholes have been fitted with motor or hand pumps, and countless wells have been dug by local people. In 1988 Zimbabwe launched a major program to provide clean water within 500 m (550 yds) of every home, and proper sanitation facilities for every village by the year 2000. Hygienic pit latrines – designed to prevent flies from getting out and contaminating food and water supplies – have also been developed, and this has had an immense impact on rural health. However, the population continues to rise, and Bulawayo, the country's second city, is chronically short of water.

A new village pump comes into action. Rural access to safe drinking water, though improving, is still uneven throughout the region. Nearly 50 percent of people in rural Malawi had access to it in 1992, but the figure was less than 10 percent in Swaziland.

South Africa has shown the benefits of working with nature. Several neighboring farmers have jointly formed conservancies – protected areas – on their land, and have appointed a trained game guard or conservation officer to patrol them. The result has been a decline in poaching, enabling small animals to multiply; this has provided adequate prey for the so-called "vermin", and reduced losses to livestock. There are other benefits too: bird and mammal species have increased, while conservation measures have helped to control erosion and the spread of alien plant species. Some farmers have profited from conservancies by introducing nature trails for ramblers and providing overnight accommodation.

Rural poverty and land degradation

Much of the severe land degradation in southern Africa stems from the forced resettlement by European settlers of African farmers and seminomadic pastoralists onto remote, marginal lands. Under increased population pressure, made worse by restrictions on rural–urban migration imposed by white governments, and population growth rates of 2 to 3 percent per annum, the land has rapidly deteriorated. In many areas where the annual rainfall is as low as 500 to 600 mm (20 to 24 in), population density exceeds 350 people per sq km (910 people per sq mi). Refugees from Mozambique have compounded the problem of overcrowding in neighboring Malawi.

Survival a priority

Overgrazing, deforestation and soil erosion are the most common problems. The progressive loss of traditional seasonal grazing land to white commercial farmers puts ever greater pressure on communal pasture in densely settled areas. The land is denuded by increased livestock numbers and frequent burning, intended to encourage new grass growth. Tree clearance is often extensive – with disastrous consequences – as the need for fuelwood outweighs the value of trees as a source of shade, fodder, fruit and medicine. This inevitably leads to further soil erosion and gullying. In addition, farmers often reduce traditional fallow periods, necessary to preserve soil fertility, in an attempt to gain additional crops from their patch of land. The result is that increasing amounts of expensive fertilizers have to be applied to maintain soil fertility and boost yields.

Most state-sponsored rural development schemes in southern Africa have tried to convert communal land to individual tenure, thereby creating a class of small African commercial farmers. The prototype of this policy was the so-called "betterment" schemes instituted in South Africa's black "homelands" in the late 1950s. These had mixed results. A minority benefited from investment in new roads, water and electricity supplies, soil conservation measures and marketing facilities. But many small farmers became poorer because they were displaced from their land or lost access to their traditional water supplies.

A pitted and sculpted landscape (*above*) in northern Natal, South Africa, shows the effects of wind and water erosion. Fragile land resources are being lost through overuse and population pressure.

No grass left (*right*) During a period of drought, too many cattle and goats have created a dustbowl outside a village in Botswana. Their trampling hooves have added to the devastation.

Throughout the region, able-bodied men leave the land for work in cities, mines and factories, sending remittances back to their families. With a shortage of labor at sowing and harvest time, yields and total harvests are often reduced. Lesotho, for example, has 100,000 mineworkers on contract in South Africa at any one time. They represent the country's principal export and, together with development aid, provide virtually all its foreign exchange.

In Lesotho, Malawi, Swaziland and Zimbabwe village woodlot schemes – either run communally or by selected individuals – have been incorporated into some rural development programs. The tree plantations provide new sources of wood for domestic use, or timber for sale. However, the planners have often been shortsighted in selecting the type of tree planted. *Eucalyptus*, imported from Australia, is the most widely planted species in Zimbabwe, and although it has the advantage of being fast-growing, it is of limited use and reduces soil fertility, while lowering the water table. In contrast, native African species of *Acacia* provide good fodder, maintain soil fertility, and make better fuelwood. Dissatisfaction and disillusionment have often led to the abandonment of such schemes.

A chain of destruction

The treeless hills of Madagascar's central plateau – deprived of protective vegetation cover – have been dramatically dissected by deep gullies as the topsoil has been washed away by the rains. The eastern half of the island was once covered in evergreen and deciduous forest, but now less than one-quarter remains. As with other developing countries around the world, Madagascar has achieved short-term economic gain in exchange for long-term agricultural and ecological loss.

The island's forests were once home to all kinds of unique plant and animal species that had evolved in isolation. Many are now extinct, their potential value to humanity lost forever. It is estimated that at the present rate of rainforest destruction around the world, some 100 species are being lost every day. Combined with the loss of other fragile ecosystems brought about by inappropriate land use, this could lead to the loss of some 20 to 25 percent of all known species by 2020.

Maintaining biodiversity is essential for human well-being and possibly even survival. The majority of the world's population still depends on natural medicines derived from plants. One species that originated in Madagascar and is now of global importance is the Rosy periwinkle (*Catharanthus roseus*). It is used to treat a range of serious illnesses, from Hodgkin's disease to diabetes, leukemia and other cancers. Unless the environments that support such plants are conserved rather than destroyed, their continued availability will become ever less certain.

Slipping through the fingers Madagascar's once forested, species-rich central plateaus have been devastated by erosion and gullying, the result of extensive deforestation.

CHALLENGES OF EXTREME DIMENSIONS

A LANDSCAPE CARVED FOR USE · DEFORESTATION AND POLLUTION · DEFENDING THE LAND

The Indian subcontinent is prone to natural disasters on a grand scale – cyclones, flooding and drought – and is beset by environmental problems, many stemming from rapid population growth. However, its peoples are resilient, responding in positive ways to the challenges that face them. For example, in the Himalayas, extensive deforestation by loggers inspired villagers to defend their trees by "hugging" them, a move that initiated a powerful conservation movement. The problems are diverse. On the arable lands of Pakistan and India intensive agriculture has led to waterlogging and salinization from over-irrigation. Other critical issues include high levels of industrial and sewage pollution in the region's overcrowded cities, in the rivers and along the coastline, and the associated health risks to the population.

COUNTRIES IN THE REGION

Bangladesh, Bhutan, India, Maldives, Nepal, Pakistan, Sri Lanka

POPULATION AND WEALTH

	Highest	Middle	Lowest
Population (millions)	853.1 (India)	19.1 (Nepal)	0.2 (Maldives)
Population increase (annual population growth rate, % 1960–90)	3.0 (Pakistan)	2.4 (Nepal)	1.9 (Sri Lanka)
Energy use (gigajoules/person)	8 (India)	2 (Bangladesh)	1 (Nepal)
Real purchasing power (US$/person)	2,120 (Sri Lanka)	870 (Nepal)	720 (Bangladesh)

ENVIRONMENTAL INDICATORS

CO₂ emissions (million tonnes carbon/annum)	230 (India)	6.8 (Nepal)	0.2 (Bhutan)
Deforestation ('000s ha/annum 1980s)	1,500 (India)	58 (Sri Lanka)	1 (Bhutan)
Artificial fertilizer use (kg/ha/annum)	113 (Sri Lanka)	77 (Bangladesh)	1 (Bhutan)
Automobiles (per 1,000 population)	7 (Sri Lanka)	2 (India)	0.4 (Bangladesh)
Access to safe drinking water (% population)	57 (India)	41 (Sri Lanka)	36 (Nepal)

MAJOR ENVIRONMENTAL PROBLEMS AND SOURCES

Air pollution: generally high, urban very high; acid rain prevalent; high greenhouse gas emissions
River pollution: medium; *sources*: agricultural, sewage
Land degradation: *types*: desertification, soil erosion, salinization, deforestation, habitat destruction; *causes*: agriculture, industry, population pressure
Resource problems: fuelwood shortage; inadequate drinking water and sanitation; coastal flooding
Population problems: population explosion; urban overcrowding; inadequate health facilities; famine
Major events: Bhopal (1984), leak of poisonous chemicals; Bangladesh (1988, 1991), major floods

A LANDSCAPE CARVED FOR USE

For over 5,000 years, people have dramatically modified the environments of the Indian subcontinent by cutting down the forests and constructing terraces and complex irrigation systems. Behind many of the changes lay the need to control water supply, both for domestic needs and for agriculture (the chief economic activity) in a region frequently struck by drought and flooding. As early as 2500 BC, the two great cities of Mohenjo Daro and Harappa – products of the lower Indus valley civilization in present-day Pakistan – were reliant for their barley and wheat crops on the fertile floodplain and waters of the Indus river.

Patchworks of irrigated agricultural plots dominate the landscape of the lower Indus plains in Pakistan's Sind province, and paddyfields cover the floodplains of the Ganges and Brahmaputra rivers in northeastern India and Bangladesh. In mountainous Nepal and Bhutan, in the Himalayan states of India, and on the hills of Sri Lanka, steeply banked rice-growing terraces have been carved out of the slopes, irrigated with rainwater run-off. The landscape of the drier parts of India, such as the southern state of Tamil Nadu, is covered with earthen "tanks", constructed to store the sparse rainfall.

The subcontinent's landscapes and economic structures – particularly those of India – changed significantly during the period of British colonial rule (1858 to 1947). The British invested heavily in irrigation canals, enabling extensive areas to be brought under cultivation for the first time. They encouraged tea growing, so that by 1871 India had over 300 tea plantations covering 12,000 ha (30,000 acres). They also helped develop modern industry – for example setting up the cotton mills of Bombay. In the 1850s the British began laying out the railroad networks of what were to become India and Pakistan, opening up the region's vast reserves of coal and rich iron-ore deposits for exploitation, and enabling agricultural produce to be transported from the interior to the ports of Bombay, Calcutta and Madras for export.

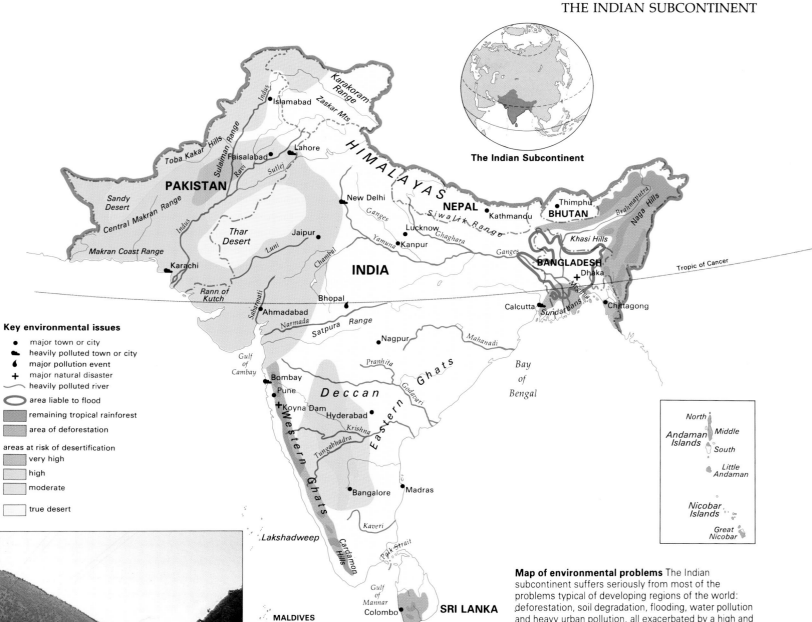

The Indian Subcontinent

Key environmental issues

- • major town or city
- ▬ heavily polluted town or city
- ⬧ major pollution event
- ✛ major natural disaster
- ∿ heavily polluted river
- ⬭ area liable to flood
- remaining tropical rainforest
- area of deforestation

areas at risk of desertification
- very high
- high
- moderate
- true desert

Map of environmental problems The Indian subcontinent suffers seriously from most of the problems typical of developing regions of the world: deforestation, soil degradation, flooding, water pollution and heavy urban pollution, all exacerbated by a high and ever-increasing population.

The drawbacks of development

In postcolonial times, the rise of industry and intensive agriculture has been paralleled by a massive increase in population. Bangladesh's population more than doubled in the 30 years from 1960 to 1990, taking it from 51.6 million to 116 million. India's increased by more than 50 percent over the same period, from 555 million to 853 million, and the region now has some of the most overcrowded and rapidly expanding cities in the world, where squatter encampments are common, and essential services – such as clean water supplies and sanitation facilities – are often completely lacking.

In rural areas, too, largescale modifications have been made to the environment

Struggling to cross the mountains, a bus tries to pass a rockfall on the Karakoram Highway between Pakistan and China. Earthquakes are frequent in the northern mountains of the region and present a constant threat to roads, bridges and dams.

since the 1960s, many of which have been affected by environmental setbacks. In the mountains of Pakistan, for example, stretches of the Karakoram Highway, built in the 1970s to link Pakistan with China across the Karakoram Mountains, are constantly washed away by raging streams and huge floodwaves – generated by the collapse of natural landslide dams – or are blocked by rockfalls, some caused by the frequent earth tremors.

Large dam-building schemes have extended irrigation, increasing agricultural yields for the expanding population, but they have also caused extensive damage to local habitats, and in one instance created a major disaster. In 1967, five years after the completion of the Koyna Dam, south of Bombay, an earthquake – probably triggered by reservoir water seeping into the rocks – killed 200 people and left thousands homeless in an area previously free of earthquake activity.

DEFORESTATION AND POLLUTION

Tropical forest once covered most of the Indian subcontinent, but today relatively little of it remains. Pressure for farmland in India has led to the destruction of more than 50,000 sq km (19,300 sq mi) by slash-and-burn cultivators. This once-sustainable system of agriculture has become a highly destructive use of land, leading to soil erosion and landslides on mountain slopes, and land degradation in lowland areas. By 1990 some 1.5 million ha (3.75 million acres) were being destroyed annually. Many of the forests that have survived largescale destruction by loggers and cultivators have been degraded by rural communities that depend on fuelwood for heating and cooking. Urban demand has increased the pressure; Delhi's firewood now comes from Madhya Pradesh, some 675 km (420 mi) away to the south, as local resources have been used up.

In Sri Lanka rainforest once covered the entire southwestern quarter of the island, but nearly all of it has now been cleared for rice and coconut plantations on the lowlands, and tea and teak on the hillsides. Similarly, the mangroves, swamps and rainforests that covered much of Bangladesh have been almost totally destroyed by an expanding population that averages a staggering 800 people per sq km (2,000 per sq mi).

The problems of deforestation are not confined to the region's tropical forests. In Nepal, the wet hill forests, which support evergreen oaks and chestnuts, are fast diminishing. Over the last 25 years Nepal has lost 30 percent of its forest cover, mainly to local fuelwood collectors who gather wood both for their domestic needs, and to keep the growing number of tourists warm at night on their mountain treks. Three-quarters of the rural population of Pakistan also depend on fuelwood. But the greatest threat to the existing forests in this part of the region comes from livestock grazing – the most widespread form of land use.

The few areas of the region that are still densely forested are under serious threat. In the Bay of Bengal, the Andaman and Nicobar Islands are largely covered by rainforest and monsoon forest. However, population growth (partly due to an influx of emigrants from the Indian main-

Draining away human effluent, an open sewer and storm drain in Peshawar, northern Pakistan. Water quality throughout this highly populated country is deteriorating, with inadequate sewage facilities being a major contributor.

land), agricultural expansion (particularly of rice) and logging have put enormous pressure on surviving areas. Bhutan in the Himalayas has retained 53 percent of its forest cover, but here the greatest threat, particularly in the south, comes from cultivation of the spice cardamom.

Price of a "Green Revolution"
In the mid 1960s the need to increase agricultural production in India, Pakistan and Bangladesh to feed the growing population led to the expansion of irrigated areas, and the introduction of new, high-yielding varieties of wheat and rice. This program of improvement launched

the so-called "Green Revolution". Food output in both Pakistan and India increased, but success was limited in Bangladesh, where the stems of the new rice variety proved too short for areas that were deeply flooded by seasonal rains.

A heavy price has been paid for the changeover from traditional farming techniques to intensive agriculture. In many areas, overirrigation has reduced the productivity of the land: India and Pakistan together have lost more than 17 million ha (42 million acres) of arable land because of waterlogging. Increased soil salinity from overirrigation and inadequate drainage has also degraded land. About one-third of Pakistan's irrigated land has been affected by salinization.

The indiscriminate use of chemical fertilizers and pesticides has brought further problems. Farmers spraying their fields

world. Sewage is perhaps the worst of the problems. Of India's 3,119 towns and cities, just eight are able to treat their sewage fully before discharging it into rivers and lakes. In Pakistan, only the modern capital Islamabad (founded in 1961) and Karachi on the coast have sewage-treatment plants (though the two plants in Karachi function only intermittently). The results are severe contamination of rivers and canals, which are frequently used for bathing, washing and drinking. For example, the river Ravi near Lahore in Pakistan sometimes flows with as much sewage as water.

Industrial effluent is also discharged directly into rivers and the sea. Along the coast of Pakistan the major pollutant is oil, which comes from Karachi's oil terminals and also from slicks that drift across the Arabian Sea from the Gulf. These are sometimes so bad that the country's fishing industry is virtually brought to a halt.

THE TRAGEDY OF BHOPAL

Just after midnight, on December 3 1984, a loud hissing sound was heard coming from the top of a tall smokestack at the American-owned Union Carbide pesticide plant in Bhopal, in northern India. The sound was caused by a leak of methyl isocyanate – a lethal chemical gas – that had been allowed to build up to very dangerous levels. Heavier than air, the cloud of gas sank from the smokestack and billowed out through the narrow streets of Bhopal's poorer districts. People awoke choking and vomiting, and often temporarily blinded. Many tried to flee in panic, but for others the gas proved fatal within minutes, leaving the streets strewn with bodies. By dawn, the chaos and carnage was so terrible that corpses were simply heaped onto funeral pyres and cremated. Over 3,000 people died, some 30,000 were seriously injured, and perhaps 300,000 were affected with relatively minor ailments.

The prime cause of the catastrophe was an inadequate management that was lax about safety precautions. In addition, a plant of this type should never have been built so close to human settlement. Legal wrangling over responsibility and compensation (originally $3 billion) has dragged on for years, and despite the "polluter pays" principle, few of the victims of Bhopal have received any compensation for their suffering. By 1991, seven years after the accident, the government had classified only 41 people as permanently disabled.

are often poisoned because they do not wear protective masks; and runoff from intensively used plots pollutes rivers and groundwater, from which drinking water is pumped directly in numerous villages. In addition, most agrochemicals have had to be imported, putting added strain on the region's developing economies.

Urban and water pollution
The combination of phenomenal urban growth – there are some 250 million urban dwellers in the region – and an almost complete absence of pollution control has led to some of the worst conditions of urban pollution and overcrowding in the

A "lucky" victim, just one of hundreds of thousands of people affected by the Bhopal disaster in December 1984, in which over 3,000 died. In all some 640,000 claims for compensation have been received so far by the Indian government.

DEFENDING THE LAND

All over the Indian subcontinent, local people – often in conjunction with non-governmental organizations – have taken initiatives to combat environmental degradation. One of the most dramatic demonstrations of "people power" took place in 1973, when villagers from the remote village of Gopeshwar in Uttar Pradesh, close to India's border with Tibet, confronted loggers who had come to cut timber for a sporting goods manufacturer. Determined to save the trees – their source of food, shelter, medicines and animal fodder – and unable to stop the loggers by persuasion, the villagers hit upon a strategy in keeping with the nonviolent philosophy of Mahatma Gandhi (1869–1948), India's revered pre-Independence leader. They decided to

put their arms around the trees and protect them with their own bodies.

This action sparked off what is possibly the best-known Third World grass roots campaign, the Chipko Movement (*chipko* in Hindi meaning "to hug"). In response, the Indian government reviewed its policy regarding forests. Today, there is no commercial logging of forests in Uttar Pradesh, and in the area where the Chipko Movement began, villagers have undertaken extensive reforestation programs on their own land. The movement's ideals have spread to other parts of India; there is even a plan to increase the extent of protected forests to cover 5 percent of the country.

Controlling the floods

One of the greatest challenges is the control of the river waters that annually inundate the floodplains. The problem is

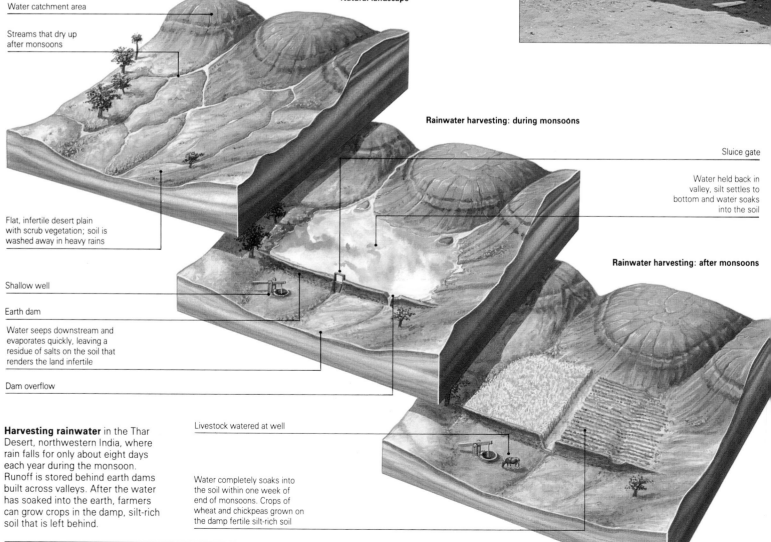

Natural landscape

Water catchment area

Streams that dry up after monsoons

Flat, infertile desert plain with scrub vegetation; soil is washed away in heavy rains

Shallow well

Earth dam

Water seeps downstream and evaporates quickly, leaving a residue of salts on the soil that renders the land infertile

Dam overflow

Rainwater harvesting: during monsoons

Sluice gate

Water held back in valley, silt settles to bottom and water soaks into the soil

Rainwater harvesting: after monsoons

Harvesting rainwater in the Thar Desert, northwestern India, where rain falls for only about eight days each year during the monsoon. Runoff is stored behind earth dams built across valleys. After the water has soaked into the earth, farmers can grow crops in the damp, silt-rich soil that is left behind.

Livestock watered at well

Water completely soaks into the soil within one week of end of monsoons. Crops of wheat and chickpeas grown on the damp fertile silt-rich soil

animal protein for most Bangladeshis.

Protecting the embankments from being breached would also be an impossible task. For example, in 1991 floods and cyclones tore down 430 km (260 mi) of earth embankments – built to protect coastal paddyfields from seawater incursion – drowning over 13,000 people and injuring half a million.

Conservation and pollution prevention

Although many of the techniques needed to solve the subcontinent's pollution problems are well known, the ability of developing countries to pay for such solutions is the major barrier to their implementation. However, efforts have been made to tackle the problems. Pakistan, for example, has embarked on a National Conservation Strategy, aided by the World Conservation Union (IUCN). The long process will ultimately provide a legislative and advisory framework to tackle the country's most pressing environmental needs.

Adequate drainage, the key to preventing salinization, has already been provided at Pakistan's historical site of Mohenjo-Daro. A ring of wells has been sunk around the site, and pumping is lowering the groundwater level to prevent salt water from penetrating and destroying the ancient bricks. Similar measures to reclaim salinized irrigation land have been put into effect with limited success in the north Indian states of Haryana, Punjab and Uttar Pradesh.

The dangers to health from overusing pesticides can be controlled by integrated pest management. This technique was used in India's northeastern Orissa state in the early 1970s, when the gall midge and other pests were severely damaging many rice crops. The strategy was three-pronged: early maturing, pest-resistant rice was introduced, enabling the harvest to be gathered before the pests were at their most prolific; pests were monitored to determine when their infestation was about to cause damage to the crops; and, perhaps most importantly, pesticides were not applied when the pests' natural enemies – parasites and animal predators – were most abundant.

SELF-HELP IN KERALA

In India the nongovernmental Council for the Advancement of Rural Technology has shown how local self-help schemes, using appropriate "intermediate technology", can help in the fight against environmental degradation. One such project was conducted in the Attapady Tribal Development Area, a 765 sq km (295 sq mi) zone in the southwestern state of Kerala that had been heavily deforested, over-grazed and over-cultivated by a rapidly growing population.

In order to reclaim the land, local people were urged to carry out soil conservation measures and reforestation. They reclaimed severely gullied slopes by digging pits, and building earth embankments and small filter dams. These reduce runoff, trap sediment normally carried away by the rains, and improve soil moisture in dry seasons. Indigenous trees have also been replanted, with a 70 percent survival rate. In addition, the Council has introduced new fuel efficient stoves to reduce the pressure of fuelwood collection. The success of the project can be attributed to two factors: the involvement of local people, and the guarantee that they will directly benefit from the improvements.

---

most severe in Bangladesh, where water brought down from the Himalayas by the Brahmaputra, Ganges, Meghna and some 250 smaller rivers is vital for growing rice: silt brought down by the rivers is crucial to the land's fertility. Heavy flooding, however, caused by monsoon rains in the mountains, brings widespread death and destruction. In 1988, Bangladesh suffered its worst-ever floods, when 80 percent of the country was inundated, killing at least 1,500 people by drowning and snakebite, and leaving 30 million homeless.

The disaster prompted several flood control studies, of which the costliest and most dramatic was put forward by a team of French engineers. Their plan was to construct, over the next 20 years, 4,000 km (2,500 mi) of embankments, averaging 4.5 m (15 ft) high, along the great rivers of the Bangladesh delta. The cost is estimated at $10 billion. Critics say the scheme relies too much on heavy engineering, while the technical difficulties would be enormous – the combined discharge of the Ganges and Brahmaputra is two-and-a-half times that of the Mississippi in the United States. Such embankments would also prevent fish, which breed in the rivers, from moving into the flooded paddyfields where they are caught for food; they provide the main source of

A Himalayan controversy

Once the last few decades, academics, foreign aid agencies and politicians have increasingly blamed deforestation of the Himalayas as the chief cause of the ever more frequent flooding of northern India and Bangladesh. According to their argument, the rapid growth of mountain-dwelling populations in India and Nepal since the 1950s has set in motion a chain reaction of environmental destruction.

Growing subsistence societies have accelerated the rate of forest clearance for fuelwood, and overgrazed their herds on the fragile mountain slopes. Once the protective vegetation cover that binds the soil has been removed, the downpours of the monsoon season quickly erode the soil, causing severe gullying and landslides. This in turn disrupts the hydrological cycle: more and more water runs down the mountains into the rivers, which flow down to flood the plains.

The rivers not only carry excess water, but also soil vital to the Himalayan people. Downstream on the plains, the massive increase in sediment clogs up irrigation channels and reservoirs used in hydroelectric generation, reducing their efficiency and useful lifespan. In the mountains the scarcity of firewood forces people to use dung as their main fuel, depriving terraced farmland of a vital fertilizer – another factor that weakens the soil and encourages erosion.

New research

This conventional theory about the cause of flooding has, however, been challenged by research being carried out with the support of several organizations in the Himalayan area, such as the United Nations University Project in Nepal and the International Centre for Integrated Mountain Development in Kathmandu (Nepal). Researchers have found little reliable data to support the idea that changes made to the land in the mountains directly affects the Ganges and Brahmaputra, and hence the plains through which they flow. They found even less evidence to back up the claims that the Himalayas have suffered from serious deforestation in recent decades.

In Nepal, for example, forests in the most densely populated area, the middle range of mountains, had been cleared by 1930 and converted to farmland; photographs claiming to show forest loss on some slopes have proved to be wrong; and it has been discovered that, because farmers look after their terraces, there are, ironically, fewer – not more – landslides in areas where the population is highest. So long as terraces are maintained, they in fact conserve the soil more effectively than forest cover.

While there is no doubting the appalling nature of the 1988 Bangladesh floods and similar catastrophes, it seems that the

Flooded out family (*above*) As here in 1989, the people of Bangladesh are accustomed to normal flooding after monsoon rains, but disastrous floods have increased in the past 40 years to about one every four years. The 1988 flood left 30 million homeless.

Cause or control? (*left*) One theory claims that increased clearance of the Himalayan forests to create terraced farmland, such as this in Nepal, has been the cause of the Bangladesh floods, but recent research indicates that well-managed terraces mitigate floods.

interplay between people and their environment may be much more complex than at first thought. Alternatively, the flooding may simply be due to exceptionally heavy rains on the plains themselves, rather than to land-use changes in the mountains. Or to the fact that the Himalayas, being a relatively young mountain chain, are readily eroded by torrential rains and landslides triggered by seismic activity. Identifying the real causes is of great importance, because if the wrong ones are targeted the remedial action will not necessarily work, and may even create as many problems as those it was intending to solve.

However, even more important than identifying the causes is the need to relieve the suffering of the people affected by the disasters. Until people living along the cyclone-swept coastline are provided with, among other things, raised homes, schools and roads, high-ground shelters, boats for escape or supplies and embankments to protect coasts, cities and farms, they will remain vulnerable to the unpredictable power of nature.

HOLDING NATURE IN CHECK

IMPRESSIONS IN THE SAND · PROBLEMS OF OVERCROWDING · ATTEMPTING TO TURN THE TIDE

China's huge and growing population lies at the heart of the region's environmental problems. In recent decades pressure on the land from farming has contributed to the southward spread of the Gobi desert. In addition, since the 1970s the country has experienced rapid urban and industrial growth, causing severe air and water pollution. The government has shown willingness to tackle these problems, but has had limited success. Attempts to curb population growth have had little impact in rural areas, where most of the population live, and efforts to prevent the desert spread with a "green wall" of trees have had mixed results. Concern for the environment has been offset by the desire for rapid economic growth, but until population growth is halted, the region's environmental problems can only grow worse.

IMPRESSIONS IN THE SAND

For centuries China has had a higher population density than most regions of the world, and it now supports more than 1 billion inhabitants. In order to provide such huge numbers with food, water, clothing and domestic goods, intense pressure has been put on the region's natural resources: the human impact on the landscape has been considerable.

Some 3,000 years ago the upper valley of the Huang river in the central north was the fertile heartland of China. Now it is largely desert, the result of climate change and inappropriate farming practices. Strong winds blew sand in from central Asia, causing the deserts (shamo) to extend southward, while overgrazing by sheep and goats on the desert fringe removed the vital protective vegetation cover. Extensive forest clearance for fuelwood over a thousand years ago further exposed the topsoil of these marginal lands, making them highly vulnerable to erosion and dust storms, which buried fertile soils and rendered them unusable. The Han agricultural area became gulleyed by rain, with topsoil washing into the Huang – or appropriately named Yellow – river.

From this heartland the Han population migrated east, settling most densely in the coastal region between the Huang and Chang rivers, and traveling south to

Choking pollution (*below*) A human traffic jam on the way to work in Baotou, in northern China, where coal-burning factories belch smoke and soot into the air, giving rise to a variety of health problems.

COUNTRIES IN THE REGION

China, Hong Kong*, Macao*, Taiwan*

POPULATION AND WEALTH

Population (millions)	1,139.1
Population increase (annual population growth rate, % 1960–90)	1.8
Energy use (gigajoules/person)	22
Real purchasing power (US$/person)	2,470

ENVIRONMENTAL INDICATORS

CO₂ emissions (million tonnes carbon/annum)	380
Municipal waste (kg/person/annum)	n/a
Nuclear waste (cumulative tonnes heavy metal)	n/a
Artificial fertilizer use (kg/ha/annum)	23
Automobiles (per 1,000 population)	0.9
Access to safe drinking water (% population)	72

MAJOR ENVIRONMENTAL PROBLEMS AND SOURCES

Air pollution: generally high, urban very high; acid rain prevalent; high greenhouse gas emissions
River/lake pollution: local/medium; *sources*: agricultural, sewage, soil erosion
Land pollution: local/medium; *sources*: industrial, agricultural
Land degradation: *types*: desertification, soil erosion, salinization, deforestation; *causes*: agriculture, industry, population pressure
Waste disposal problems: domestic; industrial
Resource problems: inadequate sanitation; land use competition; coastal flooding
Population problems: population explosion; urban overcrowding; inadequate health facilities

* Not included in figures

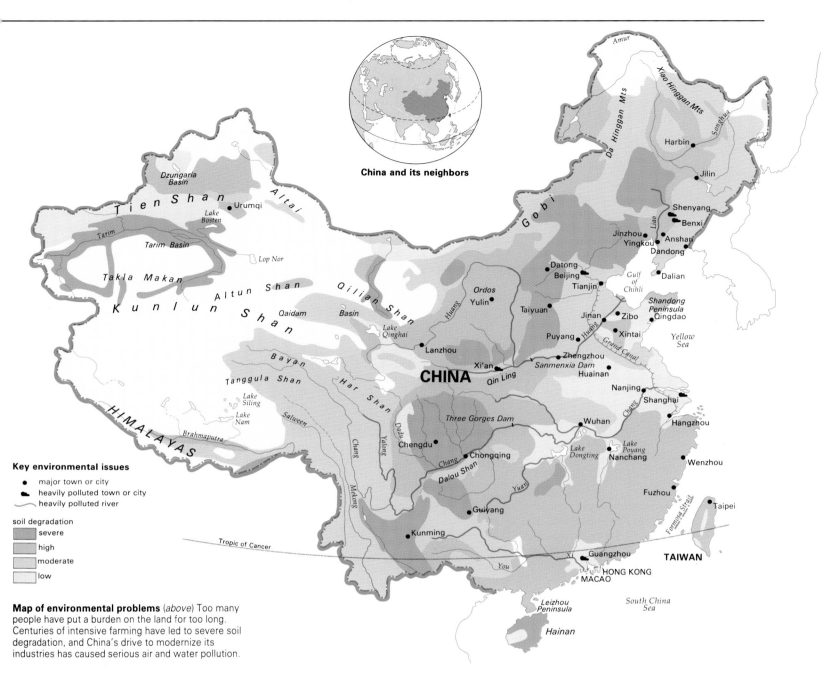

Key environmental issues

- ● major town or city
- ◢ heavily polluted town or city
- ～ heavily polluted river

soil degradation

- ▨ severe
- ▨ high
- ▨ moderate
- ▨ low

Map of environmental problems (*above*) Too many people have put a burden on the land for too long. Centuries of intensive farming have led to severe soil degradation, and China's drive to modernize its industries has caused serious air and water pollution.

---

the lush southeastern coast and tropical Hainan island. Increasing population pressure meant that all available land had to be exploited. Across the region woodland was cleared on a huge scale. (It is estimated that only 11 percent of the country is still forested.) Intricate terraces were cut into the steep hillslopes, and canals and dikes were constructed to create giant irrigation systems (one massive scheme, built in 230 BC at Dujiang on the border with Tibet in western Sichuan, is still largely intact). In some areas overirrigation led to salinization and waterlogging, greatly reducing the productivity of the land.

Modern changes

China entered a new era of environmental change in 1949, when the Chinese Communist Party gained power under the leadership of Mao Zedong (1893–1976). With economic growth a priority over the environment, the government encouraged people to view natural resources as free goods to be exploited without limit. Its target – to achieve national self-sufficiency in food and manufactured goods – put the environment under great pressure. On the Ordos plateau of northeastern China, for example, open-cast mining for coal caused widespread devastation, while in many rural areas remaining woodland was cleared to produce charcoal for iron-ore smelting.

In the early 1960s China suffered a devastating famine, which caused up to 30 million deaths. During the Cultural Revolution (1966–76) policies to collectivize agriculture by organizing family farms into large units were intensified with the aim of increasing the production of staple grains. Vulnerable land was cleared and grasslands plowed, allowing new cycles of erosion to set in, particularly on the arid plateaus of Xinjiang and

Tibet in the far west. Farming methods were centralized to an extreme degree, though local variations in soils, rainfall and climate were largely ignored, often with disastrous consequences.

Since Mao Zedong's death in 1976 local decision-making has been encouraged, easing some pressures on the environment. However, government support for rapid development has brought new problems, and urban sprawl encroaches onto some 8.5 million ha (21 million acres) of farmland every year. The arable area is under immense pressure across the country, with the average size of a family plot less than 0.083 ha (about one-fifth of an acre) – even smaller in the densely populated areas of eastern China. In addition, it is estimated that since the 1970s no less than 1,560 sq km (600 sq mi) of land has become degraded each year. By the year 2000 the country's deserts will have doubled in size since 1949.

PROBLEMS OF OVERCROWDING

With some 23 percent of the world's population to support, it is no wonder that the region has some serious pollution problems. In the mid 1980s, the capital, Beijing, with a population of about 9 million, was able to treat only about 7 percent of its daily sewage effluent. Throughout the region some sewage is used as fertilizer for farmland, but most is simply discharged into lakes, rivers and seas. Although urban authorities are well aware of the problems, an immense building program is needed to catch up with the decades of neglect.

Industrial effluent is also discharged untreated into many lakes, rivers and canals, contaminating them with high concentrations of heavy metals, oil-related wastes and chemicals that pose a serious threat to health. At least 2,400 km (1,500 mi) of China's rivers no longer support shrimp or fish, while it is estimated that the country's inshore fishing area has been reduced by one-third as a result of industrial pollution. A number of previously common marine animals have now all but disappeared from the Gulf of Chihli and the estuaries of the Huang and other rivers that flow into it.

Water demand is increasing rapidly. Groundwater reserves supplying Beijing and Tianjin have already been exhausted, and it is likely that two-thirds of the country's cities will face water shortages by the year 2000. It was hoped that the

A tide of dirty foam (*above*) Effluent from a paper mill at Chongqing, Sichuan province, is discharged directly into the Chang river. China's rivers are being poisoned by toxic industrial wastes, including heavy metals, which enter the food chain and endanger human health.

Coal makes the wheels go round (*right*) Old-fashioned coal-burning steam engines, which are still in production, are typical of China's continuing reliance on older, highly polluting technologies. Coal is also the chief commodity carried by Chinese freight trains.

Grand Canal in eastern China could be used to transfer drinking water to Beijing, but the southern end of the canal near Hangzhou is so badly polluted that all the fish have died. A massive clean-up operation would have to be implemented if the plan is ever to be carried out.

Smog and dust in the air

Air pollution is a growing problem. In Beijing and other major cities industrial fallout and boiler emissions combine with smoke from heating systems and domestic cooking to irritate the eyes and nose and cause respiratory problems. Many urban centers lie under a permanent haze of white smog. Sulfur dioxide concentrations in central Beijing are more than four times higher than World Health Organization guidelines. The city's air pollution is exacerbated by dust storms, especially in winter, when northerly winds bring sand and dust from the Gobi desert and the deteriorating grasslands of Inner Mongolia.

Benxi, near the North Korean border, is possibly the most polluted city in the world, with factories pumping out 21,300 tonnes of smoke and dust, and 87 million cu m (3,000 million cu ft) of polluting gases every year. In the late 1980s the city became so engulfed in its own smog that it vanished from satellite photographs.

Central to China's air pollution problems is the region's reliance on coal, which provides 75 percent of total energy. Of this, most is burned inefficiently, with up to 60 percent of its energy being lost. At the same time poor planning has meant that great quantities of coal have to be transported huge distances from the mines before being burned: in 1985 as much as one-third of China's extensive rail system was tied up with coal freight. Coal burning is largely responsible for the

A VANISHING WORLD

For centuries the grasslands of Inner Mongolia in the north of China were inhabited by nomadic herdsmen and their animals. The grasslands are arid and fragile, but the population was relatively small and the vast area offered sufficient pasture for their herds.

During the 20th century the situation radically altered. After 1949 the Communist government supported the migration of Han Chinese to the area from other, more crowded parts of the country; the population of Inner Mongolia is now 80 percent Han, with 60 percent living in rural areas. The new arrivals were encouraged by the state to practice familiar crop farming, in spite of the damage caused to the grasslands. Matters worsened during the Cultural Revolution when the agricultural practices of the "model village" were imposed, regardless of local conditions. Plowing was recommended in Inner Mongolia, despite its unsuitability. The resulting land degradation was severe, and soil erosion increased rapidly.

After 1976 these centralized policies were abandoned. However, the commercial incentives that replaced them brought new problems. Many Han took up sheep ranching for meat production. The Mongolians' herds and flocks have been concentrated into ever smaller areas of land, resulting in increased overgrazing. In the space of a few decades, and in the face of growing hostility between its Han Chinese and Mongolian populations, Inner Mongolia has acquired a legacy of fast deteriorating land. Some 86,000 sq km (33,200 sq mi) are now threatened with erosion, an area equal to all the land lost during the last 2,000 years.

region's very high levels of acid rain. Southern China, where the soils are naturally acidic and the warmer atmosphere converts greater quantities of sulfur dioxide into acid rain, is worst affected. Crop production has fallen as a result, and some authorities estimate that the damage in Guangdong province alone amounts to millions of dollars a year.

At present, China has very few cars. However, car ownership is increasing by as much as 20 percent a year, and new production plants are coming into operation to produce Chinese Volkswagens and Peugeots. The Chinese Environmental Protection Agency plans to test all vehicles twice a year to ensure they meet

exhaust emission standards. Nevertheless, in huge cities such as Beijing, Shanghai, Guangzhou and Chongqing, widespread car ownership will inevitably add to the major pollution problems. Hong Kong and Taiwan already have serious pollution resulting from vehicle emissions and rapid industrialization.

Deforestation's legacy

Deforestation is a key problem in terms of environmental management; it not only leads to soil erosion and the siltation of watercourses and irrigation channels, but it also increases the risk of flooding. Since the mid 1970s huge areas of forest have been cleared for commercial timber. In

hilly or mountainous areas villagers cut down trees to boost their low level of subsistence, unaware of the longterm consequences. Between 1958 and 1988 it is estimated that the volume of standing timber declined by at least 50 percent, even though the total forested area may not have changed much.

Estimates suggest that by the year 2000, timber demand from all users will exceed supply by about 50 percent. This would cause less concern if replenishment rates in the past had been higher, so that new areas were coming into production. Yet even in the early 1990s, in the face of increased demand for timber, replanting is still inadequate.

ATTEMPTING TO TURN THE TIDE

Since the end of the Cultural Revolution China's government has shown increased interest in tackling environmental problems. A series of new measures have included regulations to protect forestry and encourage energy efficiency. Fines have been imposed on factories that have exceeded legal levels of emissions and effluent discharges. However, the government has also been keen to promote rapid economic development. Whenever the two objectives clash – which they frequently do – economic growth is generally given priority.

The "Great Green Wall"

China has a long history of rulers embarking on projects of gargantuan size, and the era of environmental preservation is no exception. In 1978 the government began to deal with the threat of desertification by creating a "Great Green Wall". This extraordinary project, described as "the greatest ecological endeavor in the world", involved an immense program of tree planting to stem the spread of desert from the north. More than 9 million ha (22 million acres) have been planted with trees at various key sites along the edge of the desert, stretching all across northern China for a distance of some 7,000 km (4,350 mi).

In some areas the project has been very successful. At Yulin in Shaanxi province, for example, 720,000 ha (1.8 million acres) of forest were planted in three shelter belts designed to form a barrier against the wind, allowing dust and sand to settle. Straw and clay "checkerboards" embedded in the ground have also been used as a defense against the wind. This has enabled some 200,000 ha (500,200 acres) of farmland to be redeveloped, producing reasonable grain yields.

Despite local successes, however, the wisdom behind the Great Green Wall project is open to doubt. Critics have claimed the project is based on a mistaken idea of how desertification occurs. Much research has indicated that desert areas do not advance in a way that can be halted by a continuous linear barrier. Rather they expand over wide zones, often extending outward from isolated "blisters" beyond the main desert area.

The causes of the process also have to

be considered. Government research stations, such as at Yanchi in Ningxia Province, have come to regard overgrazing as the dominant factor in desertification, in which case a barrier of trees alone would have little effect. The Yanchi Station has substantial support from the Beijing government, and its suggestions are regarded as a blueprint for the future. The Station has promoted education programs for people living on the fringes of the desert, as well as offering financial incentives to herdsmen to keep smaller numbers of animals. Local populations have also been actively employed in efforts to try and reclaim land by leveling sand dunes and using new irrigation techniques. These new approaches may prove more effective, if less dramatic, than the Great Green Wall.

Widespread local action

Other action to protect the environment includes the building of small check dams in upland areas. These are placed across gullies to prevent valuable loess – light, dusty but rich soil – from being washed by heavy rain into rivers and reservoirs. There has also been a national campaign to encourage tree planting in all areas. The government has required everyone between the ages of 11 and 65 to plant 3 to 5 trees each year, and in 1992 it was claimed that for the first time tree planting in China exceeded logging. However,

it is doubtful that more than 30 to 40 percent of the saplings survive, particularly in arid and semiarid areas. During the 1980s the demand for timber and land clearance for farming was such that tree cutting exceeded planned levels by as much as five times.

Encouragement can be drawn, however, from the research of the Institute for the Control of the Loess Plateau in Mizhi in Shaanxi province, which has investigated planting apple and pear trees on terraced hillsides that are threatened with erosion. The fruit produced by these trees offers profit enough to persuade farmers to look after the trees and consequently a large number have survived.

During the 1980s the Chinese government tried to curb the pollution resulting from energy production. It planned to develop geothermal power, already used in Tibet to provide electricity for Lhasa. In 1984 a program to expand nuclear power, which produces low levels of carbon emissions, was also launched. By the late 1980s two nuclear power stations were under construction, near Shanghai and Hong Kong. There has been some public concern about the risk of accidents, especially in Hong Kong. Nevertheless, it is essential for China to reduce its dependence on coal if it is to avoid worsening problems of acid rain, and contributing to global increases of "greenhouse gases".

Mobilizing for the environment
(*left*) Mao Zedong's politically motivated "rustication" programs produced some practical benefits. Millions of urban workers, such as these cadres (political officers) building a break against the desert, spent weeks or years working on rural projects. Students and urban workers now spend one week each spring contributing to various projects such as planting trees.

Power on a small scale (*below*)
Water is diverted from a fast-flowing mountain stream and piped down a steep incline to turn the turbines that generate electricity for transmission to a rural community. Only small amounts of electricity are produced by these simple installations, but some 100,000 of them are in operation throughout China, generating about one-third of all the electricity used in rural areas.

VILLAGES LEAD THE WAY

Some of the most impressive developments in environmental action in China have been smallscale projects in the country's rural hinterland, rather than the immense, state-organized schemes. One of these is biogas, or local methane production, for use in cooking and lighting. Waste matter such as crop residue, pond mud and dung are placed in purpose-built chambers that have to be kept at a constant and fairly high temperature to produce methane. The process creates almost no pollutants and is particularly useful in areas far from conventional energy sources. China has several million biogas chambers, most of them located in warmer central and southern provinces.

Another important source of rural energy that creates little pollution is smallscale hydroelectricity (SHP). SHP is estimated to produce as much as one-third of rural electricity needs, offering vital energy for the pumping of water for irrigation schemes, as well as grain milling and machinery used in rural industries. There is considerable potential for more SHP schemes to be developed, particularly in the wetter southern half of the country, and for increasing domestic applications. As well as being nonpolluting, SHP and biogas also help to reduce pressure on vulnerable rural woodland, frequently the only other source of energy in remote areas.

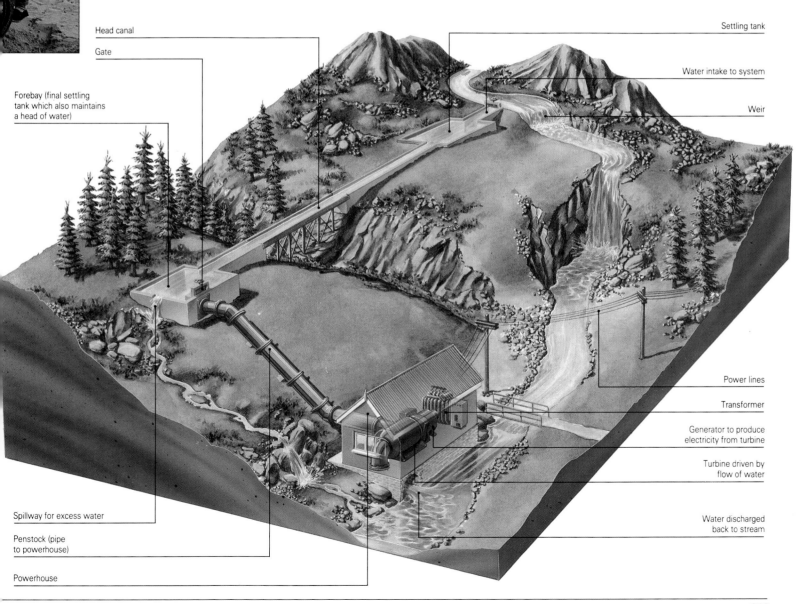

Head canal

Gate

Forebay (final settling tank which also maintains a head of water)

Settling tank

Water intake to system

Weir

Power lines

Transformer

Generator to produce electricity from turbine

Turbine driven by flow of water

Water discharged back to stream

Spillway for excess water

Penstock (pipe to powerhouse)

Powerhouse

Water control

Water control has been of vital importance in China ever since the Han first settled along the Huang river. Their success and expansion across the region was partly due to their ability to harness and control water for irrigation: mass forced labor, organized by the government, built complex irrigation systems and flood defenses. Water control has continued to preoccupy China's rulers ever since. Since 1949 the Communist leadership has tripled the amount of irrigated land – and hence agricultural output – by raising water from lakes and organizing gravity irrigation schemes fed from reservoirs.

Almost half China's arable land is now irrigated. Irrigation is essential for paddy rice, but also permits multiple cropping even in the dry season in the south. In the more arid north it is even more crucial, greatly increasing the yields of commercial crops such as cotton and soya, while also allowing rice to be grown in areas with little rainfall. In many areas, only a shortage of energy has prevented further irrigation schemes from being built. Electricity or diesel is vital for water pumping, but delivery to remote rural areas is generally a problem, and competition is often fierce among local groups vying for access to the supplies.

The hazards of overirrigation

Irrigation is not without its problems. Overwatering, or the irrigation of unsuitable land, can lead to waterlogging and salinization of the soil. Salinization occurs when dissolved salts are left behind in the soil as the irrigation water evaporates. If not flushed out these can build up to levels toxic to crops. Nearly 20 percent of China's arable land is affected by salinization, while in two arid northwestern provinces – Xinjiang and Ningxia – one-third of arable land has had to be abandoned since 1949.

Floods are a particular problem in China because of the high density of

"China's Sorrow" becomes China's power (*right*) The Huang river's frequent flooding has earned it the name "China's Sorrow." This dam in Qinghai province is the largest of several that have been built to control the Huang's waters and provide hydroelectricity.

Green at the desert's edge (*below*) An irrigation canal waters crops at Dunhuang, Gansu province, in northwestern China. Irrigation schemes have greatly increased the area of productive arable land, but overwatering can lead to longterm problems.

population in the hazardous flood plains, occupied in spite of the risks because of their fertile soils. For thousands of years there have been repeated flood disasters, particularly on the Huang and Chang rivers. In 1887 some 2.5 million people died when the Huang river burst its banks, and in 1991 floods along the Huai river – a tributary of the Huang river – drowned more than 2,000 people, as well as making a further million homeless. Despite intensive efforts to embank the Huang, much of the countryside of northern China is still at risk of moderate or severe floods.

In an effort to control flooding, and at the same time to create much needed hydroelectricity, the government decided to dam the region's major rivers. In 1960 a dam was constructed at Sanmenxia on the Huang river, but within four years the reservoir had filled with silt and the power station had to be taken out of action. Despite this, government enthusiasm for such schemes continues. A whole series of HEP projects – this time capable of dealing with silt – is planned for the Huang river. In addition, the 1991 flood disaster encouraged the government to give the go-ahead for what would be the largest HEP project in the world: the Three Gorges dam on the Chang river.

As well as preventing disastrous floods in the plains below the gorges, the massive 200 m (650 ft) high dam should be capable of producing one-sixth of China's total electricity output. However, the scheme is both highly controversial and, at $20 billion minimum, very expensive. It has met with considerable resistance from some ministries and environmentalists. The 600 km (370 mi) long reservoir would not only destroy one of China's most famous areas of natural beauty, the Chang gorges themselves, but could also increase the risks of flooding. Much of the water that reaches the Chang floodplain comes from tributaries below the planned location of the dam.

Lastly, vital food production for many rural homes would be lost. All the HEP projects involve the submersion of large areas of land, including some of the country's finest farmland, and the displacement of its workforce. The Sanmenxia dam required some 300,000 people to lose their homes. The Three Gorges project would do the same for as many as 1.7 million. The costs of water control may prove greater than the rewards.

RAPIDLY DIMINISHING ASSETS

THE ACCELERATING PACE OF CHANGE · THE ENVIRONMENT AT RISK · DEVISING STRATEGIES FOR THE FUTURE

The environment of Southeast Asia is under great pressure from population increase and rapid economic growth. Air and water quality have deteriorated as a result of urban expansion, industrialization and the intensive use of fertilizers to increase food production. Plantations of rubber and oil palm have replaced large areas of the indigenous rainforest; elsewhere it is rapidly being cleared for commercial logging and farming. Deforestation increases water runoff, leading to erosion, loss of soil nutrients and silted rivers, but so far schemes to control the scale of timber extraction and encourage replanting have met with only limited success. The forest and mangroves of Vietnam and neighboring countries in Indochina suffered widespread damage during the Vietnam war (1961–75) and later political upheavals.

THE ACCELERATING PACE OF CHANGE

The earliest, and greatest, changes to the environment of much of Southeast Asia came about with the domestication of rice between 9,000 and 5,000 years ago. Plains and river deltas were covered in complex wet-rice irrigation systems and mountain foothills were transformed by elaborate hillside terraces to grow rice. Only the region's extensive forested uplands and swampy, mangrove-fringed coasts were unaltered by farming.

The exploitation of the region's rich resources during the period of European colonialism brought further significant changes. The coastal deltas of the Irrawaddy in Burma (Myanmar) and the Mekong in Vietnam were extensively cleared and drained to increase the area of land that could be used for growing profitable cash crops. Plantations growing rubber trees, introduced from South America, and other commercial crops such as oil palm replaced large areas of the original forest in Malaysia, Indonesia and southern Vietnam. Mining also had a widespread impact on the environment. Chinese miners had first extracted tin from the river gravels of Perak and Selan-gor in western Malaysia by hand, but their efforts were intensified by the introduction of dredgers and other European mining techniques, which have left deep scars on the landscape.

Modern pressures
In recent times, population increase has put enormous strain on land and resources. The so-called "Green Revolution", which combined the planting of improved rice varieties with increased irrigation, fertilizers and pesticides, has had considerable success in raising food production to meet the growing demand – Indonesia, for example, has moved from being the world's largest rice importer to a position of self-sufficiency. However, it has also had considerable impact on the ecology of lowland farming areas and on traditional land-use patterns.

Many of the region's forested areas have been categorized as belonging to the state, thereby opening the way for governments to lease the timber rights to private companies. The indigenous forest peoples have been dispossessed and widespread ecological damage resulted from the scramble for wealth on this new frontier. Where largescale clearance has taken place for settlement, tenacious grasses (*Imperata cylindrica*) may become

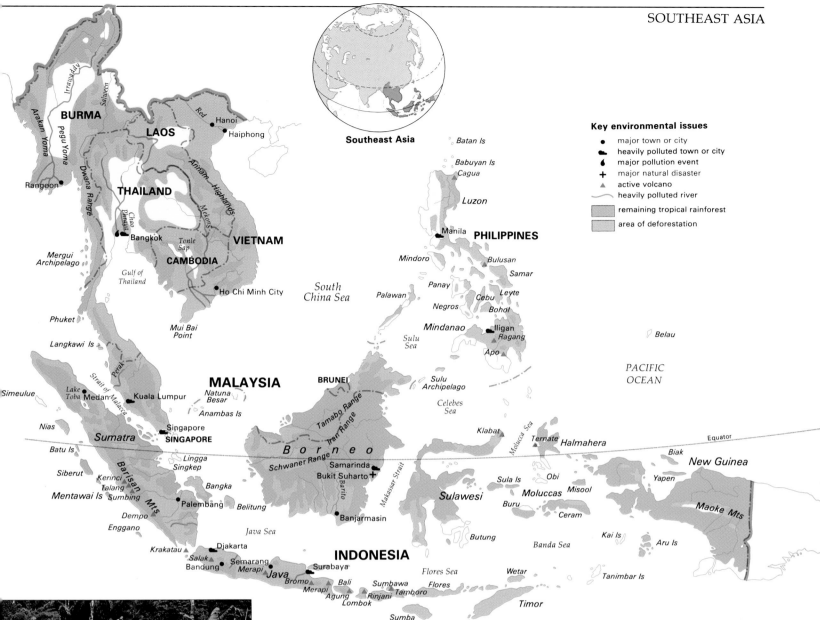

Key environmental issues

- ● major town or city
- ◣ heavily polluted town or city
- ◖ major pollution event
- ✚ major natural disaster
- ▲ active volcano
- 〰 heavily polluted river
- ▨ remaining tropical rainforest
- ▨ area of deforestation

Map of environmental problems (*above*) Southeast Asia's forests are being cleared at an alarming rate both for timber and to grow food for the region's rapidly expanding population. Erosion and soil degradation are the inevitable result.

Buried by mud (*left*) A village and its crops in Java have been struck by a landslide caused by a flash flood. Intensive logging increases the risks of landslides by leaving the soils on stripped slopes vulnerable to slippage after seasonal heavy rains.

established. These are hard to eradicate, making reforestation difficult. If soils are left bare after plowing, severe erosion is likely to follow. A pattern of droughts and flash floods – with high risk to human life – replaces the thorough absorption and slow release of rainfall in forests.

The roads cut by the loggers allow settlers to move in who often introduce farming methods and crops that are inappropriate for the fragile forest soils. Some of these new settlers are government sponsored. Indonesia has initiated several largescale resettlement schemes in the less populated islands to relieve population pressure in Java and Bali. Elsewhere, for example in Thailand and the Philippines, land hunger has fueled the migration of the rural poor into previously forested upland areas.

Rapid industrialization and growth of urban populations have brought serious problems of air and water pollution to the cities. More recently, developing industries have begun to move into the countryside to tap the large pool of unemployed rural labor, taking their pollution with them. In southern Thailand and on the west coast of peninsular Malaysia agricultural land is being turned into housing estates for workers; and the roads are constantly choked with trucks transporting industrial raw materials and finished goods. Coastal mangrove swamps – breeding grounds for fish and other aquatic life – are at risk from urban pollution, as well as from tourism and commercial prawn farming.

THE ENVIRONMENT AT RISK

International environmentalists have identified the destruction of Southeast Asia's rainforests as a major contributor to global warming, caused by the buildup of carbon dioxide in the atmosphere. Tracts of forest act as a "sink" for carbon dioxide, which is lost through largescale clearance; in addition, the burning that often accompanies clearing adds to carbon emissions, as does the breakdown of organic matter in the soil after clearance. Methane, produced by decomposing vegetation in wetlands and paddy fields and as a byproduct of digestion by ruminant animals such as cattle, is also a significant "greenhouse gas" – but is a less emotive issue than deforestation.

Although countries in the region are sensitive to their environmental image, they nevertheless defend their right to exploit their forests. However, logging companies interested in quick profits violate many of the rules drawn up for the proper management of forest and soil resources, and many ignore their obligation to replant. Logging increases the intensity of forest fires by leaving flammable detritus on the forest floor, and roads cut through the forest funnel the flames. A conflagration that destroyed 4.2 million ha (10 million acres) on the island of Borneo in 1982–83 spread in this way.

The threat to the rice supply

It is increasingly recognized that the success of the Green Revolution in boosting grain yields in the region's rice-growing lowlands has had unforeseen ecological consequences. A few new

Smokey Mountain (*right*) – a giant garbage heap in Manila – provides a living for hundreds of people. Regardless of the threats to health, many live in shacks built into the heap, which is continuously smoking due to the buildup of gases.

Deforestation of watersheds (*below*) Forest clearance has a far-reaching impact on river drainage systems. The forest acts as a sponge – once removed, less rain soaks into the ground and surface runoff increases, washing soils into the rivers.

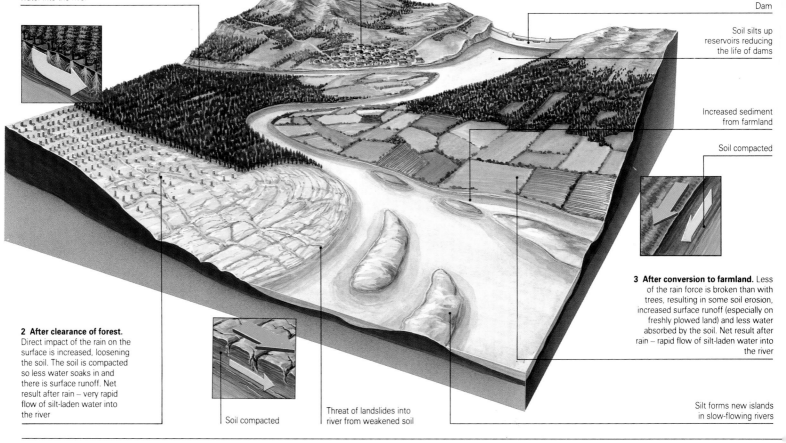

1 Natural forest. The force of the rain is broken by the trees, so there is little surface erosion and runoff. The water first soaks into the soil, which acts as a sink, and then gradually flows downslope to the river, carrying little silt with it. Net result after rain – gradual flow of clean water into the river

4 After urbanization. Rain flows from urban surfaces and through sewers to river carrying with it urban pollutants. Net result after rain – very rapid flow of pollutant laden water to the river

Dam

Soil silts up reservoirs reducing the life of dams

Increased sediment from farmland

Soil compacted

3 After conversion to farmland. Less of the rain force is broken than with trees, resulting in some soil erosion, increased surface runoff (especially on freshly plowed land) and less water absorbed by the soil. Net result after rain – rapid flow of silt-laden water into the river

2 After clearance of forest. Direct impact of the rain on the surface is increased, loosening the soil. The soil is compacted so less water soaks in and there is surface runoff. Net result after rain – very rapid flow of silt-laden water into the river

Soil compacted

Threat of landslides into river from weakened soil

Silt forms new islands in slow-flowing rivers

were made to find other measures of controlling the problem. The most successful method has been the use of a hormone that prevents the planthopper developing to sexual maturity.

Water and air pollution

Agricultural chemicals drain into water courses – one of many sources of pollution that affect water quality throughout the region. The Chao Phraya river in Thailand, which flows through the capital city of Bangkok, graphically illustrates some of the difficulties of providing clean water. Bangkok, with a population of nearly 6 million, has no sewerage system. Both sewage and domestic garbage are dumped directly in the river, along with industrial effluents. Bangkok contains at least 50,000 factories, too many for the city's hardpressed officials to monitor for failure to observe the regulations for treating waste water.

The Chao Phraya has consequently become an open sewer. The main offtake for Bangkok's water supply, 90 km (56 mi) from the mouth, is heavily polluted, with levels of health-threatening bacteria well above the limit set for drinking water, necessitating the addition of large quantities of chlorine to make it safe. The situation in the lower reaches is still worse. With an oxygen level of zero, the fish and shrimps that were still common in the estuarine waters only 10 years ago have entirely disappeared.

This pattern is repeated throughout the region. In Indonesia, the water supplies of Djakarta and Surabaya on Java are badly polluted with industrial waste, and are likely to deteriorate still further unless strong corrective measures are taken. Industrial development in the overcrowded cities of Southeast Asia exposes their inhabitants to many other serious environmental hazards. A series of chemical explosions and fires in the port area of Bangkok in 1991 gutted hundreds of squatter homes and left many people suffering from respiratory complaints and skin diseases.

The rapid growth in the number of vehicles is raising atmospheric pollution in the cities to dangerous levels. The region generally experiences low wind conditions, so pollutants are not readily dispersed. The Singapore government has attempted to cut traffic levels by imposing heavy taxes on private cars and by constructing a subway system.

high-cropping rice varieties have displaced thousands of local strains. Traditionally, farmers rotated the varieties they sowed on their plots, thereby reducing susceptibility to attack from disease and pests. Reliance on the new varieties, and the increasing practice of growing the same crop year after year, has heightened the need for nonspecific insecticides and pesticides. There has been a significant loss of genetic diversity in the region's staple food supply.

Some of the consequences have been worrying. Within seven years of the new varieties being introduced into Java, Indonesia's main rice-growing area, rice crops were devastated by a once minor pest, the brown planthopper. It was eventually realized that spraying the rice plants with insecticides destroyed the insect predators of the planthopper while it was safely maturing inside the rice stalk. A ban on many types of insecticide was imposed in Indonesia as attempts

CHEMICAL WARFARE IN VIETNAM

During the 1961–75 war in Vietnam, the United States army used high explosives and chemical defoliants (which strip plants of their leaves) to try to defeat the guerrillas in the south by destroying their croplands and the forests that gave them cover. "Carpet bombing" turned the paddy fields into flooded craters. Wide areas of dense evergreen dipterocarp forests and coastal mangroves were sprayed with the herbicide Agent Orange, which contains dioxin – a highly toxic compound. An estimated 1.7 million ha (4.2 million acres) were affected. In the coastal mangroves entire plant communities were wiped out. In the forests, trees dropped their leaves, allowing so much light to penetrate the forest floor that

bamboos and tenacious grasses took hold. Huge "Rome plows" with blades measuring 3 m (10 ft) wide were used to plow up the soil to prevent the forest from regenerating.

Many forested areas are still extremely dangerous because of unexploded bombs and other ammunition, and much of southern Vietnam remains a wasteland, despite some replanting of coastal mangroves. In a few places attempts have been made to restore the evergreen dipterocarp forest by planting exotic fast-growing species that provide shade and shelter for the young trees. However, high levels of dioxin have entered the food chain to affect wildlife and humans, where it is responsible for fetal deformities.

DEVISING STRATEGIES FOR THE FUTURE

Southeast Asian governments are under mounting pressure to safeguard the long-term future of their natural resources and environment. At the same time, they need to develop their economic and industrial base to support their growing populations. Development targets often take priority over conservation goals.

Southeast Asia's most urgent environmental requirement is to halt the present high rate of deforestation and restore the balance of forest loss. In countries such as Thailand, Vietnam and the Philippines, which have lost up to 80 percent of their tree cover, replanting is the only viable way to restore the forests. However, Indonesia and eastern Malaysia (Sarawak and Sabah on Borneo) are "forest rich" – despite the high rate of felling, the size of their indigenous forests is such that two-thirds of their land area remains forested. Theoretically, they should be able to rely on natural regeneration to stabilize their forest resources, but natural regeneration is slow and difficult to achieve.

Commercial planting – pros and cons

The wood-processing industries are central to the economies of Indonesia (the

Regrowing a flood barrier (*above*) Mangrove trees are being replanted under a new program in Java. Mangroves thrive in tidal waters and nurture a variety of plant and marine life. The roots accumulate soil and provide a crucial barrier in flood-prone coastal regions.

world's largest exporter of tropical plywood), Malaysia, Thailand and the Philippines. Thailand is already having to import timber, as does peninsular Malaysia; the Philippines may shortly have to do so. The communist countries of Vietnam and Cambodia make heavy use of timber reserves for fuelwood. In every country in the region, therefore, there is a powerful need to maintain high levels of timber supply. Most reforestation schemes consist of largescale planting of introduced species such as *Acacia mangium* or, in drier areas, *Eucalyptus camaldulensis*, which are fast-growing and higher yielding than teak and other varieties of indigenous hardwoods.

It is argued that these plantations take pressure off the remaining areas of natural forest. But the introduced trees deprive the already poor forest soils of nutrients, and their susceptibility to pests and diseases means that their longterm sustainability is far from certain. While they provide dense cover, concentration on a single species means that the complex ecosystems of the natural forest are being destroyed.

Plantation schemes also have considerable social costs. In 1985, the government of Thailand set itself the target of restoring tree cover to 40 percent of the country. This was to be achieved through an extensive program of eucalyptus planting, known as the "Green Isan" project. Left out of the account were the interests of the local forest communities, and subsequent conflicts between villagers and the plantation companies have been frequent. The plantations encroached on communal forests, and many villagers were displaced from land earmarked for planting. Large numbers settled illegally on other degraded areas of land within the forest reserves (as many as 1.2 million families, it was estimated in 1990). Others have added their numbers to the stream of migrants that flock to the cities, particularly Bangkok.

A more positive approach has been adopted in the Philippines, where local farmers in an area surrounding a pulp processing mill have been given 10 ha (25 acres) of land. On 2 ha (5 acres) they grow food crops, and on the rest they raise quick-growing pulpwood trees. They are ensured an income through supplying the pulp to the mill and comparative security of tenure, so the scheme is a popular one.

Pineapple and coconut (*left*) thrive in each other's company in the Philippines. The trees stabilize the soil in which the profitable crops are grown. Mixed land use is an increasingly common solution to balancing economic and ecological priorities.

Action for change

Even when environmental legislation exists, it is not always implemented. Environment boards or ministries in Indonesia, Thailand and Malaysia have limited powers, and are underfunded and understaffed; fines for noncompliance with regulations are low. Sometimes it takes an environmental disaster to bring about action. In 1988 a flash flood caused mud and cut logs to slide down a mountain in southern Thailand, killing 350 villagers. Illegal logging was blamed for the disaster, and the government called for a total ban on all logging. Nevertheless, evidence suggests smallscale illegal cutting is still widespread.

The growing strength of public opinion, however, may compel governments in the future to take firmer action to enforce legislation. Although most pressure groups are urban-based with an overwhelmingly middle-class, student membership, they are increasingly learning to work with rural communities to fight local issues. In 1989, for example, the Indonesian environmental group WAHLI combined with local farmers and fishermen in northern Sumatra to bring a case – which they ultimately lost – against a pulp firm and several government agencies (including the ministry responsible for imposing tougher penalties on polluters) on the grounds that effluent from a pulp mill was polluting the Asahan river and damaging their livelihoods.

COMBINING FARMING AND FORESTRY

Rather than displace local people from areas of reforested land, increasing use is being made in a number of Southeast Asian countries of schemes that allow farmers to practice a form of shifting cultivation within the plantations. This enables them to support themselves and their families while reforestation is taking place. One such is the *taungya* system, originally developed in Burma. Farmers raise annual field crops on plots between the newly planted trees until the canopy closes. They then move to a new area within the managed forest. In this way, fragile forest soils do not become degraded by overuse, and rural communities are not forced out.

Such schemes have long been practiced on the teak plantations of Java, where they are known as *tumpangsari*. One disadvantage with them was that the farmers did not feel sufficiently involved in running the schemes,

which provided no security of tenure. The Indonesian State Forest Corporation has recently been persuaded to introduce modifications to the scheme. Rows of trees are now spaced more widely to create permanent plots for crops; fruit trees as well as rice and corn may be planted. The people have a much greater stake in the entire operation, and the plantation trees are not attacked, as used to happen.

Social forestry schemes have been adopted in upland areas of the Philippines to rehabilitate land that has become severely eroded after formerly forested slopes were invaded by landless farmers. Here 25–year "stewardship" leases give farmers full control of their plots of land, on which they grow trees interspersed with crops. Secure tenure is critical, as tree cultivation requires a longterm commitment, and these schemes provide it.

The rainforest controversy

Southeast Asia's tropical rainforests are being felled at an ever-increasing rate. Between 1976 and 1980, 1.5 million ha (3.7 million acres) of forest were being cut down every year; by 1986–90 this rate had more than doubled to 3.4 million ha (8.4 million acres) a year. This attack will continue as long as conflicting economic interests tussle for control over forest and land resources. In these circumstances, sustainable management schemes have little chance of success.

The governments in the region generally divide the forests into three separate categories for management. Protected areas are set aside for the purposes of species conservation and watershed protection. Timber harvesting is carried out within productive areas of the forest. Finally, converted areas of forest land are set aside for farming, resettlement or industrial development. The most difficult problem is to prevent migrant settlers crossing the unmarked boundaries into the permanent forested areas. Only in peninsular Malaysia – where most of the rural population have been absorbed into urban industries or settled in agricultural schemes – is this avoided.

Timber harvesting should not result in deforestation. Most governments have established selective logging policies to ensure that only mature trees and certain varieties are taken. Seedlings and saplings must be replanted to allow the forest to regenerate. However, logging concessions are commonly let out to subcontractors who ignore the environmental protection laws to extract quick profits. There are too few forestry officials to police distant concessions, and very often the offenders have moved on before they can be caught; sometimes officials are bribed to turn a blind eye to violations.

The past practice in Burma and Thailand of using manual labor and elephants to harvest the forest caused little disturbance. But today the heavy machinery used may damage up to half the remaining trees, so the forest will not have regenerated when the next harvest is due. Elephants were still used in the teak forests of Burma until very recently, but since 1989 the government has been selling logging concessions to Thai firms no longer able to operate in their own country because of the ban on logging. Their heavy machinery has devastated forests in the borderlands of Burma, and they ignore the traditional obligation to replant.

Deep in the rainforest (*right*), a truck removes illegally cut trees. Patrolling for smallscale operations is difficult in remote areas where the forest is still abundant. Low-lying forests have been decimated and loggers are moving to higher altitudes.

"Don't build your home with ours" (*below*) The Penan people of Sarawak's forest have become passionate campaigners to stop largescale logging, which threatens their livelihoods. Control of resources is the central issue.

Both Malaysia and Indonesia have provided sponsorship for urgently needed largescale resettlement schemes on forest land. Such programs alleviate rural poverty and (in the case of Indonesia) also relieve population pressure in Java and Bali, which are both very densely populated. However, they have also attracted international criticism. In the past, farmers were encouraged to grow food crops, particularly rice, that were unsuited to the forest soils. Recently there has been a switch to tree-crop schemes, which are ecologically sounder but much more expensive to establish.

Clash of cultures

These resettlement schemes are also criticized for their negative impact on the forest peoples. Over hundreds of years, forest-dwellers in Southeast Asia have evolved systems for farming the forest and extracting its products without upsetting the ecological balance. In parts of West Kalimantan (the southern half of Borneo, belonging to Indonesia), centuries-old agroforestry systems have established managed forests of useful trees whose produce is gathered: tengkawang nuts and fruits such as durian and rambutan. Elsewhere in Kalimantan thorny rattan vines, used to make furniture and mats, are planted on cleared forest plots, to be harvested seven years or more later from the forest regrowth.

Yet the indigenous peoples' knowledge of the forest's ecology is ignored by government departments responsible for forest management. Officials frequently attempt to attribute blame to indigenous people for the destruction of the forests. The Indonesian government, for example, is implacably opposed to shifting cultivation, practiced by more than 250,000 families over 5.4 million ha (13 million acres) of rainforest in Kalimantan, claiming that the normal pattern of cutting and burning to create a small patch for farming is a leading cause of deforestation.

Sarawak has been the scene of particularly violent confrontations between indigenous peoples and loggers. In the most famous of these, a small group of hunter–gatherers, the Penan, who claimed the loggers were damaging their wild food and water supplies, blocked the paths of timber industry bulldozers. Their longterm chances of success seem slim, however, since many prominent Malaysian politicians are owners of extensive timber concessions in Sarawak.

COUNTING THE COSTS OF SUCCESS

ENLARGING THE LAND · THE PRICE TO THE ENVIRONMENT · CHANGING PRIORITIES

In all three countries of the region low-lying land, suitable for settlement, accounts for less than a quarter of the total land area. Human pressure on the environment is consequently very acute in these areas. During Japan and South Korea's spectacular postwar economic booms few restrictions were placed on industrial and urban development, leading to serious air and water pollution. This had damaging effects on human health and left a legacy of environmental degradation that will take a long time to clear up. Today, though a number of anti-pollution measures show that environmental awareness is generally increasing, the growing leisure industry is causing new problems in the shape of golf courses and hotel developments, both of which destroy natural habitats and endanger wildlife.

Deadly spray (*above*) A farm worker sprays pesticide onto rice plants. Using high intensity farming methods, Japan produces a rice surplus on its limited area of farmland, but at the cost of heavy chemical pollution.

COUNTRIES IN THE REGION

Japan, North Korea, South Korea

POPULATION AND WEALTH

	Japan	N Korea	S Korea
Population (millions)	123.5	21.8	42.8
Population increase (annual population growth rate, % 1960–90)	0.9	2.4	1.8
Energy use (gigajoules/person)	110	79	52
Real purchasing power (US$/person)	13,650	n/a	5,680

ENVIRONMENTAL INDICATORS

	Japan	N Korea	S Korea
CO₂ emissions (million tonnes carbon/annum)	220	20	29
Municipal waste (kg/person/annum)	344	n/a	396
Nuclear waste (cumulative tonnes heavy metal)	5,600	0	700
Artificial fertilizer use (kg/ha/annum)	433	312	422
Automobiles (per 1,000 population)	239	n/a	19
Access to safe drinking water (% population)	94	n/a	78

MAJOR ENVIRONMENTAL PROBLEMS AND SOURCES

Air pollution: locally high, urban high; acid rain prevalent; high greenhouse gas emissions
River/lake pollution: medium; *sources*: industrial, agricultural, sewage
Marine/coastal pollution: medium; *sources*: industrial, agricultural, sewage, oil
Land pollution: local; *sources*: industrial, urban/household
Population problems: urban overcrowding
Major events: Hiroshima (1945), destroyed by atom bomb; Nagasaki (1945), destroyed by atom bomb; Minamata (1940s–60s), leakage of mercury from industry; Taegu (1991), chemicals dumped in river

ENLARGING THE LAND

In both Japan and Korea the centuries-long effort to maximize the use of limited areas of low-lying land, and to extend them by reclaiming coastal and inland wetlands, continues to have a major environmental impact. The first encroachments on Japan's wetlands were made as early as the 9th century. Originally the impetus came from the need to increase the area of land used for farming, particularly rice production. Over time, marshes, ponds and lakes, as well as coastal lagoons and bays, were drained, so that today there are few lowland areas that have not been enlarged in this way.

These changes to the natural environment, however, were minimal in comparison with the largescale land reclamation schemes that have taken place since the 1950s as rapid urban and industrial growth has demanded more and more space for development. Nowhere have the pressures been greater than along the densely populated Pacific coastal belt stretching westward from Tokyo. Here the demand for land has been met by draining tidal areas to create polders; these have significantly altered the shape of the shoreline. Large artificial islands

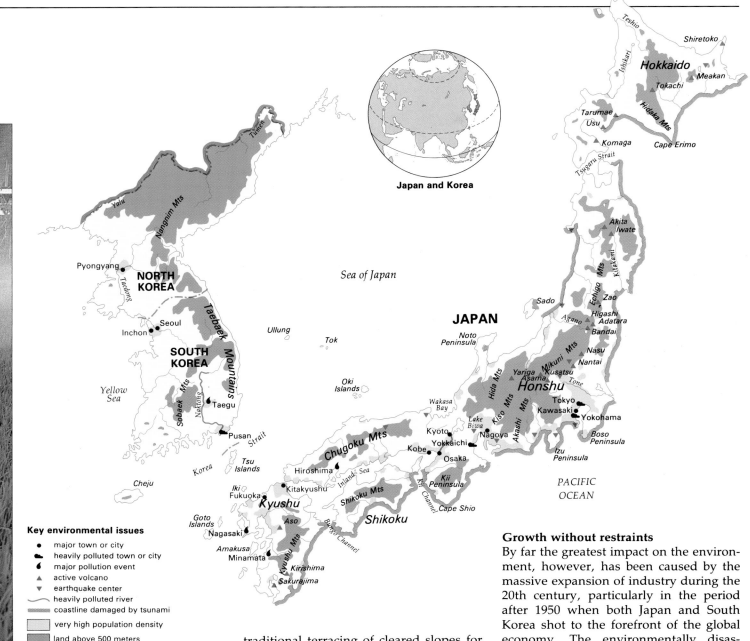

Japan and Korea

Sea of Japan

JAPAN

NORTH KOREA

SOUTH KOREA

Yellow Sea

PACIFIC OCEAN

Key environmental issues

- ● major town or city
- ◖ heavily polluted town or city
- ⬩ major pollution event
- ▲ active volcano
- ▽ earthquake center
- ⌇ heavily polluted river
- ▬ coastline damaged by tsunami
- ▢ very high population density
- ▢ land above 500 meters

Map of environmental problems (*above*)
With low-lying land for development at a premium, the areas of dense population suffer greatly from severe problems of industrial and urban pollution.

have been built for development by filling in shallow lagoons with quarried rock and other materials.

Similar largescale schemes have been undertaken in North Korea as part of the "nature remaking program" in force since 1976. Under this national plan, the reclamation of some 300,000 ha (741,000 acres) of tidal land has been targeted. To facilitate this project, an 8 km (5 mi) barrage, claimed to be the longest in the world, has been constructed across the Taedong river estuary.

A second strategy to increase the amount of land available for settlement and economic use has been to clear and terrace the forested upland slopes immediately adjacent to the lowlands. The traditional terracing of cleared slopes for paddy fields caused little environmental damage, but more recently extensive deforestation has taken place to create large sites for urban development. Tree clearance on this scale deprives the thin soils of protection and binding. This, in combination with the region's torrential rainfall, makes landslides a common occurrence: in 1990, for example, there were over 3,000 landslides in Japan, causing widespread damage to bridges, roads and railroad lines.

Outside the towns and cities, farmland – already in short supply – is being increasingly encroached upon by urban development. Farmers therefore turn to intensive agricultural methods as the only means of increasing production. The heavy and sometimes indiscriminate use of artificial fertilizers has led to the contamination of water courses by a number of chemicals, notably nitrates and phosphates. Arsenic from pesticides is known to have entered the food chain.

Growth without restraints

By far the greatest impact on the environment, however, has been caused by the massive expansion of industry during the 20th century, particularly in the period after 1950 when both Japan and South Korea shot to the forefront of the global economy. The environmentally disastrous effects of such rapid growth were aggravated by the concentration of so much of the offending development in both countries' narrow lowland areas.

The worst degradation occurred between the mid 1950s and the mid 1970s when virtually no controls were placed on industrial and urban growth. Air and water quality declined rapidly, especially in Japan. Untreated sewage and toxic industrial effluents – including polychlorinated biphenyls (PCBs) and heavy metals such as cadium and mercury – were discharged into the sea, rivers and lakes. Oil-burning industries were responsible for heavy emissions of sulfur and nitrogen dioxides into the atmosphere, while solid fossil fuels – used to heat most homes and now largely replaced by electricity and natural gas – also added to air pollution. As affluence increased more people acquired automobiles, adding to air and noise pollution, particularly in the congested cities.

THE PRICE TO THE ENVIRONMENT

The major environmental problems that the countries of the region face stem largely from the pressures created by rapid, uncontrolled industrialization and urbanization. That industrial pollutants directly affect human health is now established beyond a doubt, but it was the outbreak of two particular diseases in Japan – "Yokkaichi asthma" and "Minamata disease" – that helped to establish the connection between them in the 1950s and 1960s. The former was caused by the emission of air pollutants from a petrochemical plant in Mie prefecture on Honshu island; the latter by the discharge of effluent water containing mercury from two chemical factories in Kumamoto prefecture, Kyushu. Although these illnesses were initially diagnosed among people living in specific locations, they were subsequently found to be occurring more widely throughout the region.

With the shortage of land for building, urban areas are extremely densely populated. This has caused a variety of environmental problems, referred to by the Japanese as *kogai* (public harms). For example, vast amounts of sewage were for too long discharged largely untreated into the sea, lakes and rivers. In the relatively closed waters of Tokyo Bay and the Inland Sea, this led to a phenomenon known as "red tides", caused by the vigorous growth of a brownish red algae fed by nitrogen and phosphorus in the sewage.

Economic success has brought the advent of the throw-away consumer society, which creates ever-greater quantities of domestic garbage. The scarcity of land means that finding sites for tipping is a constant problem, and some cities are tempted to dump the waste in coastal waters – with terrible consequences for aquatic life. The sheer weight of the urban built-up area has caused land subsidence, which in coastal areas increases the likelihood of serious flooding. This problem is likely to get worse as the sea level rises due to global warming.

The costs of land reclamation

While land reclamation programs have been necessary to extend the available area and relieve pressure on crowded city space, it is now clear that the process has considerable environmental costs. The nature of the material used to create the polders is critical – in some cases seepage of toxic chemicals from domestic and industrial waste used as infilling has resulted in serious pollution of coastal waters and groundwater, contaminating the food chain in both land and marine ecosystems. Without proper precautions, deposited organic waste can lead to the

Altering the coastline (*below*) The Tokyo megalopolis sprawls over large areas of reclaimed land around Tokyo Bay. The drainage of wetlands and filling of lagoons started in the 19th century, but has accelerated in the last 40 years, destroying coastal habitats.

THE MINAMATA TRAGEDY

No single incident of environmental abuse had more profound consequences – both in its damaging effects, and in the role it played in stirring public awareness of the dangers of toxic wastes – than the Minamata tragedy. The roots of the disaster were laid in the 1940s when a factory manufacturing vinyl started to discharge mercury compounds into Minamata Bay, on the west coast of Kyushu. The first sign that something was amiss was the death of large numbers of fish; then seabirds and cats began to behave erratically. Before long local people were complaining of a range of illnesses – deafness, tremors and general weakness – and children were born with severe malformations of their limbs.

It was not until the 1960s that the link was made between what had come to be known as "Minamata disease" and the factory waste. The marine life in the bay was absorbing mercury, and the concentration of this toxic substance was increasing as it moved up through the food chain. As Japanese people are accustomed to eating large amounts of seafood, it was not long before high levels of mercury were being passed on to humans – with dire consequences. The companies responsible were eventually forced to pay damages to the victims, and the widespread publicity that the case attracted drew attention to the real human and environmental costs of Japan's eagerness for economic growth and personal affluence.

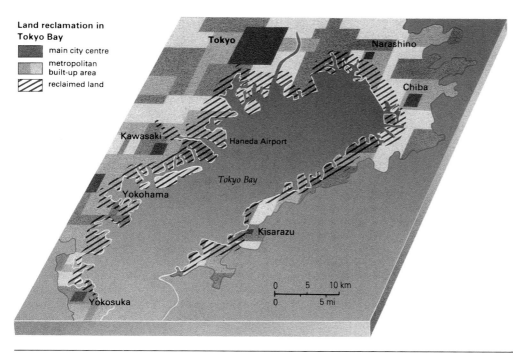

Land reclamation in **Tokyo Bay**
- ■ main city centre
- ▨ metropolitan built-up area
- ▨ reclaimed land

Tokyo
Narashino
Chiba
Kawasaki
Haneda Airport
Yokohama
Tokyo Bay
Kisarazu
Yokosuka

0   5   10 km
0   5 mi

Victims of Minamata (*above*) Some have been waiting 34 years for compensation.

A rubble playgound (*above*) A child squats in the rubble of a village in South Korea that was demolished to make way for a new car plant. In the race for economic prosperity, environmental considerations have often been overlooked.

buildup and then release of dangerous amounts of methane gas.

Earthquakes are frequent in Japan, and during the more vigorous tremors the ability of reclaimed land to support large structures – such as industrial plants and apartment blocks – is suddenly weakened. There have been incidents around Tokyo Bay, for example, in which groundfloor apartments have become basements in a matter of a few moments as earthquake vibrations have caused buildings to subside into the reclaimed land. These low-lying areas are also vulnerable to tsunamis – the large tidal waves generated by earthquakes or whipped up by the summer typhoons.

Wildlife under threat

As the remaining wetlands on the Korean peninsula and the Japanese archipelago are drained, many species of birds are deprived of breeding grounds, and vital resting places for the vast numbers of birds that migrate each year along the western Pacific "flyway" are destroyed. On Okinawa, the largest of the southern Ryukyu islands, the creation of a site for a new airport by in-filling a lagoon with rock quarried from a nearby hill eliminated one of Japan's few remaining coral reefs. Many others have been destroyed by tourist development – the islands are a favorite destination for vacationers from all over Japan. The clearance of coastal sites to build hotels or to create farmland silts up the shallow lagoons and kills the corals. The problem is exacerbated by the extraction of coral sands to meet the needs of the booming construction industry.

Although the forested mountain areas of the region have escaped most of the environmental ills that have ravaged lowland habitats, they are coming under increasing pressure. Korean forests have been badly affected by acid rain produced by industry in China and carried over by the prevailing winds. All over Japan, forested slopes have been cleared to make golf courses, which are a serious environmental hazard: the clearing of slopes increases the incidence of landslides, while the chemical fertilizers applied to the fairways and greens enter and pollute watercourses.

One clear barometer of a nation's environmental health is the state of its wildlife. Today one-fifth of Japan's endemic vertebrate species are in danger of extinction, their habitats threatened by human activity. Some 20 have already been lost for ever.

233

CHANGING PRIORITIES

All three countries in the region have been late starters to environmental management and protection. For a long time, the fairly small but vociferous environmental lobby in Japan was regarded with official disapproval – its members were perceived to be acting against the national interest by calling for constraints on industry. It was not until 1971 that a government environmental agency was set up to oversee the use and management of the environment. Born from mounting public concern, it marked the beginning of official commitment to environmental issues. Since then Japan has made considerable progress in clearing up the legacy of problems left by uncontrolled industrial expansion.

Cleaning up the environment

The public outcry over incidents such as the Minamata tragedy led to much more stringent controls over toxic emissions from Japanese factories. Cleaner – and cheaper – energy fuels were sought, and action was also taken to clean up the environment by reducing levels of atmospheric pollution caused by automobiles. Official encouragement was given to developing "lean-burn" engines, which burn fuel more efficiently and produce cleaner exhaust emissions. In 1973 the use of catalytic converters was made compulsory in all car engines, resulting in greatly reduced levels of nitrogen oxides in the atmosphere. However, after an initial sharp drop in atmospheric pollutants as a result of these measures, levels had begun to rise again by the 1990s as the number of automobiles on Japan's roads continued to increase.

Environmental awareness did not advance so rapidly in South Korea, however: here conservation and public safety still took second place to industrial and economic growth. But in 1991 at Taegu in the southeast of South Korea a factory producing printed circuit boards dumped 300 tonnes of phenol, which causes cancer and damage to the nervous system, in the Naktong river – the source of drinking water for more than 10 million people. Public outrage triggered a reaction that may prove instrumental in changing attitudes. Local government officials were arrested for trying to cover up the incident and the corporation responsible was

CHEAPER, CLEANER ENERGY

Until the early 1970s imported fossil fuels (principally coal and oil) provided more than 90 percent of Japan's energy. However, the sharp rise in oil prices during the 1970s prompted the government to rethink its energy strategies, and the subsequent search for alternative fuels, together with the introduction of energy-saving measures, have done much to reduce pollution.

Nonpolluting nuclear reactors now supply more than a quarter of Japan's electricity. However, concerns have been raised about safety: in 1991 Japan experienced the first minor accident in its nuclear program, and the threat of earthquakes means that reactors have to be carefully sited and designed. Most

power stations are now fueled by liquefied natural gas (LNG). This has the advantage of producing fewer pollutants than either coal or oil while giving more heat.

The most direct way of reducing pollution caused by energy production, however, is simply to use less of it. The steel and chemical industries have been most successful in introducing energy-saving technologies – in steelmaking the amount of energy needed to make 1 tonne of steel has been reduced by a tenth, and new chemical production techniques have cut consumption by as much as a quarter. The use of energy-saving robots in the automobile industry has achieved similar reductions.

Recycling precious waste (*above*) Bales of sorted waste metal await transportation to a factory for reprocessing. A technology has been developed for pressing 1 tonne of garbage into a 0.8 cu m (1 cu yd) block that can be coated with concrete and used in construction.

Working up steam (*left*) Geothermal energy is used to produce domestic hot water on northern Kyushu, the most southerly of Japan's four main islands. Hydroelectric power also offers a clean alternative to expensive and highly polluting fossil fuels, which must be imported.

Global responsibilities

The annual reports of Japan's environment agency have often been more concerned with discussing global issues than drawing attention to the more stubborn of its own domestic problems, such as noise and air pollution by motor vehicles. The Japanese government's recognition that all nations share the same global environment and have equal responsibilities toward it is a significant sign of changing attitudes, though the region still has a poor record of international environmental exploitation.

A great deal of the deforestation in Southeast Asia, for example, is driven by Japan's and South Korea's huge and growing demand for tropical hardwoods. Japan's hunger for shellfish has caused many Asian mangrove swamps to be cleared for shrimp farming. As factories in Japan and South Korea are improved to meet new domestic environmental standards, the old polluting manufacturing plants are often sold to poorer countries. Nevertheless, Japan's belated accession in the early 1990s to international agreements to curb whaling and drift-net fishing was an indication of a growing commitment to global conservation goals, as was its stated readiness at the Rio de Janeiro Earth Summit in 1992 to increase its aid budget for international pollution prevention and habitat protection.

forced to pay compensation of $30 million. Other factories charged with river pollution have since been closed down.

South Korea is taking steps to preserve its forest resources, though it should be noted that as the world's largest exporter of hardwood plywood it relies on imports of logs from the rapidly diminishing forests of Southeast Asia. It is proposing to help Chinese industry reduce the sulfur emissions that the prevailing winds carry across the Yellow Sea, causing the acid rain that has seriously damaged South Korean forests and agricultural land. In addition, in the early 1970s the country launched a massive village-based program to increase supplies of fuelwood. More than 300,000 ha (740,000 acres) of new stands of woods were established

and existing areas were managed to make better use of their resources. Efforts were made to improve the efficiency of traditional wood-burning stoves, and the government-sponsored Forest Research Institute developed an improved version of the traditional underfloor heating system that uses 30 percent less fuelwood.

Little is known of the level of environmental awareness and protection that exists in North Korea as the country is virtually closed off from the rest of the world. It is likely, however, that North Korea's industries are still using older – and dirtier – production methods typical of isolated communist regimes; moreover, the country's dictatorial government is unlikely to tolerate dissent on environmental issues.

Air quality in Tokyo

The population of the vast metropolitan area that stretches around Tokyo Bay has more than doubled in size since 1945. Tokyo itself has nearly 12 million people; the figure rises to more than 32 million if all the neighboring towns and cities are included. By the late 1960s Tokyo's air quality had declined dangerously; levels of the major air pollutants – sulfur, carbon and nitrogen oxides, as well as suspended particles – were among the worst in Japan, and well over the limits set down by the government as acceptable.

The major sources of pollutants were the area's high concentration of heavy industry, as well as the burning of fossil fuels to heat the homes and workplaces of the huge population. The rapid growth of automobile ownership also contributed greatly to the problem: by the late 1960s more than 1 million vehicles were choking the city's already severely congested road system.

The worst outcome of these high levels of pollution was the frequent occurrence of photochemical smog – the result of complex photochemical reactions involving the ultraviolet rays present in sunlight and the nitrogen oxides and hydrocarbons released into the atmosphere by factories and automobiles. This smog irritates the mucus membranes of the eyes and respiratory tract. It is also harmful to plants. When sulfur dioxide is included in these photochemical reactions, the result is sulfuric acid mist – one of the contributors to acid rain.

The incidence of photochemical smog over Tokyo peaked in the early 1970s. At this time it was common for people to wear face masks out of doors, and oxygen booths were set up in public places to assist people with respiratory difficulties. Up to 28,000 people in one year reported to the metropolitan government that they had suffered health trouble as a result of exposure to air pollution. By 1992 the number of such complaints was well under a hundred, and face masks and oxygen booths are a thing of the past – the antipollution measures that had been implemented for some two decades had proved effective in reducing the levels of pollutants by two-thirds.

Monitoring and control

The approach adopted by the city's administration was to set total permissible volumes for each of the atmospheric pollutants, and then to regulate individual pollution sources to maintain total levels within these limits. Factories were subject to very close monitoring. Heating and air-conditioning systems of buildings in densely built-up areas were found to give rise to high concentrations of pollutants, and so area-based heat-regulation schemes were introduced. Because these use modern equipment they are more energy efficient and less polluting. Most of the schemes are run on

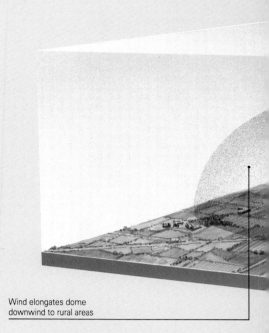

Wind elongates dome downwind to rural areas

Solar radiation from the Sun

Cold air

Cool air

Warm air

Air and pollutants rise

Normal situation
The sun heats the ground which in turn heats the surrounding air. The hot air rises into the atmosphere, carrying the pollutants with it.

Formation of thermal inversion (*above*) Pollutants are trapped in a layer of cool air that cannot rise. Surrounding hills reinforce the trap. Inversion either occurs when warm air from a high pressure system flows over cooler air – this can take days or weeks to clear – or at night when the air near the ground cools more rapidly than air above it. This normally disperses at sunrise.

Thermal inversion
Cool air is trapped below a layer of warm air, preventing the polluted air from rising.

Cold air

Warm inversion layer

Surrounding hills increase problem by trapping air

Cool air

Air pollutants trapped

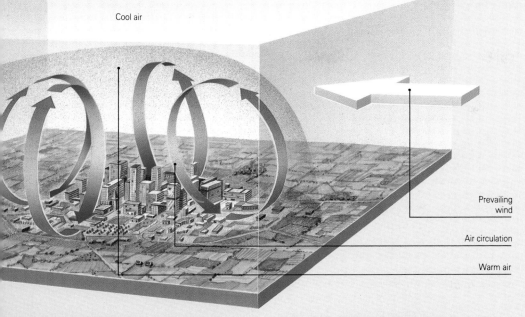

Cool air

Prevailing wind

Air circulation

Warm air

Urban heat islands and dust domes (*above*) Large cities tend to have a higher temperature than surrounding areas. This dome of heat traps suspended and gaseous pollutants, increasing urban air pollution by up to 1,000 times. Wind can stretch the dome, creating a regional heat island that extends into adjacent areas.

Health protection (*below*) During the 1970s and 1980s, face masks protected pedestrians when Tokyo's air quality was at its worst. Now they are occasionally used to prevent the spread of colds. Another device – coin-operated oxygen machines – has survived as a quick cure for hangovers.

"clean" energy such as electricity, but some exploit local energy sources such as heat from garbage incinerators.

Regulations on automobile exhaust emissions have been progressively tightened. However, levels of nitrogen oxides remain high and are creeping upward. The main causes are the ever-growing number of motor vehicles on the roads, larger engine size and an increasing proportion of diesel vehicles. Vehicles spend hours with their engines idling in traffic jams. Space to build expressways and relieve congestion is limited.

Controlling exhaust emissions is only one way to reduce the atmospheric pollution caused by rising car use. A range of measures to persuade commuters to abandon their cars and turn instead to the city's public transportation systems – among the most efficient in the world – includes the severe rationing of parking space and the imposition of high charges.

IN FRAGILE ISOLATION

Isolation kept Australasia and Oceania free from major environmental change until 1788. Since then, waves of migration, fast population growth and entry into the global economy have brought ever increasing problems to the region. In Australia and New Zealand native plants and animals declined and wilderness dwindled. Inappropriate farming methods have degraded soils, caused salinization and destroyed natural vegetation. In New Guinea and across the Pacific islands the rich tropical forests are shrinking. Radioactive contamination from nuclear tests threatens some of the islanders. People do not accept these circumstances without complaint. In Australia and New Zealand, the lobby of concerned citizens has gained the power to resolve issues, block damaging projects and attract international support.

COUNTRIES IN THE REGION

Australia, Fiji, *Kiribati, *Nauru, New Zealand, Papua New Guinea, Solomon Islands, *Tonga, *Tuvalu, *Vanuatu, *Western Samoa

POPULATION AND WEALTH

	Highest	Middle	Lowest
Population (millions)	16.9 (Australia)	3.4 (N Zealand)	0.7 (Fiji)
Population increase (annual population growth rate, % 1960–90)	2.4 (PNG)	1.7 (Australia)	1.2 (N Zealand)
Energy use (gigajoules/person)	201 (Australia)	11 (Fiji)	7 (Solomon I)
Real purchasing power (US$/person)	14,530 (Australia)	3,900 (Fiji)	1,960 (PNG)

ENVIRONMENTAL INDICATORS

CO₂ emissions (million tonnes carbon/annum)	63 (Australia)	1.5 (PNG)	0.2 (Fiji)
Deforestation ('000s ha/annum 1980s)	679 (Australia)	n/a	n/a
Artificial fertilizer use (kg/ha/annum)	709 (N Zealand)	89 (Fiji)	29 (Australia)
Automobiles (per 1,000 population)	424 (Australia)	44 (Fiji)	4 (PNG)
Access to safe drinking water (% population)	90 (Australia)	75 (Solomon I)	34 (PNG)

MAJOR ENVIRONMENTAL PROBLEMS AND SOURCES

Air pollution: urban high
Marine/coastal pollution: medium; *sources:* industrial, agricultural, sewage
Land degradation: *types:* desertification, soil erosion, salinization, deforestation, coastal degradation, habitat destruction; *causes:* agriculture, industry, population pressure
Resource problems: inadequate drinking water; inadequate sanitation; coastal flooding

* Figures for these small island states not included

NEWCOMERS AND NEW THREATS

The Aboriginal inhabitants of Australia lived a nomadic life of hunting and gathering over territories limited by the availability of freshwater. There was no cultivation or grazing, and only small-scale changes were made to the natural habitat, such as burning off the scrub vegetation to prevent major conflagrations. The peoples of New Guinea and the Pacific islands practiced smallscale farming – they cultivated yams and taro, and raised pigs and chickens – but the needs of their communities did not place great pressure on the environment. The Maoris, who migrated from Polynesia to New Zealand some 1,250 years ago, cultivated sweet potatoes and fished: because of their isolation there were no native mammals on the islands.

The impact of Europeans
Into this isolated and relatively tranquil paradise came the first European settlers some 200 years ago. They introduced crops, animals, pests, weeds, diseases, new methods of land use and advanced technology. The environmental impact was enormous.

In Australia, farms and ranches spread rapidly through the eastern third of the country, altering natural landscapes that had never been touched. Extensive irrigation projects turned the dry habitats along the Murray and other rivers into orchards and pastures. Much of the forested land in the east was cleared, some of it to supply timber, and industrial exploitation of the continent's mineral resources began. Changes were greatest in the coastal areas where the bulk of the settlers were concentrated.

Cattle and sheep farming were quickly established as the mainstay of Australian farming. However, the native grasses were fragile and could not withstand the trampling of hard-hoofed animals. Sustained grazing simply wiped them out. But grasses introduced to take their place were poorly adapted to the continent's hot, dry summers. They could provide only partial land cover, exposing the soil to erosion from wind and rain. Introduced rabbits multiplied extremely rapidly in the wild. They competed with livestock for food, and depleted the grass cover still further. In Queensland, the

grazing potential of pasture land was reduced by half between 1895 and 1970.

Introduced plants also wreaked havoc on the environment. Originally introduced as a hedge plant from the Americas, the prickly pear cactus outcompeted the native vegetation. Vast stretches of open plains and water holes in the tropical north were invaded and turned into impenetrable thickets by the giant plant. By 1920, the prickly pear cactus had become established across 25 million ha (65 million acres) of light woodland and pastures. Not until scientists found and introduced its natural predator, the Argentinian moth *Cactoblastis cactorum*, was the prickly pear brought under control. No effective biological control has been found for other introduced ornamental plants that now pose a threat.

New Zealand has suffered similar problems. The greater part of its subtropical

and temperate beech forests have been cleared, and a landscape very similar in many places to that of Britain was created by the introduction of alien species. European grasses displaced the native tussock grassland on the uplands, while Red deer, introduced for sport, competed for grazing with native birds and reptiles. In the drive to develop the country for farming, little regard was taken for the native vegetation or wildlife, or for the territorial rights of the Maori people.

Vulnerable islands

The limited land area of the Pacific islands makes them particularly vulnerable to environmental disturbance. Competition and predation by introduced plants and animals have wiped out the tiny, isolated populations of a number of endemic species. Increased human populations on the islands now add to the pressure on natural habitats by demanding ever more space for farms, roads and settlements. Only fragments of many of the old forests now remain.

Papua New Guinea's mountains and impenetrable forests were one reason why it remained largely unexplored and underdeveloped until 1945. It still has huge areas of wilderness. Environmental damage there has not been so severe, but high population growth, increasing deforestation, soil erosion, and the growth of tourism make it a land trembling on the edge of change.

Smoke from a bush fire (*above*) hangs over farmland in Victoria, Australia. Prolonged drought makes fire a constant hazard. In 1983, winds whipped fires from Adelaide to Melbourne, destroying seven towns and several thousand homes.

Map of environmental problems (*right*) Large areas of Australia's semiarid grasslands are threatened with desertification – a consequence of their conversion to grazing land, which has stripped them of cover. Deforestation threatens New Zealand and the tiny Pacific islands, including the rainforests of Papua New Guinea.

Key environmental issues

- ●   major town or city
- ☢   former nuclear test site
- heavily polluted river
- area of salt scald
- remaining tropical rainforest
- area of deforestation

areas at risk of desertification
- very high
- high
- moderate
- true desert

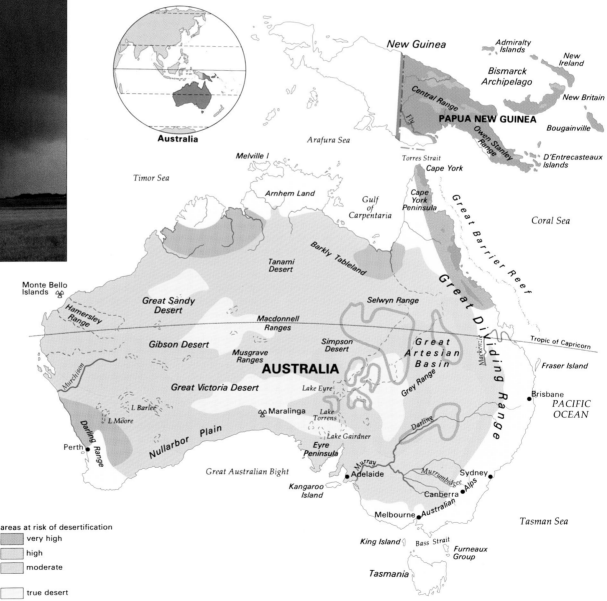

THE MISUSE OF LAND

Much of Australia is arid, with variable rainfall and sparse vegetation. Soils are vulnerable to erosion by wind and rain, especially if the land is used too intensively. After the 1970s falling prices for agricultural products encouraged wheat and livestock farmers to expand cultivation and increase livestock in marginal areas in order to recoup losses. Salinization and soil erosion have increased, and in the early 1980s dust storms of whipped-up topsoil choked several major cities, including Melbourne. A major inquiry into land degradation in 1984 found that well over half the areas used for cropping and pasture needed some form of soil conservation.

Salinization owes much to the naturally high salt content of many soils, together with high rates of surface evaporation in the hot, dry conditions. But the clearance of natural vegetation for cultivation and grazing has made the problem much worse. Before the natural vegetation was cleared, the deep roots of native trees and shrubs tapped enough moisture from the soil to keep water tables low. However, the new crops and grasses that have taken their place have much shallower roots, and as a result water tables have risen, bringing deep, salty water near the surface. The highly saline water from the raised water tables now feeds numerous springs used to irrigate the land; some 800,000 ha (1.2 million acres) of prime agricultural land have been adversely affected. In addition, excessive grazing over some 3.7 million ha (9.3 million acres) in low rainfall areas has exposed saline soils and created salt crusts or "scalds" on the surface.

Just as serious has been the heavy salinization and consequent loss of 156,000 ha (385,000 acres) of intensively irrigated land along major rivers, such as

the Murray, which runs along the boundary of Victoria and New South Wales; the Glenelg river in southwestern Victoria; and the Blackwood river in Western Australia. As water drains from the irrigated areas back into the rivers, salinity increases downstream. The salt content of the Murray river has become so high that South Australia has threatened legal action against the upstream states of Victoria and New South Wales if they introduce new irrigation projects. Adelaide and other cities in South Australia rely on the Murray for 50 percent of their domestic water.

Conflicting priorities

The increasing affluence of Australians, together with greater mobility and increased leisure time, has put ever greater pressure on the scenic areas surrounding large cities: for example, the Dandenong

NUCLEAR TESTING – A DEADLY EXERCISE

Three of the world's nuclear powers – France, the United States and Britain – have conducted over 250 nuclear weapons tests in the Australian desert and the Pacific islands since 1946. Although atmospheric testing was outlawed in 1963, France continued to test devices underground on their uninhabited Pacific island territories.

The tests not only caused immediate devastation, but also longterm radioactive contamination of land, air and water in the surrounding areas. By the early 1990s, an area of some 375 sq km (145 sq mi) at Maralinga in the South Australian desert – used for British tests in the 1950s – was still too contaminated for human habitation. With levels of radiation up to 300 times the accepted limit, the Aboriginal groups who lived there before the testing have been unable to return. The cost of cleaning up the soil could amount to US $500 million. Bird colonies on the US-owned islands of Bikini Atoll and Eniwetok Atoll in the Marshall Islands, where testing was also carried out in the 1950s and 1960s, suffered massive breeding failure after contamination, and islanders absorbed dangerous levels of radiation through the food chain.

Jacques Cousteau, the famous French marine explorer, visited the Mururoa Atoll, still being used for French tests, in 1987. He found extensive damage and deep fissures in the coral and surrounding sea bed. Some of the water samples taken in the area showed contamination by radioactive isotopes, which meant that the underground testing chambers were leaking into the ocean. Despite such evidence, together with the efforts of international pressure groups such as Greenpeace and the decision of regional governments to declare the Pacific a nuclear-free zone, the French government remains committed to testing in the Pacific region.

Radioactive debris (*left*) discarded after British nuclear tests litters Christmas Island.

Killed by salt (*above*) Dead trees, encrusted with white salt, stand in Lake Yoting, Western Australia. Throughout the arid country, sudden heavy rainfalls flood parched land; the water evaporates rapidly under the hot sun, leaving a residue of salt.

Range east of Melbourne and the Blue Mountains west of Sydney. These areas have become battlegrounds for the conflicting interests of conservationists and developers. On one side is the demand for greater protection of natural parks and nature reserves; on the other, the creation of more recreational amenities, the spread of freeways, and quarrying in prominent sites.

Other development projects that have sparked off disputes with environmentalists include logging and the associated woodchip industries in the Daintree tropical rainforest, Queensland, uranium mining in Kakadu National Park, Northern Territory, and scrub clearance for

farming in the semiarid Little Desert, Victoria. A hydroelectric scheme based on the King, Gordon and Franklin rivers that threatened a unique wetland habitat in southwestern Tasmania became a *cause célèbre* in the 1980s. Construction sites were blockaded by protesters, and the issue eventually brought down the state government in 1982.

New Zealand also has its share of fiercely contested environmental issues, many of them concerning hydroelectric schemes in national parks. Conservationists also protest against the planting of fast-growing introduced *radiata* pine in commercial plantations instead of the native kauri trees. As in Australia, the land rights of the indigenous people loom large over all environmental issues. The Treaty of Waitangi of 1840 between the British and the Maori is used as a firm legal basis for negotiating rights and the

sharing of resources, such as fisheries and land. Unlike in Australia, the declining agricultural market has led to the contraction of pasture land, which is reverting to natural vegetation. Farmers are concentrating on more profitable, land-intensive operations such as fruit farming.

Island deforestation

In Papua New Guinea and across much of Oceania, high rates of deforestation are having devastating environmental consequences. Clear felling for timber and slash-and-burn for farming expose the fragile forest soils to heavy tropical rainstorms, leading to severe erosion and the leaching away of nutrients. In mountainous areas, the erosion of topsoil from cleared slopes leads to siltation in rivers downstream, as happened on the island of Bougainville, PNG, when forests were felled to allow for copper mining.

THE ENVIRONMENTAL BATTLE

An estimated 300,000 to 400,000 people belong to conservation groups in Australia – three times the total membership of the political parties. This unusually high number is sustained by nationally emotive issues such as the preservation of native animals and plants, nuclear testing, rainforest conservation and controversy over damaging developments such as the Franklin river hydroelectric scheme. Political parties ignore the environmental vote at their cost.

On certain development projects, public opposition in Australia has gathered enough weight to influence and even halt plans. In Adelaide and Melbourne environmentalist pressure groups have forced the abandonment of freeway development to link up the outer suburban sprawl with the center. Continuing public opposition has repeatedly prevented the Australian authorities from building a high-temperature incinerator for hazardous chemical waste. Environmentalists claim that such incinerators produce toxic residues and emit harmful toxic gases.

Changing attitudes in government have been revealed by recent reversals of policy. One of the most celebrated concerned mining proposals for Kakadu National Park. The mining company Broken Hill Proprietary had originally been given rights to mine gold, platinum, palladium and other valuable minerals in a 2,250 sq km (870 sq mi) section of the park (the bulk of the park is already protected from mining). But in 1989, after opposition from conservation groups and Aboriginal rights groups, the concession was reduced by the Australian government to just 37 sq km (14 sq mi), with no mining to begin until an official investigation into the environmental consequences had taken place.

Fighting wind, salt and fire

Soil erosion can be ameliorated by promoting farming practices that put the needs of the environment before economic gain. Farmers have been urged to abandon areas highly vulnerable to erosion, and to practice the rotation of crops with fallow periods. This allows the land to build up moisture and also provides the soil with protective vegetation cover. Other methods of soil conservation include tree planting to prevent wind erosion, plowing along the contours of the land rather than up-and-down slopes to reduce water erosion, and the restoration of natural vegetation. However, farmers have been reluctant to adopt new practices without financial compensation for production losses.

Soil salinity can also be tackled effectively, though never entirely eliminated. Along the Murray river, a proposal has been made to build a pipeline that will carry the saline water discharged from irrigated fields directly to the ocean, thereby reducing the salinity of the river. The amount of water discharged could be lessened if the water being applied to the fields was regulated to take into consideration the local soil types, and adjusted minutely to daily and seasonal weather conditions. At present, most irrigators flood the land every few days regardless of location or season. At present, about half the water being used need not be applied.

Bushfires are another hazard in dryland Australia: every year they destroy houses, livestock, crops and human life. In 1983, for example, uncontrolled bushfires in one area killed 71 people and 300 livestock. To try and prevent such incidents, strict regulations over the use of barbecues and incinerators are brought into force on a daily basis in the dry season. Although these have helped to reduce the frequency of bushfires, the risk of further fires increases as city suburbs continue to expand into the bush.

Concern in New Zealand and Oceania

In New Zealand environmental awareness is equally strong, particularly over issues concerning the protection of the country's national parks and reserves. For example, plans to develop hydroelectric

power that would have altered Lake Manapouri in the Fiordland National Park of South Island were effectively scotched by a petition of 300,000 names to parliament – an impressive achievement for a country of only 2.5 million people. Wider environmental campaigns that have gained popularity in the country include strong opposition to nuclear testing in the Pacific.

In Oceania, environmental concern has only just begun to make itself heard as a public issue. In 1985, the South Pacific Nature Conservation and Protected Areas Conference approved goals and guidelines for the islands, but so far there has been little effective action. In 1990, a conservation strategy was established for the bird life of the South Pacific, which put forward a number of management guidelines for the many threatened bird species across the islands.

Cleaner commuting (*above*) A monorail system serves the 3.5 million residents of Sydney – an attempt to alleviate pollution and congestion caused by heavy commuter traffic. The monorail is clean, quiet and takes up less space than a freeway.

Protected laboratory (*below*) A mobile biology lab observes Emperor penguins in Antarctica. Scientific research is carried out at international base camps that are monitored by Greenpeace to ensure good environmental practices are followed.

PROTECTING ANTARCTICA FOR THE FUTURE

Attempts to turn the whole of Antarctica into a World Park, preserving its ecological value as a wilderness forever, moved a step closer in 1991. The Antarctic Treaty nations agreed to a ban on mining and offshore exploration for at least the next 50 years. The agreement also included measures to control pollution and waste disposal.

Such protection is vital if Antarctica's harsh, vast, but fragile environment – supporting a diversity of marine and bird life – is to be preserved. Disturbance and pollution are already problems around the continent's scientific bases. A mere footprint on delicate lichens, mosses and algae can take years to repair, and waste materials from the bases do not decompose in the low temperatures. Any industrialscale exploitation of the continent's mineral wealth would cause immense damage.

The 1991 protocol set up a committee of scientists to assess the effectiveness of the new measures. Any new activity in Antarctica – particularly tourism – will be subject to close scrutiny and an environmental impact assessment. As tourism becomes increasingly popular, it poses an ever-greater threat to the area's fragile ecosystems. Proposals have also been made to improve the management of protected areas such as sites of specific scientific interest, historic monuments (such as the camps of Antarctic explorers), and specific marine environments.

The Great Barrier Reef

The Great Barrier Reef is a spectacular assemblage of coral reefs stretching for about 2,000 km (1,240 mi) off the Queensland coast on the continental shelf of Australia. The largest coral ecosystem in the world, its 2,500 reefs, islands and shoals cover some 270,000 sq km (104,000 sq mi). Many are visible only at low tide, others are capped by wooded islands, while a few prominent islands – such as Whitsunday and Magnetic Islands – stand over 1,000 m (3,300 ft) above sea level.

The Great Barrier Reef is one of the most diverse ecosystems on earth. It is home to some 400 different types of coral, 1,500 species of fish and 4,000 species of shellfish. It supports 240 bird species, and its waters are frequented by rare turtles, dolphins, whales and the dugong – an endangered marine mammal.

The hospitable environment (*left*) of the Great Barrier Reef is perfect for the growth of coral, which needs sunlight; deep waters (more than 70 m or 230 ft) are too dark and cold. The reefs support a spectacular variety of marine life that attracts divers.

Reef walkers (*below*) on Heron Island. Trampling feet can cause great damage to corals, and tourists also like to snap off pieces as souvenirs. Although the Great Barrier Reef Marine Park authority tries to control access to the reef, areas above the low water mark are difficult to safeguard.

Threats to the Reef

Human activities pose three main threats to the reef's unique collection of marine life. First, rivers flowing toward it contain increasing loads of sediment and pollution. This, along with offshore sewage discharge, has impaired coral growth by making the seawater cloudy (the colonies of polyps that make up corals need sunlight in order to develop their fantastically shaped skeletons). Second, the pressure of millions of tourists attracted by the beauty of the coral and marine life has damaged the natural habitats. The third threat – offshore oil drilling and mineral exploitation – first came to public attention in 1975. It provoked a public outcry that led to the establishment of the Great Barrier Reef Marine Park.

The park authority has the power to regulate and prohibit activity on and around the reef. The management plan divides the park into seven zones of increasing control. In the General Use Zone most activities, including all types of fishing, coral collecting and cruise ship tours, are allowed to take place freely (though coral collecting requires a permit). More restricted Buffer Zones come between this and the highly restricted Scientific Zone, where permits are needed for the only activities possible, traditional fishing and scientific research. In the exclusive innermost Preservation Zone only research is allowed.

The park authority aims to promote the maximum human enjoyment and use at a level commensurate with the conservation of the ecosystem. It is a delicate balance that is likely to be put under stress by the increasing level of tourism, especially to reefs above low water mark that are not adequately protected by the park legislation.

Thorny invasion

There is one other menace, however, that threatens to undermine all the conservation efforts. Ironically, it is an invasion by one of the reef's inhabitants, the Crown-of-Thorns starfish. The starfish's digestive juices break down coral, and its population is spreading out of control. Perhaps a third of the Great Barrier Reef has been invaded, and some scientists believe that the corals could break up entirely by the year 2040. The reason for the sudden explosion of the starfish population is unknown, though there is evidence that overfishing of snapper and emperor fish has reduced the number of predators that eat the young starfish.

Thinking globally, acting locally

Nonviolent direct action is the hallmark of the Greenpeace environmental protection group. Since 1969 its members have been drawing media and public attention to some of the worst instances of pollution and wildlife exploitation in the world. The waters of the South Pacific and Antarctica have been the scene of some of their most ardent campaigning.

Since the beginning of the 20th century 99 percent of the Blue whales, 97 percent of the Humpbacks and 80 percent of the Fin whales in these waters have been killed by whaling fleets. Greenpeace has led the campaign to put an end to the slaughter, bringing world attention to the fact that the Japanese have continued to hunt the Minke whale in contravention of an international moratorium forbidding it. The activists have spent weeks at a time hunting down the whaling fleets. Once found, they place their boats directly between the whales and the harpoons to bring an end to the hunt.

In 1985 Greenpeace suffered a serious blow when their flagship, the *Rainbow Warrior*, was blown up in Auckland harbor, New Zealand, by French security forces just before it was due to sail for Mururoa Atoll to protest against French nuclear weapons testing. One crew member died in the attack. Since then the group has rebuilt its fleet, and continues to patrol the seas.

In 1992 Norway (a member of the International Whaling Commission) stated its intention to resume the hunting of Minke whales in Arctic waters, claiming that their numbers had recovered sufficiently. But without the sustained campaign against whaling by Greenpeace, and protection of the whales by international initiatives, such an increase in numbers would never have been possible.

A game of cat-and-mouse in the South Seas. Greenpeace activists pull alongside a Japanese factory ship processing harpooned whales. The protest kept the Japanese whalers from meeting their target of Minke whales.

GLOSSARY

Acid rain (Also known as acid deposition.) PRECIPITATION as rain, snow or fog of dilute acid solutions formed by the mixing in the atmosphere of industrial gases, especially sulfur dioxide and nitrogen oxides, with naturally occurring oxygen and water vapor. Can also refer to the dry deposition of acidic PARTICULATE MATTER.

Acidification The process by which soils and surface water become more acid as a result of ACID RAIN.

Agrochemical Any artificial or manufactured chemical used in agriculture to control pests, diseases and weeds.

Agroforestry Any system of farming that incorporates forestry with crop and livestock production. It allows a diversity of products to be grown without exhausting any particular component of the ENVIRONMENT.

Air pollution The presence of gases and PARTICULATE MATTER in the air in high enough concentrations to harm humans, other animals, vegetation or materials. Such pollutants are introduced into the atmosphere principally as a result of human activity.

Algae Simple plants that do not have true stems, leaves or roots but possess chlorophyll and so are capable of photosynthesis (the ability to convert carbon dioxide to sugars, using sunlight as energy). They are found in rivers, lakes, ponds, oceans and other surface waters, and in damp soil.

Algal bloom Population explosion of ALGAE in surface waters due to an increase in plant NUTRIENTS such as nitrates or phosphates. The decomposition of the algae by BACTERIA consumes large quantities of DISSOLVED OXYGEN, killing fish and other forms of aquatic life.

Alternative energy Energy obtained from sources other than FOSSIL FUELS or NUCLEAR POWER, which are generally pollution free and use RENEWABLE RESOURCES; for example BIOGAS, GEOTHERMAL ENERGY, HYDROELECTRIC POWER, SOLAR ENERGY, TIDAL POWER and WIND POWER.

Ambient quality control A method of controlling POLLUTION that establishes the maximum concentrations of air or water pollution the ENVIRONMENT can sustain and allocates pollution quotas to businesses accordingly. In some countries these quotas can be traded. Standards vary from country to country, and controls are not easily enforced because of the difficulty of tracing EMISSIONS, particularly gases, to their source.

Aquifer An underground layer of permeable rock, sand or gravel that absorbs and holds GROUNDWATER.

Bacteria Single cell organisms that are neither plant nor animal. They are the smallest and most primitive form of life. In SOIL, they break down dead organic matter into substances that dissolve in water and are used as NUTRIENTS by plants.

Biodegradable Material that can be broken down into simpler substances by BACTERIA or other DECOMPOSERS. Products made of organic materials such as paper, woollens, leather and wood are biodegradable; many plastics are not.

Biodiversity The number of different species of plants and animals found in a given area. The greater the number of species, the more stable and robust the ECOSYSTEM is.

Biogas A fuel in the form of a METHANE-rich gas derived from rotting organic matter such as animal manure, human excreta or crop residues that can be used for heating and to produce steam for generating electricity.

Biological pest control The control of pests in farming by using natural predators, parasites or disease-causing BACTERIA and viruses rather than chemical PESTICIDES.

Carrying capacity The optimum population of a particular species that a given area of land, or HABITAT, can support under a given set of environmental conditions.

Catalytic converter A device fitted to the exhaust systems or mufflers of motor vehicles that is able to convert harmful carbon monoxide, nitrogen oxides and hydrocarbons into less harmful carbon dioxide, nitrogen and water.

Chlorofluorocarbons (CFCs) Organic compounds made up of atoms of carbon, chlorine and fluorine. Gaseous CFCs used as aerosol propellants, refrigerant gases and solvent cleaners are known to cause depletion of the OZONE LAYER.

CITES The Convention on the International Trade in Endangered Species of Wild Flora and Fauna – the international agreement that prohibits trade in the rarest endangered species and controls the legal trade in others. Signatory countries have to supply data to the WORLD CONSERVATION UNION, which monitors the trade.

Clear cutting (felling) A method of timber harvesting in which all the trees in an area of natural or managed forest are removed in a single cutting.

Compaction Caused when SOILS become closely packed and compressed, for example by the movement of heavy machinery. The amount of air spaces in the soil are reduced, so that water is unable to drain down through it, thus increasing RUNOFF and EROSION.

Coniferous forest A forest of mainly coniferous, or cone-bearing trees, frequently with evergreen needle-shaped leaves. The timber they produce is known as softwood.

Conservation The use, management and protection of NATURAL RESOURCES so that they are not degraded, depleted or wasted and are available on a sustainable basis for use by present and future generations.

Contour plowing The plowing of land for farming in furrows that follow the contours of the land. Such a system helps retain water on the slope and prevent EROSION.

Coppicing A method of forest or woodland management in which mature trees are cut off at the base to encourage the abundant regrowth of side shoots from the stump.

Coral reef A barrier of limestone rock at or near the surface of warm tropical seas that has been formed by the accumulation of the skeletons of millions of coral polyps.

Crop rotation A method of farming in which the same field or area is planted with different crops from year to year or season to season to reduce the depletion of SOIL NUTRIENTS. For example, LEGUMES often follow a cereal crop because they help restore nitrogen levels in the soil.

Dam A structure built across a river to hold back water to form a lake or RESERVOIR.

"Dead" lake/river An area of water in which DISSOLVED OXYGEN levels have fallen so far as a result of ACIDIFICATION, EUTROPHICATION or high levels of POLLUTION that few or no living things are able to survive.

Decomposer Any organism such as fungi or BACTERIA that feeds on dead organic matter, releasing NUTRIENTS into the ENVIRONMENT.

Deciduous forest A forest consisting of trees that drop their leaves in the winter or dry season.

Deforestation The felling and clearing of forested land, which is then converted to other uses.

Desertification The alteration of arable or pasture land in arid or semiarid regions to desert-like conditions. It is usually caused by a combination of OVERGRAZING, SOIL EROSION, prolonged DROUGHT and climate change.

Developed country Any country characterized by having high standards of living and a sophisticated economy, particularly in comparison to DEVELOPING COUNTRIES. A number of indicators can be used to measure a country's wealth and material well-being: for example, the gross national product, the per capita consumption of energy, the number of doctors per head of the population and the average life expectancy.

Developing countries Any country that in relation to the DEVELOPED COUNTRIES is characterized by low standards of living and national wealth. Sometimes called Third World countries, they include most of Africa, Asia and Central and South America.

Dissolved oxygen The amount of oxygen gas dissolved in a certain amount of water at a particular pressure and temperature. It is often expressed as a concentration in parts of oxygen per million parts of water.

Domestication The process by which wild animals and plants are brought under human control and are bred to possess special characteristics that enhance their usefulness for human exploitation, particularly food production.

Domestic waste The solid waste or garbage generated by households.

Dredging The process of removing silt from rivers and harbors to keep shipping channels open.

Drip irrigation The delivery of small amounts of IRRIGATION water to the roots of plants by the use of small pipes or tubes from which water flows slowly and does not cause WATERLOGGING.

Drought An extended period in which rainfall is substantially lower than average and the water supply is insufficient to meet demand.

Dust dome The dome of heated air that in certain climatic conditions forms above an urban area and traps pollutants, especially suspended PARTICULATE MATTER, leading to SMOG.

Ecology The study of the interactions of living organisms with each other and with their ENVIRONMENT; the study of the structure and functions of nature.

Ecosystem A community of plants and animals and the ENVIRONMENT in which they live.

Effluent Any liquid waste discharged into the ENVIRONMENT as a byproduct of industry, agriculture or sewage treatment.

Emission A substance discharged into the air in the form of gases and PARTICULATE MATTER, as from automobile engines and industrial smokestacks.

Energy use The total amount of energy used at a variety of levels from the individual up to the global level, expressed in gigajoules per person at the level under consideration.

Environment The external conditions – climate, geology, other living things – that influence the life of individual organisms or ECOSYSTEMS; the surroundings in which all animal and plants live and interact with each other.

Environmental degradation The process of depleting or destroying a RENEWABLE RESOURCE such as SOIL, forests, pasture or wildlife by using it at a faster rate than it can be replenished.

Erosion The process by which exposed land surfaces are broken down into smaller particles and carried away by water, wind or ice.

Eutrophication The process by which a HABITAT, usually aquatic, becomes enriched with NUTRIENTS such as nitrates and phosphates. This can lead to ALGAL BLOOM.

Exotic species Any plant or animal species that has been introduced to an area in which it does not normally live.

Fallow Farmland that is tilled but left uncultivated for a period of time to allow it to regain its fertility.

Famine An acute shortage of food, leading to widespread malnutrition and starvation, in a particular area. Crop failures due to DROUGHT or flood are rarely the sole cause of famine, but are compounded by other factors such as OVERGRAZING, refugee problems, or war.

Fertilizers Organic or inorganic plant NUTRIENTS that are spread on the land to improve its ability to grow crops, trees or other vegetation.

Floodplain A stretch of low, flat ground on one or both sides of a river channel formed by sediments deposited by the river when it is in flood.

Food and Agriculture Organization (FAO) A specialized agency of the United Nations created in 1945 to combat worldwide malnutrition and hunger by coordinating food, farming, forestry and fisheries development programs.

Food chain The term that describes the process by which energy in the form of food is passed from one living organism to another. Plants (primary producers) are eaten by plant-eating animals, or herbivores (primary consumers), which in turn are eaten by meat-eaters, or carnivores (secondary consumers), and so on up the chain.

Fossil fuels Fuels such as oil, coal, peat and natural gas, formed beneath the Earth's surface under conditions of heat and pressure from organisms that died millions of years ago.

Friends of the Earth An environmental pressure group working at international, national and local level.

Fuelwood Wood that is used as a fuel usually for household heating or cooking, but sometimes by

industry. It may be burned as charcoal, which gives off greater heat.

Garigue A type of SCLEROPHYLLOUS vegetation of the Mediterranean region, in which aromatic shrubs and scattered trees predominate. See also MAQUIS.

Genetic engineering The manipulation of genetic material by humans to produce plants and animals with new characteristics or new combinations of characteristics that are useful to humans, such as disease-resistant crop plants. See also HYBRID.

Geothermal energy An ALTERNATIVE ENERGY source that is derived from heat sources within the Earth's interior. They can be tapped to provide power for electricity or domestic heating, either by harnessing steam or hot water rising to the surface through cracks in the rock, or by drilling holes to hot rocks close to the surface and injecting water to create steam.

Gigajoule A unit of energy equivalent to one billion joules.

Global warming The increase in the average temperature of the Earth that is believed to be caused by the GREENHOUSE EFFECT.

Greenhouse effect The effect of certain gases in the atmosphere, such as carbon dioxide and METHANE, in absorbing solar heat radiated back from the surface of the Earth and preventing its escape into space. Without these gases the Earth would be too cold for living things, but the burning of FOSSIL FUELS for industry and transportation has caused atmospheric levels of these gases to increase, and this is believed to be a cause of GLOBAL WARMING.

Greenpeace An environmental pressure group working internationally and nationally that specializes in active campaigns against specific groups causing environmental damage.

Green Revolution The introduction of high-yielding varieties of cereals (especially rice and wheat) and modern farming techniques to increase agricultural production in DEVELOPING COUNTRIES. It began in the early 1960s.

Groundwater Water that has percolated into the ground from the Earth's surface, filling pores, cracks and fissures. An impermeable layer of rock prevents it from moving deeper so that the lower levels become saturated. The upper limit of saturation is known as the WATER TABLE.

Gullying A severe form of SOIL EROSION that occurs near the bottom of slopes when water RUNOFF removes soil and soft rock to erode a deep channel or gully.

Habitat The locality within which a particular plant or animal naturally lives.

Half-life The time required for half the atoms of a RADIOACTIVE substance to disintegrate, and radioactivity levels are considered to have fallen by one half. Times vary enormously: iodene-131 has an 8-day half-life, but plutonium-239 and uranium-238, the main nuclear fuels, have half-lives reckoned in hundreds of thousands of years.

Hazardous waste Material of any kind – solid, liquid or gaseous – that poses an immediate or longterm risk to health and the ENVIRONMENT. See also TOXIC WASTE.

Heavy metal Any metal such as mercury, cadmium and lead with a high atomic weight. Widely used in small quantities in industry, they concentrate in SOILS and water and are highly poisonous to living things. Once they enter the FOOD CHAIN, they accumulate in organs such as the brain, liver and kidneys of animals, with longterm toxic effect.

Herbicide Any chemical used by farmers to control weeds.

Humus A complex mixture of decaying organic matter (e.g. leaf litter, dead plants and animals and animal feces) and inorganic compounds found in the SOIL. This insoluble material helps retain water and water-soluble NUTRIENTS so they can be taken up by plant roots.

Hunter–gatherer Any individual who subsists by hunting wild animals and gathering the berries and fruits from wild plants.

Hybrid An animal or plant that is the offspring of two species that would not normally interbreed. Hybrid crops are often developed for agriculture because they give higher yields and are more resistant to disease.

Hydroelectric power (HEP) Electrical energy generated by the force of falling or flowing water, which is used to spin a turbine. See also SMALLSCALE HYDROELECTRIC POWER.

Hydrological cycle The continuous process whereby the Earth's fixed supply of water circulates around the oceans, the atmosphere, the SOIL and the rocks beneath through evaporation, TRANSPIRATION, PRECIPITATION, RUNOFF and the movement of GROUNDWATER.

Igneous rock Rock formed when magma, molten material within the Earth's crust, cools and solidifies.

Incineration The disposal of waste by burning it in a furnace under controlled conditions. TOXIC WASTE can be destroyed by incineration at very high temperatures.

Industrialization The process by which a country or region moves from an agricultural economy to an industrial one.

Insecticide Any chemical used to control insect pests on crops or other plants.

Intensive agriculture A method of farming that uses high inputs of capital and labor to maximize production from a small area of land.

Intercropping A method of farming in which two or more crops are grown on an area of land side by side.

International Whaling Commission An international organization that seeks to monitor and control the hunting of large whales so that their future is assured.

Invader An EXOTIC SPECIES that, in the absence of natural predators, is able to flourish, crowding out competing native species and causing imbalance to the ENVIRONMENT. The removal of one species from a HABITAT by, for example, grazing or DEFORESTATION, may lead to other species moving in to invade it.

Irrigation The artificial application of water to the SOIL, for example by sprinklers or through canals, to enable crops to be grown successfully.

IUCN (International Union for Conservation of Nature and Natural Resources) *see* **World Conservation Union**.

Land degradation The process of depleting or destroying the SOIL by using its NUTRIENTS and minerals at a faster rate than natural process are able to restore them. As land becomes sterile, it becomes liable to SOIL EROSION and DESERTIFICATION.

Landfill site A site such as a former quarry, disused mineworking or gravel pit that is used to dispose of waste by burying it between layers of soil or building rubble.

Leaching The process by which water washes NUTRIENTS and minerals downward from one layer of SOIL to another, or into streams.

Legume/leguminous plant Any plant of the family Leguminosae, including many food and forage crops, which enrich SOIL through their ability to fix nitrogen (convert it to nitrates) in their roots. See NITROGEN CYCLE.

Mangrove A dense forest of shrubs and trees growing on tidal coastal mudflats and estuaries throughout the tropics. Many plants have aerial roots.

Maquis SCLEROPHYLLOUS vegetation of the Mediterranean coast, consisting of small trees such as olive, fig and hermes oak and aromatic shrubs, created by centuries of human activity in the form of grazing and burning. See also GARIGUE.

Marginal land Farmland that yields a very low productive return for the use to which it is put.

Mechanization The introduction of machines to replace human labor and animal power in agriculture and other human activities.

Methane A gas produced by decomposing organic matter that burns without releasing pollutants and can be used as an ALTERNATIVE ENERGY source. Excessive methane production from vast amounts of animal manure is believed to contribute to the GREENHOUSE EFFECT.

Monoculture The cultivation of one type of crop on the same area of land year after year. See also CROP ROTATION, POLYCULTURE.

Municipal waste Normally refers to the solid waste from homes and businesses collected by or for the local authority. It excludes industrial waste.

Napalm A thick liquid that burns very easily and is used in firebombs and flamethrowers. Its main use is in warfare, but it may also be used to clear wide areas of vegetation.

Natural resources All the resources that are produced by the Earth's natural processes including mineral deposits, FOSSIL FUELS, SOIL, air, water, plants and animals, and are used by people for agriculture, industry and other purposes.

Nitrogen cycle The natural circulation of nitrogen through the air, SOIL and living organisms. Atmospheric nitrogen is converted by soil BACTERIA, certain ALGAE and organisms in the root nodules of LEGUMINOUS PLANTS into organic nitrogen compounds (nitrates). These are absorbed by green plants and synthesized into more complex compounds. When the plants (or animals feeding on them) die they are broken down again by DECOMPOSERS so that nitrogen gas is released into the atmosphere.

Nonrenewable resource NATURAL RESOURCES that are present in the Earth's makeup in finite amounts and cannot be replaced once reserves are exhausted.

Nuclear power The term used to describe the electricity generated from the heat energy released when atoms, usually of uranium-238 or plutonium-239, are split. This is called a nuclear reaction.

Nuclear reactor The part of a NUCLEAR POWER station where the nuclear reaction takes place.

Nuclear waste *see* **Radioactive waste**

Nutrient Any substance or compound, derived from the ENVIRONMENT, that contributes to the survival, growth and reproduction of plants or animals.

Old-growth forest *see* **Primary forest**

Open-cast mining The mining of NATURAL RESOURCES such as iron ore or coal from immediately below the surface of the land. Narrow layers of SOIL or rocks may have to removed first. Such mining creates large holes in the ground that are sometimes used as LANDFILLS.

Overgrazing The result of grazing too many animals on an area of pasture so that the CARRYING CAPACITY is reduced. The stripping away of vegetation exposes the SOIL to EROSION by wind and water, with risk of DESERTIFICATION.

Ozone A poisonous gas that exists naturally in the upper atmosphere (see OZONE LAYER) but is also formed when certain pollutants such as hydrocarbons and nitrogen oxides react with sunlight, and is a major contributor to SMOG. Even in small concentrations it causes respiratory problems and hinders plant growth.

Ozone hole *see* **Ozone layer**

Ozone layer A band of enriched oxygen or OZONE found in the upper atmosphere. It absorbs harmful ultraviolet radiation from the Sun. The heat this creates provides a cap for the earth's weather systems. OZONE in the upper atmosphere is being depleted by the build-up of CFCs, and seasonal "holes" have been detected in the layer over Antarctica. There is evidence that there is a seasonal drop over the Arctic as well.

Paddyfield A flooded field in which rice is planted and cultivated. The fields are later drained as the rice grains ripen.

Particulate matter Solid particles or liquid droplets suspended or carried in the air.

Permafrost Ground that is permanently frozen. The layer of soil above the permafrost melts in summer but water is unable to drain through it, leading to WATERLOGGING.

Pesticide Any chemical substance used to control the pests that can damage crops, such as insects and rodents. Often used as a general term for HERBICIDES, INSECTICIDES and fungicides.

Photovoltaic cell A device that converts SOLAR ENERGY directly into electrical energy. Also known as a solar cell.

pH Measurement on a scale of 0–14 of the acidity or alkalinity of air, water or SOIL. Neutral levels are shown as pH7, less than pH7 are acidic and more than pH7 are alkaline.

Phytoplankton Small, drifting plants, mostly ALGAE, that are found in marine and freshwater ECOSYSTEMS.

Polder An area of level land at or below sea level that has been reclaimed from the sea or a lake. It is normally used for agriculture.

"Polluter pays principle" The idea that the person or business that causes POLLUTION should pay for cleaning it up and for its prevention in the future, and compensate for any damage caused.

Pollution The contamination of the ENVIRONMENT with substances that are harmful to it.

Polyculture A complex form of INTERCROPPING in which a large number of different crops maturing at different times are cultivated together.

Precipitation Moisture that reaches the Earth from the atmosphere, whether as rain, snow, sleet, hail, fog or mist.

Primary air pollutant A pollutant that has been added directly to the air by natural events (such as a volcanic eruption) or human activities and which occurs in concentrations harmful to living things.

Primary forest Any original, virgin forest that has not been cut and may contain massive trees that can be hundreds or even thousands of years old. Also known as OLDGROWTH FOREST.

Qanat An underground tunnel system of ancient origin dug to transport water long distances for IRRIGATION in desert areas.

Radioactivity The radiation emitted from atomic nuclei. This is greatest when the atom is split, as in a NUCLEAR REACTOR. There are three different types – alpha, beta and gamma – which have very different properties. Prolonged exposure to radioactive material can cause damage to living tissue, leading to cancers, destruction of bone marrow and, ultimately, death.

Radioactive waste The waste products from NUCLEAR POWER plants, medicine, research, atomic weapons or other processes that have involved nuclear reactions.

Rainforest Usually known as tropical rainforest and found in the equatorial belt where there is heavy rain and no marked dry season. Growth is very lush and rapid. Rainforests probably contain half of all the world's plant and animal species.

Real purchasing power A measure of wealth that takes account of the goods and services that a person can buy with the wealth he or she possesses.

Recycling Collecting and reprocessing a NATURAL RESOURCE so it can be used again.

Reforestation The replanting with trees of an area that was formerly forested.

Renewable resource A NATURAL RESOURCE that is normally replenished through natural processes. Examples are oxygen in the air and water in lakes. A renewable resource may become NONRENEWABLE if it is used up at a quicker rate than it is replenished; for example, forest trees or fish stocks in the oceans.

Reservoir A large body of water created in the landscape, usually by building a DAM, for storage for drinking or IRRIGATION water or for generating HYDROELECTRIC POWER.

Rill erosion SOIL EROSION caused when surface water flowing at high velocities over the ground becomes channelled into many miniature streams. See also GULLYING.

Runoff Water produced by rainfall or melting snow that flows across the land surface into streams and rivers. Delayed runoff is water that soaks into the ground and later emerges on the surface as springs.

Salinization The accumulation of soluble salts near or at the surface of the SOIL. This occurs naturally in arid and semiarid areas through evaporation, but can also result from the incorrect application of water for IRRIGATION. Eventually the land becomes worthless for cultivation as plants cannot cope with the high levels of salt.

Saltwater intrusion The movement of saltwater into freshwater AQUIFERS in coastal and inland areas that occurs when GROUNDWATER is withdrawn faster than it is recharged by PRECIPITATION.

Sclerophyllous The term used to describe woody plants with small leathery evergreen leaves that are found in hot, dry areas such as the Mediterranean.

Scrubber A device that reduces the amount of pollutants in the form of PARTICULATE MATTER from the gases, especially sulfur dioxide, given off by industrial plants in the burning of FOSSIL FUELS before they enter the atmosphere.

Secondary air pollutant A harmful chemical formed in the atmosphere when two or more pollutants react together or with the natural components of the atmosphere. See for example OZONE.

Sheet erosion SOIL EROSION caused when water runs in a thin sheet or film down slopes bare of vegetation, carrying soil with it.

Shelter belt A belt of trees planted in windswept areas to reduce SOIL EROSION by the wind and protect crops.

Shifting cultivation A method of farming prevalent in tropical areas in which a piece of land is cleared and cultivated until its fertility is diminished. The land is then left to restore itself naturally.

Siltation The process by which the bed of a river, lake or RESERVOIR becomes clogged with silt (eroded rock particles and SOIL) that has been carried by the river from upstream and deposited.

Sierra Club An American organization devoted to the study and protection of the Earth's scenic and ecological resources.

Slash-and-burn farming A method of farming in tropical areas where vegetation cover is cut and burned to FERTILIZE the land before crops are planted. Often a feature of SHIFTING CULTIVATION.

Sludge The semi-solid mixture of organic matter laden with BACTERIA, viruses, HEAVY METALS, synthetic organic chemicals and solid chemicals that is removed from waste water at a sewage treatment plant.

Smallscale hydroelectric power A HYDROELECTRIC POWER station that generates less than 100kW. It is a lowcost solution to providing electrical power to smallscale industries in remote rural areas. Also known as a microhydro.

Smog Originally referring to a thick mixture of smoke and fog, the term is now used to describe the haze in many cities caused by a variety of pollutants. Photochemical smog is a complex mixture of air pollutants produced in the lower atmosphere by the reaction of hydrocarbons and nitrogen oxides under the influence of sunlight. OZONE is one of the harmful components.

Soil The unconsolidated, weathered layer of material that lies at and immediately below the Earth's surface and is able to support plant growth. It consists of mineral material and living and dead organic matter (HUMUS).

Soil conservation Any one of a number of methods that may be used to reduce SOIL EROSION, to prevent the depletion of soil NUTRIENTS and to restore nutrients already lost by EROSION, LEACHING and excessive crop harvesting.

Soil erosion The loss of the top layers of SOIL from an area as a result of EROSION by wind or water. In the case of water it may be caused by RILL EROSION, SHEET EROSION or GULLYING.

Solar energy The radiant energy produced by the Sun, which powers all the Earth's natural processes. It can be captured and used to provide domestic heating or converted to produce electrical energy. See PHOTOVOLTAIC CELL.

Surface water PRECIPITATION that does not enter the ground or return to the atmosphere but runs over the ground and collects in streams, lakes, rivers and the sea. See RUNOFF.

Sustainability The concept of using the Earth's NATURAL RESOURCES to improve people's lives without diminishing the ability of the Earth to support life today and in the future.

Sustainable agriculture The use of methods that allow farming to be continued year after year without diminishing the productivity of the land or exhausting NATURAL RESOURCES.

Temperature inversion The situation that occurs in particular climatic conditions when a layer of cool, dense air becomes trapped under a layer of warmer, less dense air. Air pollutants in the surface layer are unable to disperse and can build up to levels harmful to living things.

Terrace A level area of land cut out of a slope on which crops are grown.

Thirty Percent Club The group of countries that have agreed to cut their EMISSIONS of sulfur dioxide by 30 percent (based on levels in 1985) by the mid 1990s.

Tidal energy An ALTERNATIVE ENERGY source in which tidal movement (the fall between high and low tide) is used to turn a turbine to generate electricity.

Topsoil The upper horizon or layer of the SOIL, which is usually rich in NUTRIENTS and HUMUS.

Toxic waste Any form of HAZARDOUS WASTE capable of causing death, serious injury or illness.

Transpiration The transference of water, drawn up from the SOIL through the roots and stems of living plants, to the atmosphere as water vapor through pores in their leaves.

Tundra Land lying in the very cold northern regions of Europe, Asia and Canada, where the winters are long and cold and the ground beneath the surface is permanently frozen, limiting vegetation. See PERMAFROST.

United Nations Environment Program (UNEP) An agency of the United Nations established in 1972 to coordinate international concern and activities for protecting the ENVIRONMENT.

Unleaded gasoline Gasoline to which little or no lead has been added. Lead is added to gasoline to enhance engine performance, but it is highly damaging to all living organisms and in humans is known to cause brain damage, particularly in the young. Most industrial nations have now applied legislation to reduce the level of lead in automobile EMISSIONS.

Urbanization The transformation of a population from rural to urban status; the process of city formation and growth.

Volatile compound A liquid or solid that vaporizes at a relatively low temperature; for example, gasoline.

Water erosion The wearing away of rocks and soil by running water.

Waterlogging Descriptive of land that has become completely saturated with water.

Watershed An imaginary line dividing the headwaters of two separate river systems, also known as a water-parting or divide. The term watershed may also be used to describe the whole area, or basin, drained by a river and its tributaries.

Water table The uppermost level of underground rock that is permanently saturated with GROUNDWATER.

Wetland Any area of low-lying land where the WATER TABLE is at or near the surface for most of the year, resulting in a flooded or WATERLOGGED landscape.

Wind erosion The removal of loose SOIL and fine rock fragments by the wind, or the wearing away of rocks and other materials by tiny particles carried in the wind.

Wind power The use of the wind to drive machinery (as in windmills or wind pumps) or to generate electricity by harnessing the movement of air to turn aerofoil-shaped blades or turbines. Several such devices together are known as a wind farm.

Woodlot An area of land reserved for the cultivation of trees.

World Conservation Union (WCU) An international scientific organization for CONSERVATION that brings together government and nongovernment organizations. Among many of its projects it monitors CITES. Formerly the International Union for the Conservation of Nature and Natural Resources (IUCN).

World Health Organization (WHO) An agency of the United Nations, established in 1948, responsible for coordinating international health activities.

Further reading

Allaby, M. *Green Facts: The Greenhouse Effect and Other Key Issues* (Hamlyn, London, 1989)

Collins, M. (ed.) *The Last Rain Forests* (Mitchell Beazley, London, in association with The World Conservation Union, 1990)

Daniel, Joseph E. (ed.) *1992 Earth Journal – Environmental Almanac and Resource Directory* (Buzzworm Books, Boulder, CO, USA, 1991)

Lean, G., Hinrichsen, D. and Markham, A. *Atlas of the Environment* (Arrow Books, London, 1990)

May, J. *The Greenpeace Book of Antarctica* (Dorling Kindersley, London, 1988)

McCormick, J. *Acid Earth: the Global Threat of Acid Pollution* (Earthscan, London, 1989)

Rennie, J. K. *Population, Resources and Development: a Guidebook* (IUCN, Gland, UK, 1988)

Seager, Joni *The State of the Earth* (Unwin Hyman, London, 1990)

The Organization for Economic Cooperation and Development *Environmental Indicators* (OECD, Paris, 1991)

Tyler Miller Jr., G. *Environmental Science – An Introduction* and *Resource Conservation and Management* (Wadsworth, Belmont, California, 1988 and 1990)

United Nations Environment Programme *The Greenhouse Gases, The Ozone Layer* and *Urban Air Pollution* (UNEP/GEMS, Nairobi, Kenya, 1987 and 1990)

World Resources Institute *The 1992 Information Please Environmental Almanac* (Houghton Mifflin, Boston, 1992)

Acknowledgments

Picture credits
Key to abbreviations: APA Andes Press Agency, London, UK; **BCL** Bruce Coleman Limited, Middlesex, UK; **C** Colorific!, London, UK; **COP** Christine Osborne Pictures, London, UK; **E** Explorer, Paris, France; **EPL** Environmental Picture Library, London, UK; **FLPA** Frank Lane Picture Agency, Suffolk, UK; **HL** The Hutchison Library, London, UK; **IB** The Image Bank, London, UK; **IP** Impact Photos, London, UK; **KF** Knudsens Fotosenter, Norway; **M** Magnum Photos Limited, London, UK; **PP** Panos Pictures, London, UK; **SAP** South American Pictures, Suffolk, UK; **SP** Still Pictures, London, UK; **SPL** Science Photo Library, London, UK; **Z** Zefa Picture Library, London, UK

t=top; c= center; b=bottom; l=left; r=right

1 PP/Ron Giling 2 OSF/Kim Westerskov 3 Rob Badger 4 M/Peter Marlow 6–7 Rob Badger 8–9 EPL/Micheal McKinnon 10–11t IB/Guido Alberto Rossi 10–11b PEP/Sean Avery 12 PEP/Jonathan Scott 12–13 IB/Joseph B Brignolo 14 PP/R Berriedale 15 Greenpeace Communications Limited/Gulley 16 IP/Christophe Bluntzer 16–17 NHPA/Anthony Bannister 18 OSF/Jack Dermid 20–21 IB/Guido Alberto Rossi 20 SPL/Nasa GSFC 22–23 PR/Shirley Richards 23 HL/Crispin Hughes 24 IB/Flip Chalfant 24–25 EPL/Robert Brook 26–27 IP/Roger Scruton 27 EPL/I Lilly 28 PP/Marc French 29 M/Steve McCurry 30 BCL/C B & D W Frith 30–31 BCL/Frances Furlong 31 OSF/Waina Cheng 32 Ecoscene/Adrian Morgan 32–33 Intermediate Technology 34–35 OSF/David Cayless 36 PP/Ron Giling 36–37 Ecoscene/Chinch Gryniewicz 38 PP/Jeremy Hartley 38–39 PP/N Cooper and J Hammond 40–41t SPL/Peter Menzel 40–41b Sue Cunningham 44–45 PR/Norman R Lightfoot 46–47 Fred Bruemmer 47 Fred Bruemmer 48 IB/Steve Dunwell 49 Fred Bruemmer 50 SPL/Nasa, Goddard Institute for Space Studies 50–51 PEP/J MacKinnon 52–53 PR/Lowell Georgia 53 OSF/Mike Birkhead 54 PR/Tom McHugh 56–57 OSF/Breck P Kent 58–59 PR/Spencer Grant 61 Rob Badger 62–63 OSF/Frank Huber 63 PR/Tom and Pat Leeson 64 Rob Badger 64–65 PR/David R Frazier 66 M/Paul Fusco 66–67 M/Paul Fusco 68–69 IB/Steve Proehl 71 SAP/Tony Morrison 73 PEP/Doug Perrine 74–75 OSF/Michael Fogden 75 APA/Carlos Reyes 76 FSP/Hoagland/LN 76–77 EPL/Matt Sampson 78–79 PP/M Hanney 80–81 SAP/Robert Francis 81 SAP/Tony Morrison 83 Sue Cunningham 84 PEP/Richard Matthews 84–85 M/Rio Branco 86–87 SP/Mark Edwards 88–89 KF/Andre 90 KF/Liasson 91 OSF/Tom Ulrich 92 OSF/Fredrick Ehrenst rom 94 Ecoscene/Harwood 96–97 NHPA/David Woodfall 98–99 Swift Picture Library/Mike Read 99 OSF/Martyn Colbeck 100 OSF/Mike Birkhead 102–103 SPL/James Holmes/Zedcor 104–105 E/Christian Delu 106 E/Francis Jalain 107 E/Jose Dupont 108–109 E/Louis Salou 109 E/Henri Guillon 110–111 FSP/Patrick Aventurier 112–113 E/F Jourdan 114–115 Greenpeace Communications Limited/Hoffman 116 NHPA/Picture Box 116–117 NHPA/Picture Box 118 NHPA/Picture Box 120 NHPA/Picture Box 120–121 SP/Mark Edwards 122–123 Z/Schipol Air 124 IB/David W Hamilton 126 M/Harry Gruyaert 126–127 OSF/Konrad Wothe 128–129 OSF/David & Sue Cayless 130 Graham Bateman 130–131 E/Philippe Roy 132 FLPA/Silvestris/Horst Wole 134–135 FSP/Luigi Tazzari 136 E/Walter Geiersperger 137 Katz Pictures/Richard Baker 139 FSP/Germain Rey 140–141 E/Patrick Broquet 142 FLPA/Silvestris/Norbert Schwirtz 144 Ecoscene/Chinch Gryniewicz 146–147 Ecoscene/Sally Morgan 147 OSF/Hans-Dieter Schlag 148–149 M/James Nachtwey 150–151 FSP/Noel Quidu 152–153 E/EWA Guillou/Alain 154–155 Ecoscene/Miessler 156 M/James Nachtwey

157 M/James Nachtwey 158–159 M/James Nachtwey 160–161 OSF/Doug Allan 164–165 M/Fred Mayer 165 M/Fred Mayer 166–167 FSP/Gilles Saussier 168–169 PEP/Hans Christian Heap 170–171 COP 171 IP/David Stewart-Smith 172 E/Elie Bernager 173t COP 174–175 EPL/Micheal McKinnon 176–177 E/Christine Delpal 178–179 COP 179 COP 180 PEP/John Lythgoe 181 HL 182–183 E/Gerard Boutin 183 PP/Katri Burri 184–185 M/Stuart Franklin 186–187 OSF/Edward Parker 188–189 PP/Bruce Paton 190 HL 192 Friends of the Earth 192–193 FSP/Micozzi 194–195 PEP/Jonathan Scott 196–197 BCL/Jennifer Fry 198 BCL/Norman Myers 198–199 BCL/Gerald Cubitt 200 NHPA/Anthony Bannister 201 PP/David Reed 202 BCL/Gerald Cubitt 202–203 PEP/J R Bracegirdle 204–205 BCL/Konrad Wothe 206–207 EPL/Jimmy Holmes 208–209 EPL/Jimmy Holmes 209 M/Raghu Rai 210–211 E/Laurent Giraudon 212 PEP/John Waters, Bernadette Spiegel 213 PP/Trygve Bolstad 214 M/Hiroja Kubota 216 M/Bruno Barbey 216–217 E/Yann Layma 218–219 S & R Greenhill 220 E/Rosi Baumgartner 220–221 M/Hiroja Kubota 222–223 BCL/Alain Compost 224–225 PP/Ron Giling 226 BCL/Alain Compost 226–227 COP 228 EPL/Nigel Dickinson 228–229 NHPA/Morten Strange 230–231 EPL/Jimmy Holmes 232 M/W Eugene Smith 233 M/Rene Burri 234–235 E/Katia Krafft 235 C/Michael Yamashita 237 M/Stuart Franklin 238–239 OSF/Michael Fogden 240 PEP/Richard Beales 240–241 OSF/Babs & Bert Wells 242–243 Ecoscene/Whittle 243 E/Hiroynki Matsumoto 244 NHPA/Bill Wood 245 OSF/Kathie Atkinson 246–247 Greenpeace Communications Ltd/Culley

Editorial, research and administrative assistance
Jo Ahier, Nick Allen, Dr Mike Bradshaw, Mike Brown, Joanna Chisholm, Pamela Egan, Roger Few, Reina Foster-de Wit, Claire Gabbey, Matthew Kneale, Hilary McGlynn, Jo Rapley, Eelin Thomas, Claire Turner

Artists
The Maltings Partnership, Derby, England

Cartography
Pauline Morrow, Sarah Rhodes
Maps drafted by Euromap, Pangbourne

Index and Glossary
Barbara James and John Baines

Production
Clive Sparling

Typesetting
Brian Blackmore, Niki Moores

Color origination
Scantrans pte Ltd, Singapore

INDEX

Page numbers in **bold** refer to extended treatment of topic; in *italic* to illustration or map